APPLIED CHEMISTRY AND CHEMICAL ENGINEERING

Volume 3

Interdisciplinary Approaches to Theory
and Modeling with Applications

APPLIED CHEMISTRY AND CHEMICAL ENGINEERING

Volume 3

Interdisciplinary Approaches to Theory and Modeling with Applications

Edited by

A. K. Haghi, PhD
Lionello Pogliani, PhD
Francisco Torrens, PhD
Devrim Balköse, PhD
Omari V. Mukbaniani, DSc
Andrew G. Mercader, PhD

Apple Academic Press Inc.
3333 Mistwell Crescent
Oakville, ON L6L 0A2 Canada

Apple Academic Press Inc.
9 Spinnaker Way
Waretown, NJ 08758 USA

© 2018 by Apple Academic Press, Inc.

First issued in paperback 2021

No claim to original U.S. Government works

ISBN-13: 978-1-77463-116-4 (pbk)
ISBN-13: 978-1-77188-566-9 (hbk)

Library and Archives Canada Cataloguing in Publication

Applied chemistry and chemical engineering / edited by A.K. Haghi, PhD, Devrim Balköse, PhD, Omari V. Mukbaniani, DSc, Andrew G. Mercader, PhD.
Includes bibliographical references and indexes.
Contents: Volume 1. Mathematical and analytical techniques --Volume 2. Principles, methodology, and evaluation methods --Volume 3. Interdisciplinary approaches to theory and modeling with applications --Volume 4. Experimental techniques and methodical developments --Volume 5. Research methodologies in modern chemistry and applied science.
Issued in print and electronic formats.
ISBN 978-1-77188-515-7 (v. 1 : hardcover).--ISBN 978-1-77188-558-4 (v. 2 : hardcover).--ISBN 978-1-77188-566-9 (v. 3 : hardcover).--ISBN 978-1-77188-587-4 (v. 4 : hardcover).--ISBN 978-1-77188-593-5 (v. 5 : hardcover).--ISBN 978-1-77188-594-2 (set : hardcover).
ISBN 978-1-315-36562-6 (v. 1 : PDF).--ISBN 978-1-315-20736-0 (v. 2 : PDF).-- ISBN 978-1-315-20734-6 (v. 3 : PDF).--ISBN 978-1-315-20763-6 (v. 4 : PDF).-- ISBN 978-1-315-19761-6 (v. 5 : PDF)
1. Chemistry, Technical. 2. Chemical engineering. I. Haghi, A. K., editor
TP145.A67 2017   660   C2017-906062-7   C2017-906063-5

Library of Congress Cataloging-in-Publication Data

Names: Haghi, A. K., editor.
Title: Applied chemistry and chemical engineering / editors, A.K. Haghi, PhD [and 3 others].
Description: Toronto ; New Jersey : Apple Academic Press, 2018- | Includes bibliographical references and index.
Identifiers: LCCN 2017041946 (print) | LCCN 2017042598 (ebook) | ISBN 9781315365626 (ebook) | ISBN 9781771885157 (hardcover : v. 1 : alk. paper)
Subjects: LCSH: Chemical engineering. | Chemistry, Technical.
Classification: LCC TP155 (ebook) | LCC TP155 .A67 2018 (print) | DDC 660--dc23
LC record available at https://lccn.loc.gov/2017041946

Apple Academic Press also publishes its books in a variety of electronic formats. Some content that appears in print may not be available in electronic format. For information about Apple Academic Press products, visit our website at **www.appleacademicpress.com** and the CRC Press website at **www.crcpress.com**

ABOUT THE EDITORS

A. K. Haghi, PhD

A. K. Haghi, PhD, holds a BSc in Urban and Environmental Engineering from the University of North Carolina (USA), an MSc in Mechanical Engineering from North Carolina A&T State University (USA), a DEA in applied mechanics, acoustics and materials from the Université de Technologie de Compiègne (France), and a PhD in engineering sciences from the Université de Franche-Comté (France). He is the author and editor of 165 books, as well as of 1000 published papers in various journals and conference proceedings. Dr. Haghi has received several grants, consulted for a number of major corporations, and is a frequent speaker to national and international audiences. Since 1983, he served as professor at several universities. He is currently Editor-in-Chief of the *International Journal of Chemoinformatics and Chemical Engineering* and the *Polymers Research Journal* and on the editorial boards of many international journals. He is also a member of the Canadian Research and Development Center of Sciences and Cultures (CRDCSC), Montreal, Quebec, Canada.

Lionello Pogliani, PhD

Lionello Pogliani, PhD, was Professor of Physical Chemistry at the University of Calabria, Italy. He studied Chemistry at Firenze University, Italy, and received his postdoctoral training at the Department of Molecular Biology of the C. E. A. (Centre d'Etudes Atomiques) of Saclay, France, the Physical Chemistry Institute of the Technical and Free University of Berlin, and the Pharmaceutical Department of the University of California, San Francisco, CA. Dr. Pogliani has coauthored an experimental work that was awarded the GM Neural Trauma Research Award. He spent his sabbatical years at the Centro de Química-Física Molecular of the Technical University of Lisbon, Portugal, and at the Department of Physical Chemistry of the Faculty of Pharmacy of the University of Valencia-Burjassot, Spain. He has contributed nearly 200 papers in the experimental, theoretical, and didactical fields of physical chemistry, including chapters in specialized books. He has also presented at more than 40 symposiums. He also published a book on the numbers 0, 1, 2, and 3. He is a member of the International Academy of

Mathematical Chemistry. He retired in 2011 and is part-time teammate at the University of Valencia-Burjassot, Spain.

Francisco Torrens, PhD

Francisco Torrens, PhD, is lecturer in physical chemistry at the Universitat de València in Spain. His scientific accomplishments include the first implementation at a Spanish university of a program for the elucidation of crystallographic structures and the construction of the first computational chemistry program adapted to a vector facility supercomputer. He has written many articles published in professional journals and has acted as a reviewer as well. He has handled 26 research projects, has published two books and over 350 articles, and has made numerous presentations.

Devrim Balköse, PhD

Devrim Balköse, PhD, is currently a faculty member in the Chemical Engineering Department at the Izmir Institute of Technology, Izmir, Turkey. She graduated from the Middle East Technical University in Ankara, Turkey, with a degree in Chemical Engineering. She received her MS and PhD degrees from Ege University, Izmir, Turkey, in 1974 and 1977, respectively. She became Associate Professor in Macromolecular Chemistry in 1983 and Professor in process and reactor engineering in 1990. She worked as Research Assistant, Assistant Professor, Associate Professor, and Professor between 1970 and 2000 at Ege University. She was the Head of the Chemical Engineering Department at the Izmir Institute of Technology, Izmir, Turkey, between 2000 and 2009. Her research interests are in polymer reaction engineering, polymer foams and films, adsorbent development, and moisture sorption. Her research projects are on nanosized zinc borate production, ZnO polymer composites, zinc borate lubricants, antistatic additives, and metal soaps.

Omari V. Mukbaniani, DSc

Omari Vasilii Mukbaniani, DSc, is Professor and Head of the Macromolecular Chemistry Department of Iv. Javakhishvili Tbilisi State University, Tbilisi, Georgia. He is also the Director of the Institute of Macromolecular Chemistry and Polymeric Materials. He is a member of the Academy of Natural Sciences of the Georgian Republic. For several years he was a member of the advisory board of the *Journal Proceedings of Iv. Javakhishvili Tbilisi State*

University (Chemical Series) and contributing editor of the journal *Polymer News* and the *Polymers Research Journal*. He is a member of editorial board of the *Journal of Chemistry and Chemical Technology*. His research interests include polymer chemistry, polymeric materials, and chemistry of organosilicon compounds. He is an author more than 420 publications, 13 books, four monographs, and 10 inventions. He created in the 2007s the "International Caucasian Symposium on Polymers & Advanced Materials," ICSP, which takes place every other two years in Georgia.

Andrew G. Mercader, PhD

Andrew G. Mercader, PhD, studied Physical Chemistry at the Faculty of Chemistry of La Plata National University (UNLP), Buenos Aires, Argentina, from 1995–2001. Afterwards he joined Shell Argentina to work as Luboil, Asphalts and Distillation Process Technologist, as well as Safeguarding and Project Technologist. His PhD work on the development and applications of QSAR/QSPR theory was performed at the Theoretical and Applied Research Institute located at La Plata National University (INIFTA). He received a post-doctoral scholarship to work on theoretical-experimental studies of biflavonoids at IBIMOL (ex PRALIB), Faculty of Pharmacy and Biochemistry, University of Buenos Aires (UBA). He is currently a member of the Scientific Researcher Career in the Argentina National Research Council, at INIFTA.

Applied Chemistry and Chemical Engineering, 5 Volumes

**Applied Chemistry and Chemical Engineering,
Volume 1: Mathematical and Analytical Techniques**
Editors: A. K. Haghi, PhD, Devrim Balköse, PhD, Omari V. Mukbaniani, DSc, and
Andrew G. Mercader, PhD

**Applied Chemistry and Chemical Engineering,
Volume 2: Principles, Methodology, and Evaluation Methods**
Editors: A. K. Haghi, PhD, Lionello Pogliani, PhD, Devrim Balköse, PhD,
Omari V. Mukbaniani, DSc, and Andrew G. Mercader, PhD

**Applied Chemistry and Chemical Engineering,
Volume 3: Interdisciplinary Approaches to Theory and Modeling with
Applications**
Editors: A. K. Haghi, PhD, Lionello Pogliani, PhD, Francisco Torrens, PhD,
Devrim Balköse, PhD, Omari V. Mukbaniani, DSc, and Andrew G. Mercader, PhD

**Applied Chemistry and Chemical Engineering,
Volume 4: Experimental Techniques and Methodical Developments**
Editors: A. K. Haghi, PhD, Lionello Pogliani, PhD, Eduardo A. Castro, PhD,
Devrim Balköse, PhD, Omari V. Mukbaniani, PhD, and Chin Hua Chia, PhD

**Applied Chemistry and Chemical Engineering,
Volume 5: Research Methodologies in Modern Chemistry and Applied Science**
Editors: A. K. Haghi, PhD, Ana Cristina Faria Ribeiro, PhD,
Lionello Pogliani, PhD, Devrim Balköse, PhD, Francisco Torrens, PhD,
and Omari V. Mukbaniani, PhD

CONTENTS

LIST OF CONTRIBUTORS

Gloria Castellano
Departamento de Ciencias Experimentales y Matemáticas, Facultad de Veterinaria y Ciencias Experimentales, Universidad Católica de Valencia San Vicente Mártir, Guillem de Castro-94, E-46001 València, Spain

Mahdi Ghavami
Department of Mechanical Engineering, Isfahan University of Technology, Isfahan, Iran

Shahriar Ghammamy
Department of Chemistry, Faculty of Science, Imam Khomeini International University, Ghazvin, Iran. E-mail: shghamami@yahoo.com

V. I. Kodolov
Doctor of Science in Chemistry, Professor, Head of Department of Chemistry and Chemical Engineering, Izhevsk State Agricultural Academy, Izhevsk, Udmurtskaja Respublika, Russia. E-mail: kodol@istu.ru

G. A. Korablev
Doctor of Science in Chemistry, Professor, Izhevsk State Agricultural Academy, Izhevsk, Udmurtskaja Respublika, Russia. E-mail: korablevga@mail.ru

Shima Maghsoodlou
University of Guilan, Rasht, Iran

E. K. Molchanov
Institute of Mechanics, Ural Division, Russian Academy of Sciences, Izhevsk, Russia

N. N. Novykh
Doctor of Veterinary Sciences, Professor, Head of the Department of Anatomy and Biology, Izhevsk State Agricultural Academy, Izhevsk, Udmurtskaja Respublika, Russia

S. Poreskandar
University of Guilan, Rasht, Iran

Francisco Torrens
Institut Universitari de Ciència Molecular, Universitat de València, Edifici d'Instituts de Paterna, PO Box 22085, E-46071 València, Spain

A. V. Vakhrushev
Institute of Mechanics, Ural Division, Russian Academy of Sciences, Izhevsk, Russia; Kalashnikov Izhevsk State Technical University, Izhevsk, Russia. E-mail: vakhrushev-a@yandex.ru

Yu. G. Vasiliev
Doctor of Science in Medicine, Head of Department of Physiology and Animal Sanitation, Izhevsk State Agricultural Academy, Izhevsk, Udmurtskaja Respublika, Russia; Professor of Department of Department of Histology, Cytology and Embryology, Izhevsk State Technical University, Izhevsk, Udmurtskaja Respublika, Russia. E-mail: devugen@mail.ru

G. E. Zaikov
Doctor of Science in Chemistry, Professor of Emmanuel Institute of Biochemical Physics, RAS, Moscow, Russia. E-mail: chembio@sky.chph.ras.ru

LIST OF ABBREVIATIONS

AC	alternating current
ADMET	absorption, distribution, metabolism, excretion, and toxicity
ANN	artificial neural network
ANOVA	analysis of variance
BC	boundary condition
BD	big data
BER	base-excision repair
BTX	botulinum toxin
CGRP	calcitonin gene-related peptide
DC	direct current
DEL	double electrical layer
DRMs	DNA repair mechanisms
EC	equipartition conjecture
ECAP	equal-channel angular pressing
ECD	electrocodeposition
EDL	electric double layer
EHD	electrohydrodynamics
FDM	finite-difference method
FISABIO	Foundation for Health and Biomedical Research Promotion of Valencia Region
HTS	high-throughput screening
IBC	isolated boundary condition
IC	initial condition
IDs	imaging databanks
IE	information entropy
IR	infrared
MALDI-TOF MS	matrix-assisted laser desorption–ionization time-of-flight mass spectrometry
MD	molecular dynamics
MI	medical imaging
MIDs	medical-imaging databanks
MMEC	matrix composite electrochemical coatings
NA	non-Archimedean

NER	nucleotide-excision repair
NGS	next-generation sequencing
NPs	nanoparticles
ODE	ordinary differential equations
OSs	oxidation states
PAIDS	paralyzed academic investigator's disease syndrome
PBC	periodic boundary condition
PDT	photodynamic therapy
PMIE	principle of maximum information entropy
PT	periodic table
PTE	periodic table of the elements
RCE	rotating cylindrical electrode
RDE	rotating disc electrode
SB	synthetic biology
SPD	severe plastic deformation
TEM	transmission electron microscopy
TMDs	transition metal dichalcogenides
VR	Valencia region

PREFACE

Involving two or more academic subjects, interdisciplinary studies aim to blend together broad perspectives, knowledge, skills, and epistemology in an educational setting. By focusing on topics or questions too broad for a single discipline to cover, these studies strive to draw connections between seemingly different fields.

Understanding the mathematical modeling is fundamental to the successful career of a researcher in chemical engineering. This book reviews, introduces, and develops the mathematical model that is most frequently encountered in the sophisticated chemical engineering domain. This volume bridges the gap between classical analysis and modern applications.

This volume, the third of the 5-volume **Applied Chemistry and Chemical Engineering,** provides a collection of models illustrating the power and richness of the mathematical sciences in supplying insight into the operation of important real-world systems. It fills a gap within modeling texts, focusing on applications across a broad range of disciplines.

The first part of the book discusses the general components of the modeling process and highlights the potential of modeling in production of nanofibers. These chapters discuss the general components of the modeling process and the evolutionary nature of successful model building in electrospinning process. Electrospinning is the most versatile technique for the preparation of continuous nanofibers obtained from numerous materials. This section of book summarizes the state of the art in electrospinning as well as provides updates on theoretical aspects and applications.

The second part provides a rich compendium of case studies, each one complete with examples, exercises, and projects.

This volume covers a wide range of topics in mathematical modeling, computational science, and applied mathematics. It presents a wealth of new results in the development of modeling theories and methods, advancing diverse areas of applications and promoting interdisciplinary interactions between mathematicians, scientists, engineers, and representatives from other disciplines.

Applied Chemistry and Chemical Engineering, 5-Volume Set includes the following volumes:

- Applied Chemistry and Chemical Engineering,
  Volume 1: Mathematical and Analytical Techniques
- Applied Chemistry and Chemical Engineering,
  Volume 2: Principles, Methodology, and Evaluation Methods
- Applied Chemistry and Chemical Engineering,
  Volume 3: Interdisciplinary Approaches to Theory and Modeling with
  Applications
- Applied Chemistry and Chemical Engineering,
  Volume 4: Experimental Techniques and Methodical Developments
- Applied Chemistry and Chemical Engineering,
  Volume 5: Research Methodologies in Modern Chemistry and
  Applied Science.

PART I
Electrospun Nanofibers

SYNOPSIS (Chapters 1–13)

Sh. Maghsoodlou and S. Poreskandar

University of Guilan, Rasht, Iran

Electrospinning is the most versatile technique for the preparation of continuous nanofibers obtained from numerous materials. This section of the book summarizes the state of the art in electrospinning as well as updates on theoretical aspects and applications.

CHAPTER 1

AN OVERVIEW ON ELECTROSPUN NANOFIBERS: INFLUENCES OF VARIOUS PARAMETERS AND APPLICATIONS

SHIMA MAGHSOODLOU* and S. PORESKANDAR

Textile Engineering, University of Guilan, Rasht, Iran
Corresponding author. E-mail: sh.maghsoodlou@gmail.com

CONTENTS

ABSTRACT

Electrospinning is an appropriate method for nanofibers production. Understanding the influence of various parameters became important for producing suitable fibers with special applications. Also, the fine electro-spun nanofibers make them useful in a wide range of innovative applications. For example, nanofibers with small pore size and high surface area are suitable for biomedical applications such as tissue engineering scaffolds. Thus, it is desirable to control morphology, porosity, and specially, diameter of producing fibers for selecting applications. The aim of this chapter is to review the applications and the effects of these parameters. In addition, some commonly used techniques to align the fibers are discussed in this chapter.

1.1 INTRODUCTION

Nanotechnology is taken one of the expanding technologies interested by many scientists in everywhere in recent years.[1-3] Nobel Prize winner for Physics in 1965, Richard Feynman, came up with the brilliant concept of the nano when he said "there is plenty of room at the bottom" during a conference of the American Physical Society, in 1959.[4] Research in nanotechnology are directed toward realizing and creating improved materials, devices, and systems that exploit these new properties.[5]

One-dimensional structures, with nanoscale diameter, due to the unique properties such as a high area to surface and high porosity, have broadly attracted the attention of used in varied applications.[6-8]

A fiber is of course first of all a geometric shape, a one-dimensional object, having a certain diameter, a given axial ratio, and a certain length that can often approach infinity. Nanofibers have a complex morphology or even topology that does not rely on mechanical means with all their restrictions in terms of fiber diameters, fiber structures, and hierarchical structures.[9]

Electrospinning compares favorably with other methods of nanofiber manufacture, such as drawing, template synthesis, phase separation, and self-assembly, have attracted a great deal of interest as a novel technique.[10,11] A summaries of comparison between these methods are shown in Table 1.1.

This process, which is likewise known as electrostatic spinning, is perhaps the most versatile process,[12] because it is a fascinating, very uncomplicated, inexpensive, and powerful tool used to prepare polymeric fibers with nanoscale diameter.[13,14] An electrospinning process is shown in Figure 1.1.

TABLE 1.1 Comparison of Common Methods for Producing Nanofibers.

Repeatability	Controllability	Simplicity	Scalability	Technology	Technique
No	Yes	Yes	No	Laboratory scale	Drawing
Yes	Yes	Yes	Yes	Laboratory scale	Template synthesis
No	Yes	Yes	No	Laboratory scale	Phase separation
No	No	Yes	No	Laboratory scale	Self-assembly
Yes	Yes	Yes	Yes	Industrial process	Electrospinning

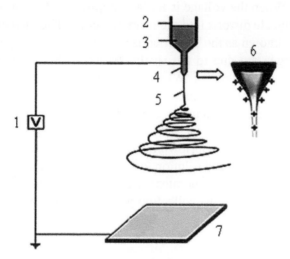

FIGURE 1.1 Sections of electrospinning process: (1) high voltage, (2) polymer solution, (3) syringe, (4) needle, (5) whipping instability, (6) Taylor cone, and (7) collector.

An ordinary electrospinning setup contains three main parts[15–17]:

- a high power supply voltage;
- a syringe with a needle and a pump; and
- a collector.

Employing electrostatic forces to deform materials in the liquid state goes back many centuries,[18,19] but the origin of electrospinning technique goes back to 100 years.[20,21]

Many researchers work on electrospinning set-up and effective factor on this technique. Rapid, reliable, and inexpensive characterization of polymers are reported between 1934 and 1944.[18,22]

The works of Henry Martin *Formhals* (August 2, 1908–May 12, 1981) provided a better understanding of electrospinning process. Several research groups such as Dr. Darrell Reneker and his research group further showed interest in electrospinning with a series of papers published starting in early to mid-1990s and continuing till date. This renewed interest spread quickly and many secondary academic groups became interested in the field of the electrospinning.[1,23,24]

The electrospinning process is grounded upon the simple concept that creates nanofibers through an electrically charged jet of polymer solution or polymer melt. When the voltage is initially applied to the solution fluid, the droplet at the nozzle distorts into the form of a cone. The final conical shape has come to be known as the Taylor cone.[25]

These changes are due to the rivalry between the increasing solution charge and its surface tension. When the applied voltage is sufficient, the electrostatic force in the polymer and solvent molecules can have enough charge to overcome surface tension, a stream is turned out from the tip of the Taylor cone.[11,26–29]

The solution is drawn as a jet toward an oppositely charged collecting plate, which will cause the charged solution to speed up toward the collector.[25,30]

The solvent gradually evaporates, and a charged, solid polymer fiber is allowed to accumulate on the collection plate.[28,29,31] This process is summarized in Figure 1.2.

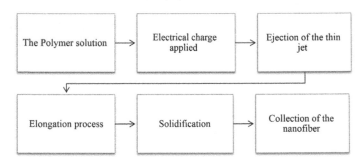

FIGURE 1.2   Steps of electrospinning process.

1.2   VARIOUS PARAMETERS FOR CONTROLLING ELECTROSPINNING PROCESS

The most significant challenge in this process is to attain uniform nanofibers consistently and reproducibly.[1,32–34] In addition, the mechanics of this process deserving a specific attention and necessary to predictive tools or way for better understanding and optimization and controlling process.[35] Also, fiber diameter is an important characteristic for electrospinning, because of its direct influence on the properties of the produced webs.[23,34] Depending on several solution parameters, different results can be obtained using the same polymer and electrospinning setup.[16] Many parameters effect fiber formation. These factors that are studied to have a primary effect on the formation of uniform fibers are the process parameters, environmental parameters and solution parameters.[1,36–39] Processing parameters play a major character in the electrospinning technique because the electrospinning technique is governed by the external electric field produced by the applied voltage caused by charges on the jet surface. It can be divided in two sections: processing parameters and type of collector that is summarized in Figures 1.3 and 1.4.

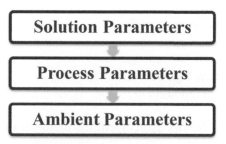

FIGURE 1.3   Effective parameters on the morphology of electrospun nanofibers.

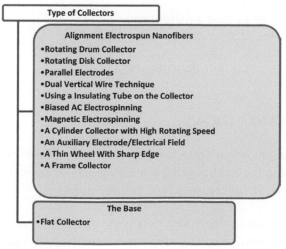

FIGURE 1.4   Type of collectors.

In addition, the processing conditions and external parameters also have a significant result on the diameter and morphology of the nanofibers. The electric field between the needle tip and the target can be controlled through these varying parameters[36]:

- The applied voltage
- The distance between needle tip and target
- The shape of the collector
- The diameter of the needle.

Solution polymer parameters affect forming the nanofibers, causing defects in the fibers in the pattern of beads and junction by their low concentration or viscosity, low conductivity, and by their decrease of molecular weight. Ambient parameters play a role in the properties of the polymeric solution and so affect the morphology of the nanofibers.[24] The solution must also have these feathers to prevent the gate from collapsing into droplets before the solvent has evaporated.[40]

- A surface tension, low enough
- A charge density, high enough
- A viscosity, high enough.

In addition, major factors that control the diameter of the fibers are[21]:

- concentration of polymer in their solution;
- type of solvent used;
- conductivity of their solution; and
- feeding rate of the solution.

In the next part, we discussed about these parameters.

1.2.1   CONCENTRATION

The concentrations of polymer solution play a significant role in the fiber formation during the electrospinning technique. Four critical concentrations from low to high should be remarked in these points[22,32,41,42]:

- As the concentration is low, polymeric micro- or nanoparticles will be produced. Now, electrospray occurs instead of electrospinning owing to the low viscosity and high surface tensions of the solution.

- As the concentration is little higher, a mixture of beads and fibers will be received.
- When the concentration is suitable, smooth nanofibers can be obtained.
- If the concentration is high, not nanoscale fibers, helix-shaped micror-ibbons will be obtained.

Normally, by increasing concentrates solution, the fiber diameter will increase if the solution concentration is suitable for electrospinning. Also, solution viscosity can be also tuned by setting the solution concentration.[32] In the electrospinning technique, a minimum solution concentration is needed for fiber formation. It has been found that at low concentration, beads and fibers are obtained. When the solution concentration increases, the shapes of beads changes and finally uniform fibers are finding. There is an optimum solution concentration for the electrospinning technique. (At low concentrations, beads are formed and at higher concentrations continuous fibers are banned.) Researchers have tried to determine a relationship between solution concentration and fiber diameter. They found a power-law relationship. Increasing concentration of the solution causes increase in fiber diameter with gelatin in electrospinning. Solution surface tension and viscosity also play important roles in determining the range of concentrations from continuous fibers which can be obtained in electrospinning.[37]

Polymer concentration determines the spinnability of a solution, that is to say, whether a fiber form or not. The solution must have a high enough polymer concentration for chain entanglements to occur; however, the solution cannot be either too dilute or too centralized. On the other hand, polymer concentration influences both viscosity and surface tension of the solution technique. If the solution is to dilute, then the polymer fiber will give away up into droplets before reaching the collector (due to the effects of surface tension). If the solution is too concentrated then fibers cannot be made (due to the high viscosity), because it is difficult to control solution flow rate through capillaries. Thus, an optimum range of polymer concentrations is called for.[16]

The molecular properties of polymer play a vital part in controlling fiber initiation and stabilization. Several investigators have attempted to establish optimum ranges for concentration and molecular weight to insure stable fiber formation. At any molecular weight, the effect of concentration on the breakdown of the solution jet can be distinguished by two critical concentrations, C_i and C_f. Below C_i, only beads may be developed (due to insufficient chain entanglements in the solution). Above C_i, a combination of beads and fibers is observed. When the concentration is increased above C_f, complete fibers are created. C_i is typically about the entanglement concentration C_e, at which chain entanglements in the solution become significant. Hence, $\tilde{C}_i$ is a transition concentration at which fibers begin to come forth from the beads and C_f is the concentration at which a fibrous structure is stabilized. Shenoy et al. have developed the solution entanglement number, $(n_e)_{sol}$, for depicting the transition points for fiber initiation and complete fiber formation[18]:

$$(n_e)_{sol} = \frac{\varphi_p M_w}{M_e} \tag{1.1}$$

Lyons et al. have reported that fiber diameter varies exponentially with a molecular weight for melt electrospun polypropylene. The fiber diameter, normalized with difference to the Berry number, is plotted. A power-law relationship is observed between D and $[\eta]C$ as follows[18]:

$$D(nm) = 18.6([\eta]C)^{1.11} \tag{1.2}$$

Various investigators have also correlated the fiber diameter with a normalized concentration defined as[18]

$$C_e = \frac{\rho M_e}{M_w} \tag{1.3}$$

1.2.2   SOLUTION VISCOSITY

One of the most significant parameters that influence the diameter and the morphology of the fiber is the viscosity of the resolution, which is indirectly affected by polymer characteristics such as molecular weight and concentration. When a polymer with a higher molecular weight is dissolved in a solvent, the viscosity of the polymer solution is higher than a solution of the same polymer with a lower molecular weight. Similarly, the viscosity of a polymer solution increases with an increased concentration of polymer in that solution. It is obvious that the viscosity of solution and polymer chain entanglements have a direct relationship. It has been proven that continuous and smooth fibers cannot become in low viscosity, whereas high viscosity results in the hard ejection of jets from solution, namely there is a need of suitable viscosity for electrospinning. The viscosity range of different polymer solution at electrospinning is different. It is important that viscosity, polymer concentration, and polymer molecular weight are linked to one another. For solution with low viscosity, surface tension is the dominant factor and just beads or beaded fiber formed. If the solution is of suitable viscosity, continuous fibers can be made. Also, the shape of the beads changes from spherical to elliptical when the viscosity of solution varies from low to high. Higher viscosity also results in larger diameter fibers and smaller deposition areas.[32,36,37]

Taken together, these studies suggest that there exist polymer-specific, optimal viscosity values for electrospinning and this property possesses a remarkable influence on the morphology of the fibers.[37]

1.2.3 MOLECULAR WEIGHT

Molecular weight of polymer also has an important impression on the morphology of electrospun fiber. In principle, molecular weight reflects entangling polymer chains in solutions, namely the solution viscosity. Keep the concentration fixed, lowering the molecular weight of polymer trends to form beads rather than smooth fiber. By increasing the molecular weight, smooth fiber will be obtained. Further by increasing molecular weight, microribbon will be received. Also, the authors establish that as the molecular weight is high, some patterned fibers can also be obtained at low concentration.[32]

It has been remarked too that a low molecular weight solution form beads rather than fibers. A high molecular weight solution gives fibers with larger average diameters. Chain entanglement plays an important role in technique electrospinning. It has been discovered that high molecular weights are not always essential for the electrospinning technique if enough intermolecular interactions can provide a substitute for the interchange connectivity got through chain entanglements.[37] In addition, this parameter plays a vital role in controlling fiber beginning and stabilization. Several investigators have tried to demonstrate the ideal ranges for concentration and molecular weight to insure stable fiber formation. The molecular weight of the polymer has a significant purpose of proving the structure in the electrospun polymer. At a constant concentration, the structure changes from beads, to beaded fibers, to complete fibers, and to flat ribbons as the molecular weight is increased.[18]

1.2.4 SURFACE TENSION

Surface tension is important in electrospinning. Different solvents may give different surface tensions. With the concentration fixed, reducing the surface tension of the solution, beaded fibers can convert into smooth fibers. The surface tension and solution viscosity can be adjusted by varying the mass ratio of solvents mix and fiber morphologies. Surface tension decides the upper and lower boundaries of the electrospinning window if all other conditions specified.[32]

A lower surface tension of the spinning solution helps electrospinning to occur at a lower electric field.[15,36,37] However, not necessarily a lower surface tension of a solvent will always be more suitable for electrospinning. Also, this parameter determines the upper and lower boundaries of the electrospinning window if all other variables are held constant.[15,37] Adding a surfactant to the polymer solution also changes the surface tension. If all other variables are held constant, surface tension decides the upper and lower bounds of the electrospinning technique.[36]

It is caused by the attraction between the molecules in a liquid. In the most of liquid, each molecule is attracted equally in all directions by neighboring liquid molecules, resulting in a net force of zero. At the surface of the liquid, the molecules are subjected to a net inward force to balancing only by resisting liquid to compression. The net effecting causes the surface area to reduce it until controlling the possible lowest ratio of surface area to volume. In electrospinning, the charges on the polymer solution must be high enough to overwhelm the surface tension of the solution. As the electrician jet speeds up from the needle to the aim, the polymer jet is stretched. Then, surface tension of the solution may cause the jet to break up into droplets. In addition, if there is a lower concentration of polymer molecules, the surface tension causes beaded fibers formed.[36]

1.2.5 CONDUCTIVITY/SURFACE CHARGE DENSITY

Solution conductivity is mainly decided by the polymer type, solvent sort, and the salt. Usually, natural polymers are polyelectrolyte in nature, subjecting to higher tension under the electric field, resulting in the poor fiber formation. Also, the electrical conductivity of the solution can be tuned by adding the ionic salts. With the aid of ionic salts, nanofibers with small diameter can be produced. Sometimes high solution conductivity can be also accomplished by using organic acid as the solvent. An increase in the solution conductivity favors forming thinner fibers.[32] Also, solutions with high conductivity will cause a greater charge carrying capacity than solutions with low conductivity. Therefore, the fiber jet of conducive solutions will be subjected to a greater tensile force in the mean of an electric field than will a fiber jet from a solution with a low conductivity.[16] The minimum voltage for electrospinning to occur can also be cut back if the conductivity of the polymer solution is increased. Also, higher solution conductivity results in greater bending instability and produces a larger deposition area of collecting fibers. It has been reported that the size of the ions in the solution has an important impact on the electrospun fiber diameter besides the charges carried by the jet. Ions with a smaller atomic radius have a higher charge density. Thus, a higher mobility under an external electric field is present.[36] However, conductivity solution is unstable in the presence of strong electric fields, which results in a dramatic bending instability as well as a broad diameter distribution. Electrospun nanofibers with the smallest fiber diameter can be taken with the highest electrical conductivity and it has been found, there is a drop in the size of the fibers because of the increased electrical conductivity. The ions increase the charge-holding capacity of the jet with it subjecting it to higher tension. Thus, the fiber-forming ability of the gelatin is less compared to the synthetic ones. For example, Zong et al. have proved the effect of ions by adding ionic salt on the morphology and diameter of electrospun fibers.[37]

Stanger et al. found that an increase in charge density results in a reduction in the mass deposition rate and initial jet diameter during the electrospinning. In addition, a theory was proposed where they correlated reducing the curvature diameter of the Taylor cone with increasing charge density. Decreasing total electrostatic forces cause a smaller effective area. Similarly, other researchers have described the different behavior of the Taylor cone compared to the Taylor's observation of ionic liquids.[36]

1.2.6 SOLVENT VOLATILITY

Selecting a suitable solvent or solvent as the carrier of a particular polymer is fundamental for optimizing electrospinning. On the other hand, it is critical in determining the critical minimum solution concentration to allow the transition from electrospraying to electrospinning, by significantly affecting solution spinnability and the morphology of the electrospun fibers.[42] In addition, choice of solvent is too critical about whether the fibers are forming, as well as influencing fiber porosity. For enough solvent evaporation to happen between the capillary tip and the collector a volatile solvent must be used. As the fiber jet travels through the air toward the collector, a phase separation occurs before the solid polymer fibers deposited, a technique that is influenced by the volatility of the solvent. For volatile solvents, the region close to the fiber surface can be saturated with solvent in the vapor phase, which further limits penetrating nonsolvent. This can hinder skin formation leading to developing a porous surface morphology.[16]

1.2.7 FLUID CHARGING

In electrospinning, generation of charging within the fluid, usually occurs under contract with and flow across an electrode held at high (positive or negative) potential, referred to as induction charging. Depending on the nature of the fluid and polarity of the applied potential, free electrons, ions, or ion pairs may be produced as charge carriers in the fluid; the generation of charge carriers can be sensitive to solution impurities. Forming ions or ion pairs by induction, result in forming an electrical double layer. Without flow, the double layer thickness is found by the ion mobility in the fluid; in the presence of flow, ions may be convected away from the electrode and the double layer continually renewed. Charging of the fluid in electrospinning is typically field-limited, with the breakdown field strength in dry air being between flat plates.[43]

1.2.8 PERMITTIVITY

As with conductivity, the permittivity of a solvent has an important influence on the electrospinning technique and fiber morphology. However, not much discussion has been published around these effects. There is a method for deciding the permittivity of an electrospinning solution by measuring the complex resistance of a small cylindrical volume of the fluid. Bead formation and the diameter of the resultant electrospun fibers can be diminished by using a solution with a higher permittivity. With bending instability and the traversed jet in the path, electrospinning jet increase with higher permittivity, which results in a reduction in the diameter of the fiber and a larger fiber deposition area. Solvents such as dimethylformamide can be utilized to increase the permittivity of polymer solutions.[36]

1.2.9 ELECTRICAL VOLTAGE

One of the major parameters which affect the fiber diameter to a remarkable extent is the applied electric potential. In general, a higher applied voltage ejects more fluid in a jet, resulting in a larger fiber diameter.[36,44]

If the applied electric potential is higher, a greater amount of charge will cause the jet to speed up faster, and more solution will be drawn out from the tip of the needle. At a critical voltage, the Taylor cone is no longer seen. The jet imminent directly from the nozzle with increasing applied voltage. The resultant electrical field between the needle and the target increases as well, which contributes to greater stretching of the solution because of the larger columbic force between the surface charges. An increase in the applied voltage therefore leads to a lessening in the diameter of the electrospun nanofibers. The drier fibers can be generated if the voltage is increased because of the faster evaporation of the solvent that results. A lower voltage leads to a weaker electrical field, which reduces speeding up the jet and increases the flight time of the electrospinning jet, thus, producing thinner fibers. A voltage close to the minimum critical voltage needed for the onset of electrospinning might be efficient for getting thinner fibers. Higher voltages are related to a greater of bead formation, possibly because of the increased instability of the jet as the Taylor cone recedes into the syringe needle with the increased potential. The shape of the beads transforms from a spindle to a spherical shape with increased voltage, and sometimes the beads will get together to form thicker fibers because of the increased density of the beads on the other hand.[36,45,46]

Reneker and Chun have proved there is not much effect of electrical field on the diameter of electrospun PEO nanofibers. Several groups suggested that higher voltages simplified form large diameter fiber. For example, Zhang et al. explored the effect of voltage on morphologies and fiber diameter distribution with polyvinyl alcohol or a water solution as a model. Several groups suggested that higher voltages can increase the electrostatic repulsive force on the charged jet, favoring the narrowing of fiber diameter. For example, Yuan et al. analyzed the effect voltage on morphologies and fiber alignment with polysulfone, dimethylacetamide, or acetone as a model. Beside those phenomena, some groups also established that higher voltage offers the greatest chance of bead formation. Thus, we can find that voltage does the influence fiber diameter, but the meanings vary with the polymer solution concentration and on the distance between the tip and the collector.[22,32]

1.2.10 FLOW RATE

The flow rate decides solution available for electrospinning. Keeping a stable Taylor cone needs a minimum solution flow rate for a given voltage and electrode gap. On the other hands, at low flow rates, the Taylor cone recedes into the needle, and the jet originates from the liquid surface within the needle. In contrast, if the solution flow rate is greater than the electrospinning rate, it causes solution droplets to come from the needle tip because of lack of time for electrospinning the complete droplet to be an electrician. It has been viewed that the diameter of the fiber and the size of the bead both increase with an increased flow rate.[16,37] Also, at high flow rates, significant numbers of bead defects were noticeable, because of the inability of fibers to dry before progressing to the collector. Incomplete fiber drying also leads to forming ribbon-like (or flattened) fibers compared to fibers with a circular cross-section.[16] A summary of researchers' work is shown in Figure 1.5.[16,37]

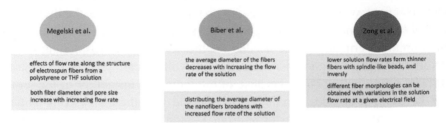

FIGURE 1.5   Summary of researchers' work on the flow rate parameter.

1.2.11   NEEDLE DIAMETER

The diameter of the needle has an effect on the electrospinning technique. A smaller needle diameter has been found to reduce clogging at the tip of the needle as well as the number of beads in the collected nanofibers because of the lower exposure of the solution to the atmosphere during electrospinning. In addition, using smaller diameter needles means the diameter of the electrospun nanofibers can also be smaller. The jet-flying time of the solution between the needle and the collector plate can also be increased if the needle diameter is cut because the surface tension of the droplet is increased and the jet acceleration decreased. Few studies worked in this field are summarized in Figure 1.6.[36]

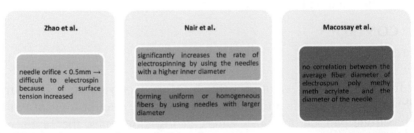

FIGURE 1.6   Summary of researchers' study.

1.2.12 *DISTANCE BETWEEN THE COLLECTOR AND THE TIP OF THE SYRINGE*

Varying the distance between the needle and the target causes a modification in the behavior of the electrospun jet and the morphology of the resultant nanofibers. Shortening the distance between the two electrodes causes an increase in the electrical field strength between the needle and the target and speeds up the electrospinning technique, therefore, reducing the time available for evaporation. It has been also reported spreading the diameter of the nanofibers becomes narrower when the space between the two electrodes is increased. Conversely, in other cases, it was considered as the average diameter of the fiber increases with increased distances because of the decreased strength of the electric field.[36] As a brief summary, it has been proven that the distance between the collector and the tip of the syringe can also affect the fiber diameter and morphology. In brief, if it is too short, the fiber will not have adequate time to solidify before reaching the collector, because dryness from the solvent is an important parameter on electrospun fiber. If the distance is too long, bead fiber can be obtained. It has been reported that flatter fibers can be brought out at closer distances. However, spinning distance is not important effect on fiber morphology.[16,37] Also, many researchers studied this parameter effects on fibers morphology and diameters as shown in Figure 1.7.[16]

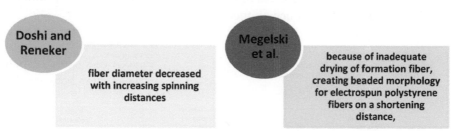

FIGURE 1.7   Summary of researchers' results on this parameter.

1.3   CONTROL ALIGNED FORMATION FIBERS

Aligned fibers have found importance in many engineering applications, such as tissue engineering, sensors, nanocomposites, filters, electronic devices. Some commonly used techniques to align the fibers are discussed in the following sections.[21] There are various ways to control aligning the depositing fiber. One way is to use different collector like a rotating wheel instead of foil sheet collector.[47] Recently, it was decided that the nature and type of the collector influences significantly the morphological and the physical characteristics of spun fibers. The density of the fibers per unit area of the collector and fiber arrangement is affected by the degree of charge dissipation on fiber deposition. The most commonly used targets are the conductive metal plate that results in collection of randomly oriented fibers in the nonwoven form. The use of metal and conductive collectors helped dissipate the charges and reduced the repulsion among the fibers. Therefore, the fibers collected are smooth and densely compacted. However, the fibers collected on the nonconductive collectors do not fritter away the charges which repel one another. In addition, the fibers can also be collected on specially designed collector to get aligned fibers or arrays of fibers.[21] In the next parts, we are discussing about the types of collectors.

1.3.1 TYPES OF COLLECTORS

In the former stages of electrospinning, researchers have used a needle and a flat collector plate as electrodes. However, with developing electrospinning technology, many electrode arrangements have been tried as a means of changing the electric field and getting wanted nanofiber morphologies. Collector electrode arrangements have all been ground out as ways, summarized in Figure 1.8, to produce aligned fibers.[36]

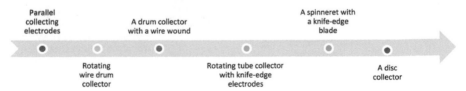

FIGURE 1.8   A summary of electrode arrangements.

One important of the electrospinning technique is the type of collector used. In electrospinning technique, a collector serves as a conductive substrate where the nanofibers are collected. Aluminum foil is used as a collector, but because of difficulty in transferring of collecting fibers and with the need for aligning fibers for various applications, other collectors are also common types of collectors today as summarized in Figure 1.9.[37]

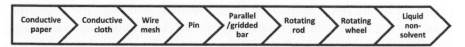

FIGURE 1.9   A summarized on other common types of collectors.

In addition, the fiber alignment is found out by the type of the target or collector and its rotation speed. The created nanofibers are deposited in the collector as a random mass because of the bending instability of the charged jet. Nowadays, several types of collectors, such as a rotating drum or a rotating wheel-like bobbin or metal frame, have been utilized for getting aligned nanofibers.[37] In the next section, we discussed briefly about these types of collectors.

1.3.2   FLAT COLLECTOR

During the electrospinning technique, collectors usually acted as the conductive substrate to collect the charged fibers. As a usual, a foil is used as a collector.[32,48] In addition, this is the most widely utilized methods of fiber collection. The collector can be either a solid metal, foil, or screen. Other materials can also be localized between the capillary and the collector.[47] However, it is difficult to transfer the collected nanofibers to other substrates for several applications. With the need of fibers transferring, diverse collectors have been developed including[32]:

- Wire mesh
- Pin or grids
- Parallel or gridded bar
- Rotating rods or wheel
- Liquid bath.

1.3.3   ROTATING DRUM COLLECTOR

This method normally used to collect aligned arrays of fibers. Also, the diameter of the fiber can be controlled and tailored based on the rotational speed of the drum. The cylindrical drum is rotating at high speeds (a few 1000 rpm) and of orienting the fibers circumferentially. Ideally, the linear rate of the rotating drum should match the evaporation rate of the solvent, such that the fibers deposited and held up on the surface of the drum. The alignment of the fibers is induced by the rotating drum and the degree of alignment improves with the rotational velocity. At rotational speeds slower than the fiber take-up speed, randomly oriented fibers are obtained along the drum. At higher speeds, a centrifugal force is prepared near the vicinity of the circumference of the rotating drum, which elongates the fibers before being collected on the drum. However, at much higher speed, the take-up velocity breaks the depositing fiber jet and continuous fibers are not taken (Fig. 1.10).[21,49,50]

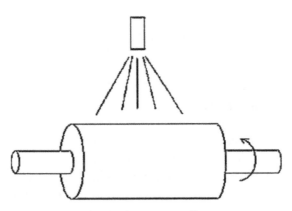

FIGURE 1.10 Rotating collector for electrospun nanofibers.

1.3.4 ROTATING DISK COLLECTOR

The rotating disk collector is a variation setup of the rotating drum collector and is practiced to obtain unaxially aligned fibers. The advantage of using a rotating disk collector over a drum collector is that most of the fibers deposited on the sharp-edged disk and are collected as aligned patterned nanofibers. The jet travels in a cone and inverse conical path with the utilization of the rotating disk collector as opposed to a conical path got when using a drum collector. During the first level, the jet follows the usual envelope cone path which is because of the instabilities influencing the jet. At a point above the disk, the diameter of the loop decreases as the conical form of the jet starts to shrink. This results in the inverted cone appearance, with the top of the cone resting on the disk. The electrical field applied concentrated on the tapered edge of the magnetic disc. Therefore, the charged polymer jet pulled toward the edge of the wheel, which explains the inverted conical form of the jet at the disk edge. The fibers that attracted to the edge of the disk are wound round the perimeter of the disk owing to the tangential force acting on the fibers produced from rotating the disk. This force further stretches the fibers and reduces their diameter. The quality of fiber alignment got using the disk is more beneficial than the rotating drum. However, only a small quantity of aligned fibers can be obtained, since there is just a small area at the tip of the disk.[21,51]

1.3.5   PARALLEL ELECTRODES

There has already been several groups produced well-aligned nanofibers by using two grounded parallel electrodes, such as aluminum strips with a 1-cm gap used in Yi Xin's group and the metal frame in Dersch's group et al. This apparatus used in this method is uncomplicated. The same as the rotating drum method, it operates by varying the collectors. The drawbacks of both these two methods are taught[47]:

- They can only produce aligned fiber in a small area.
- Fibers fabricated by this method cannot be conveniently transferred to different types of substrates.

The advantage of utilizing this technique lies in the simplicity of the set-up and the ease of collecting single fibers for mechanical testing. Good alignment has been contracting with this technique. The air gap between electrodes creates residual electrostatic repulsion between spun fibers, which helps align fibers. Two nonconductive strips of materials are placed along a straight line and an aluminum foil is placed on each of the strips and connected to the ground. This technique enables fibers to be deposited at the end of the strips so the fibers cling to the strips in an alternate fashion and collected as aligned arrays of fibers. A similar technique by Teo and Ramak-rishna used double-edge steel blades on a line to collect aligned arrays of fibers. The fibers were deposited at the gap between the electrodes, however, few fibers were found to deposit along the blades. It was solved by applying a negative voltage between the blades, resulting in the deposition of fibers between the blades.[21,52]

1.3.6   DUAL VERTICAL WIRE TECHNIQUE

The dual vertical wire technique used in Surawut's study is a variance of the parallel electrode technique, comprised two stainless steel wires, used as the secondary target, and the grounded aluminum foil, used as the primary target. The two stainless steel wires are mounted vertically in parallel to each other along a center line between the top of the needle and the grounded aluminum foil. Both the needle and the foil were tilted about 45° from a vertical baseline. They also tried to utilize the secondary electrodes alone, but much smaller amounts of aligned fibers were good. This was because most of the fibers would instead deposit randomly around the first wire electrode. Based on this observation, both the primary and secondary electrodes were essential for making good-aligned fibers for the present set-up. The mechanism for the depositing fibers to extend across the wire electrodes is similar to the parallel electrodes described previously. Both aligned fibers between the parallel vertical wires and a randomly aligned fiber mat on the aluminum foil could be reached.[47,53]

1.3.7   USING AN INSULATING TUBE ON THE COLLECTOR

Yang et al. produced a large area of oriented fibers by putting an insulating tube on the target for a long time. The changed electrical field made the jet bends around the pipe, so the oriented fibers could be received with a suitable tube. Based on different height and diameter of the pipe, there were three kinds of collection of aligned fibers formed[47,54]:

- Only a round mat within the tube area
- A round belt outside the tube and a round mat within the tube area
- Only a ring belt outside the tube area.

Still, since there is only one electrode as the target, the repelling force is not large enough on the tube to keep the jet falling around the tube all the time. In this font, the jet is not coaxial with the tube and results in disordered fibers.[47]

1.3.8   BIASED AC ELECTROSPINNING

Biased AC electrospinning is a new method used by Sarkar et al. It employs a combination of DC and AC potentials. The aim of this study is lessening the inherent instability of the fiber itself, compared to all techniques relying on lessening the fiber instability by using external forces on the fibers during electrospinning. By introducing a DC biased AC potential instead of either a pure AC or DC potential, alternating positively and negatively charged regions in the fiber result in a reduction of electrostatic repulsion and increase in fiber stability and stability can improve electrospun fiber quality.[47]

1.3.9   MAGNETIC ELECTROSPINNING

In this method, the polymer solution is magnetized by adding a few magnetic nanoparticles. The magnetic field stretches the fibers across the gap to make a parallel array as they land on the magnets. When the fibers fall down, the parts of the fibers close to the magnets are attracted to the surface of the magnets, finally the fibers land on the two magnets and suspend over the gap. This method is fundamentally different from all previously discussed methods in preparing aligned fibers, because the driving force is the magnetic field in it, while electrostatic interaction plays the function as driving force in the other methods. In addition, this method has several advantages[47,55]:

• The magnetic field can be manipulated accurately.
• The resultant nanofibers can be transferred on to any substrate with full retention.
• The area of the aligned fibers is large compared to other techniques.

1.3.10   A CYLINDER COLLECTOR WITH HIGH ROTATING SPEED

It has been suggested that by rotating a cylinder collector at a high-speed up to thousands of rpm, electrospun nanofibers could be oriented circumferentially. Researchers from Virginia Commonwealth University have utilized this technique summarized in Figure 1.11.[44,56]

FIGURE 1.11   The speed of this parameter for aligned fibers.

When a linear velocity of rotating cylinder surface matches that of evaporated jet depositions, fibers are taken up on the surface of the cylinder tightly
circumferentially, resulting in a fair alignment. Such a speed can be predicted
as an alignment speed. If the surface velocity of the cylinder is slower than
the alignment speed, randomly deposited fibers will be collected, as it is the
fast chaos motions of jets control the final deposition manner. On the other
hand, there must be a limit rotating speed above which continuous fibers
cannot be collected from the over fast take-up speed will stop the fiber jet.
The reasons a perfect alignment is difficult to achieve can be applied to the
fact that the chaotic motions of polymer jets are not probable to be consistent
and are less controllable.[44]

1.3.11   A THIN WHEEL WITH SHARP EDGE

A significant advancement in collecting aligned electrospun nanofibers has
been recently constructed. The tip-like edge substantially concentrates the
electrical field so the as-spun nanofibers are almost all attracted to and can
be continuously wound on the bobbin edge of the revolving wheel. It was
explained that before getting to the electrically grounded target, the nanofibers keep enough residual charges to repel each other. As a result, once a
nanofiber is attached to the wheel tip, it will exert a repulsive force on the
next fiber attracted to the tip. This repulsion is from one another results in a
detachment between the deposited nanofibers. The variation in the separation distances is because of varying repulsive forces related to nanofiber
diameters and residual charges.[44]

1.3.12   A FRAME COLLECTOR

To make an individual nanofiber for experimental characterizations, we
recently developed another approach to fiber alignment by simply placing
a rectangular frame under the spinning jet. In addition, different frame
materials result in different fiber alignments (i.e., aluminum frame favors
better fiber alignments than a wooden form). More investigation is under-
going to understand the alignment characteristics in varying the configuration and size of frame rods, the distance between the frame rods, and the
inclination angle of a single frame. These will be useful in deciding how
many positions would be best suitable making a polygonal multiframe
structure.[44]

1.4 THE APPLICATIONS OF NANOFIBERS IN VARIOUS SCIENCE RESEARCH

The fine electrospun nanofibers make them useful in a wide range of innovative applications.[57,58] Also new applications have been explored for these fibers continuously. Main application fields are shown in Figure 1.12[41,59,60]:

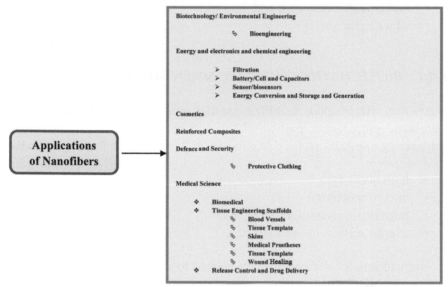

FIGURE 1.12   Potential applications of electrospun fibers.

For selected applications, it is desirable to control not only the fiber diameter but also the internal morphology. Porous fibers are of interest for applications such as filtration or prepare nanotubes by fiber templates.[22,31,33,61] Besides, small pore size and high surface area inherent in nanofiber has implications in biomedical applications such as scaffoldings for tissue growth.[22,42] Also, researchers have spun a fiber from a compound naturally present in the blood. This nanofiber can be used as forms of medical applications such as bandages or sutures that ultimately dissolve into the body. This nanofiber minimizes infection rate, blood loss, and is also taken up by the body.[62] One of the most significant applications of nanofibers is to be used as reinforcements in composite developments. With these reinforcements, the composite materials can offer superior properties such as high modulus and strength to weight ratios, which cannot be achieved by other engineered monolithic materials alone. Information on the fabrication and structure–property relationship characterization of such nanocomposites is believed to be utilitarian. Such continuous carbon nanofiber composite also has possible applications as filters for[44]:

- separation of small particles from gas or liquid;
- supports for high-temperature catalysts;
- heat-management materials in aircraft and semiconductor devices;
- rechargeable batteries; and
- super capacitors.

1.4.1 BIOTECHNOLOGY/ENVIRONMENTAL ENGINEERING

Nonwoven electrospun nanofiber meshes are an excellent material for membrane preparation, particularly in biotechnology and environmental engineering applications for this reason[8]:

- high porosity;
- interconnectivity;
- microscale interstitial space; and
- a large surface-to-volume ratio.

Biomacromolecules or cells can be tied to the nanofiber membrane for these applications[8,63]:

- in protein purification and waste-water treatment (affinity membranes);
- enzymatic catalysis or synthesis (membrane bioreactors); and
- chemical analysis and diagnostics (biosensors).

Electrospun nanofibers can form an effective size-exclusion membrane for particulate removal from wastewater. Affinity membranes are a broad class of membranes that selectively captures specific target molecules by immobilizing a specific capturing agent onto the membrane surface. In biotechnology, affinity membranes have applications in protein purification and toxin removal from bioproducts. In the environmental industry, affinity membranes have applications in organic waste removal and heavy metal removal in water treatment. To be used as affinity membranes, electrospun nanofibers must be surface functionalized with ligands. Mostly, the ligand molecules should be covalently attached to the membrane to prevent leaching of the ligands. Also, water pollution is now becoming a critical global issue. One important class of inorganic pollutant of great physiological significance is heavy metals, for example, Hg, Pb, Cu, and Cd. Distributing these metals in the environment is mainly applied to the release of metal containing waste-water from industries. For example, copper smelters may release high quantities of Cd, one of the most mobile and toxic among the trace elements, into nearby waterways. It is impossible to eliminate some classes of environmental contaminants, such as metals, by conventional water purification methods. Affinity membranes will play a critical role in wastewater treatment to remove (or recycle) heavy metal ions in the future. Polymer nanofibers functioned with a ceramic nanomaterial, mentioned below, could be suitable materials for fabrication of affinity membranes for water industry applications[8,64]:

- hydrated alumina hydroxide;
- alumina hydroxide;
- iron oxides.

The polymer nanofiber membrane acts as a bearer of the reactive nanomaterial that can attract toxic heavy metal ions, such As, Cr, and Pb, by adsorption or chemisorption and electrostatic attraction mechanisms. Again, affinity membranes provide an alternative access for removing organic molecules from wastewater.[8]

1.4.2   BIOENGINEERING

From biological viewpoint, almost entirely the human tissues and organs are deposited in nanofibrous forms or structures. Some examples include the following[44]:

- bone
- dentin
- collagen
- cartilage
- skin

All of them are characterized by well-organized fibrous structures realigning in nanometer scale. Current research in electrospun polymer nanofibers has focused one of their major applications on bioengineering. We can easily find their promising potential in various biomedical fields.[44]

1.4.2.1 MEDICAL SCIENCE

Nanofibers are used in medical applications, which include, drug and gene delivery, artificial blood vessels, artificial organs, and medical face masks. For example, carbon-fiber hollow nanotubes, smaller than blood cells, have the potential to transport drugs into blood cells.[62]

1.4.2.1.1 Biomedical Application

Biomedical field is one of the important application areas among others, using the technique of electrospinning like[17,40]:

- filtration material;
- protective material;
- electrical applications;
- optical applications;
- sensors; and
- nanofiber-reinforced composites;

Current medical practice is based almost on treatment regimes. However, it is envisaged that medicine in the future will be based heavily on early detection and prevention before disease expression. With nanotechnology, new treatment will emerge that will significantly reduce medical costs. With recent developments in electrospinning, both synthetic and natural polymers can be produced as nanofibers with diameters ranging from decades to hundreds of nanometers with controlled morphology. The potential of these electrospun nanofibers in human healthcare applications is promising, for example[8]:

- in tissue or organ repair and regeneration;
- as vectors to deliver drugs and therapeutics;
- as biocompatible and biodegradable medical implant devices;
- in medical diagnostics and instrumentation;
- as protective fabrics against environmental and infectious agents in hospitals and general surroundings; and
- in cosmetic and dental applications.

Tissue or organ repair and positive feedback are new avenues for potential treatment, avoiding the need for donor tissues and organs in transplantation and reconstructive surgery. In this advance, a scaffold is usually needed that can be fabricated from either natural or synthetic polymers by many techniques including electrospinning and phase separation. An animal model is utilized to study the biocompatibility of the scaffold in a biological system before the scaffold is introduced into patients for tissue-regeneration applications. Nanofibers scaffolds are suited to tissue engineering. These can be made up and shaped to fill anatomical defects. Its architecture can be designed to supply the mechanical properties necessary to support cell growth, growth differentiation, and motility. Also, it can be organized to provide growth factors, drugs, therapeutics, and genes to stimulate tissue regeneration. An inherent property of nanofibers is that they mimic extracellular matrix of tissues and organs. The ECM is a complex composite of fibrous proteins such as collagen and fibronectin, glycoproteins, proteoglycans, soluble proteins such as growth factors, and other bioactive molecules that support cell adhesion and growth. One of the aims is to create electrospun polymer nanofiber scaffolds for engineering blood vessels, nerves, skin, and bone. In the pharmaceutical and cosmetic industry, nanofibers are promising tools for controlled these aims[8,40,62]:

- delivery of drugs;
- therapeutics;
- molecular medicines; and
- body-care supplements.

1.4.2.1.2   Tissue Engineering Scaffolds

Successful tissue engineering needs synthetic scaffolds to bear similar chemical compositions, morphological, and surface functional groups to their natural counterparts. Natural scaffolds for tissue growth are three-dimensional networks of nanometer-sized fibers made of several proteins. Nonwoven membranes of electrospun nanofibers are well-known for their interconnected, 3D porous structures and large surface areas, which provide a class of ideal materials to mimic the natural ECM needed for tissue engineering. The electrospun nanofibrous support was treated with the cell solution and the nanofiber cell was cultured in a rotating bioreactor to create the cartilage which controlled compressive strength similar to natural cartilage. The tissue-engineered cartilages could be applied in treating cartilage-degenerative diseases. The scaffold was applied as biomimic ECM, enzyme, gene, and medicine to revive skin, cartilage, blood vessel, and nerve. The scaffold was helpful in biocompatibility, mechanical property, porosity, degradability in the human physical structure. The electrospun nanofibers showed moderate porosity, excellent mechanical property, and biocompatibility, which could be utilized to repair blood vessels, skin, and nervous tissue.[1,65] Tissue engineering is an emerging interdisciplinary and multidisciplinary research study. It involves the utilization of living cells, manipulated through their extracellular environment or genetically to develop biological substitutes for implantation into the body or to foster remodeling of tissues in some active manners. The purpose of tissue engineering is to renovate, replace, say, or improve the function of a particular tissue or organ. For a functional scaffold, a few basic needs have to be satisfied (Fig. 1.13):

- A scaffold should control a high degree of porosity, with a suitable pore size distribution.
- A large surface area is needed.
- Biodegradability is often needed, with the degradation rate matching the rate of neotissue formation.
- The scaffold must control the needed structural integrity to prevent the pores of the scaffold from collapsing during neotissue formation, with the suitable mechanical properties.
- The scaffold should be nontoxic to cells and biocompatible, positively interacting with the cells to promote cell adhesion, growth, migration, and distinguished cell function.

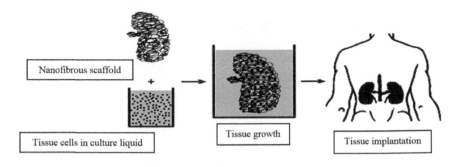

FIGURE 1.13 Principle of tissue engineering.

Among all biomedical materials under evaluation, electrospun nanofibrous scaffolds have presented great performances in cell attachment, increase, and penetration.[66] One of the most promising potential applications is tissue scaffolding. The nonwoven electrospun mat has a high surface area and a high porosity. It contains an empty space between the fibers that is approximately the size of cells. The mechanical property, the topographical layout, and the surface chemistry in the nonwoven mat may have a direct effect on cell growth and migration.[47] Ultrafine fibers of biodegradable polymers produced by electrospinning have found potential applications in tissue engineering because of their high surface area to volume ratios and high porosity of the fibers. However, the flexibility of seeding stem cells and human cells on the fibers makes electrospun materials most suited for tissue engineering applications. The fibers produced can be used systematically to design the structures that they perform not only mimic the properties of ECM, but also control high strength and high toughness. For instance, nonwoven fabrics show isotropic properties and support neotissue formation. These mats resemble the ECM matrix and can be applied as a skin-scaffold and wound-dressing materials, where the materials are needed to be more elastic than stiff. Many natural polymers like collagen, starch, chitin, and chitosan, and synthetic biodegradable polymers like PCL, PLA, PLGA have been widely investigated for potential applications in developing tissue scaffolds. These results confirm that electrospinning of natural or synthetic polymers for tissue engineering applications are promising.[21] Tissue engineering is one of the most exciting interdisciplinary and multidisciplinary research fields today, and there has been exponential growth in the number of research publications in this area in recent years. It involves the utilization of living cells, manipulated through their extracellular environment or genetically to develop biological substitutes for implantation into the body or to foster remodeling of tissues in some active manners. The purpose is to repair, replace, maintain, or increase the use of a particular tissue or organ. The core technologies intrinsic to this effort can be organized into three fields[6,67]:

- cell technology;
- scaffold frame technology; and
- technologies for in vivo integration.

The scaffold frame technology focuses on these objectives[68]:

- designing;
- manufacturing;
- characterizing three-dimensional scaffolds for cell seeding; and
- in vitro or in vivo culturing.

1.4.2.1.3   Blood Vessels

Blood vessels vary in sizes, mechanical and biochemical properties, cellar content, and ultrastructural organization, depending on their location and specific role. It is needed that the vascular grafts engineered should have wanted characteristics. Blood vessel replacement, a fine blood vessel (diameter < 6 mm), has stayed a great challenge. Because the electrospun nanofiber mats can give good support during the initial development of vascular smooth muscle cells, smooth film combining with electrospun nanofiber mat could form a good 3D scaffold for blood vessel tissue engineering.[6,66]

1.4.2.1.4   Muscles

Collagen nanofibers were first applied to assess the feasibility of culturing smooth muscle cell. The cell growth on the collagen nanofibers was promoted and the cells were easily integrated into the nanofiber network after 7 days of seeding. Smooth muscle cells also adhered and proliferated well on another polymer nanofiber mats blended with collagen, incorporating collagen into nanofibers was observed to improve fiber elasticity and tensile strength, and increase the cell adhesion. The fiber-surface wet ability influences cell attachment. The alignment of nanofibers can induce cell orientation and promote skeletal muscle cell morphologenes is and aligned formation.[6]

1.4.2.1.5   Medical Prostheses

Polymer nanofibers fabricated by electrospinning have been offered for several soft tissue prosthesis applications such as blood vessel, vascular, breast, etc. In addition, electrospun biocompatible polymer nanofibers can also be deposited as a slender, porous film onto a hard tissue prosthetic device designed to be implanted into the human body. This coating film with a fibrous structure works as an interface between the prosthetic device and the host tissues. It is anticipated to reduce efficiently the stiffness mismatch at the tissue or interphase and from here prevents the device failure after the implantation.[44,69]

1.4.2.1.6   Tissue Template

For treating tissues or organs in malfunction in a human body, one of the challenges in the area of tissue engineering or biomaterials is the design of ideal scaffolds or synthetic matrices. They can mimic the structure and biological functions of the natural ECM. Human cells can attach and organize well around fibers with diameters smaller than those of the cellular phones. Nanoscale fibrous scaffolds can provide an ideal template for cells to seed, migrate, and produce. A successful regeneration of biological tissues and organs calls for developing fibrous structures with fiber architectures useful for cell deposition and cell growth. Of particular interest in tissue engineering is creating reproducible and biocompatible three-dimensional scaffolds for cell growth resulting in biometrics composites for various tissue repair and replacement processes. Recently, people have begun to pay attention to making such scaffolds with synthetic polymers or biodegradable polymer nanofibers. It is believed that converting biopolymers into fibers and networks that mimic native structures will eventually improve the usefulness of these materials as large diameter fibers do not mimic the morphological characteristics of the native fibrils.[44,62]

1.4.2.1.7 Wound Healing

Wound healing is a native technique of regenerating dermal and epidermal tissues. When an individual is wounded, a set of complex biochemical actions take place in a closely orchestrated cascade to repair the harm. These events can be sorted into four groups:

- inflammatory;
- proliferative;
- remodeling phases; and
- epithelialization.

Ordinarily, the body cannot heal a deep dermal injury. In full thickness burn or deep ulcers, there is no origin of cells remaining for regeneration, except from the wound edges. Dressings for the wound healing role to protect the wound, exude extra body fluids from the wound area, decontaminate the exogenous microorganism, improve the appearance, and sometimes speed up the healing technique. For these functions, a wound-dressing material should provide a physical barrier to a wound, but be permeable to moisture and oxygen. For a full thickness dermal injury, when an "artificial dermal layer" adhesion and integration consisting of a 3D tissue scaffold with well-cultured dermal fibroblasts will aid the epithelialization. As shown in Figure 1.14, nanofiber membrane is a good wound-dressing candidate because of its unique properties like[6]:

- the porous membrane structure and
- well-interconnected pores.

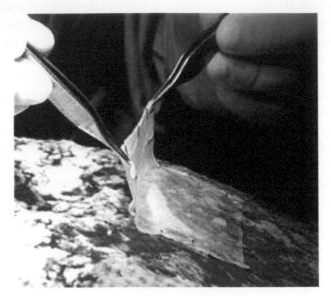

FIGURE 1.14 Nanofiber mats for medical dressing.

They are important for exuding fluid from the wound. The small pores and high specific surface area not only inhibit the exogenous microorganism invasions but also assist the control of fluid drainage. In addition, the electrospinning provides a simple path to add drugs into the nanofibers for any possible medical treatment and antibacterial purposes.[6]

For wound healing, an ideal dressing should have certain features:

- hemostatic ability;
- efficiency as bacterial barrier;
- absorption ability of excess exudates (wound fluid);
- suitable water vapor transmission rate;
- enough gaseous exchange ability;
- ability to conform to the contour of the wound area;
- functional adhesion;
- painless to patient;
- ease of removal; and
- low cost.

Current efforts using nanofibrous membranes as a medical dressing are still in its early childhood, but electrospun materials meet most of the needs outlined for wound-healing polymer. Because their microfibrous and nanofibrous provide the nonwoven textile with desirable properties,[37] polymer nanofibers can also be utilized for the treatment of wounds or burns of a human skin, as well as designed for hemostatic devices with some unique characteristics. Fine fibers of biodegradable polymers can spray/spun on to the injured location of the skin to make a fibrous mat dressing. They let wounds heal by encouraging forming a normal skin development and remove form scar tissue, which would occur in a traditional treatment. Nonwoven nanofibrous membrane mats for wound dressing usually have pore sizes ranging from 500 nm to 1 mm, small enough to protect the wound from bacterial penetration by aerosol particle capturing mechanisms. High surface area of 5–100 m^2/g is efficient for fluid absorption and dermal delivery.[44,62] The electrospun nanofibers have been utilized in treating wounds or burns of human skin because of their high porosity which allows gas exchange and a fibrous structure that protects wounds from infection and dehydration. Nonwoven electrospun nanofibrous membranes for wound dressing usually have pore sizes in the range of 500–1000 mm which is low enough to protect the wound from bacterial penetration. High surface area of electrospun nanofibers is efficient for fluid absorption and dermal delivery. Chong invented a composite containing a semipermeable barrier and a scaffold filter layer of skin cells in wound healing by electrospinning.[1] Electrospinning could create scaffold with more homogeneity besides meeting other needs like oxygen permeation and protection of wound from infection and dehydration for use as a wound-dressing materials. Many other synthetic and natural polymers, like carboxyethyl, chitosan, or PVA; collagen or chitosan; silk fibroin, have been electrician to advise them for wound-dressing applications.[17]

1.4.2.1.8 Release Control

Controlled release is an effective technique of delivering drugs in medical therapy. It can balance the following features:

- delivery kinetics;
- minimize the toxicity;
- side effects; and
- improve patient convenience.

In a controlled release system, the active substance is loaded into a carrier or device first, and then releases at a predictable rate in vivo when governed by an injected or noninjected route. As a potential drug delivery carrier, electrospun nanofibers have showed many advantages. The drug loading is easy to implement by electrospinning technique, and the high applied voltage used in the electrospinning technique had little influence on the drug activity. The high specific surface area and short diffusion passage length give the nanofiber drug system higher overall release rate than the bulk material (e.g., film). The release profile can be finely controlled by modulating of nanofiber morphology, porosity, and composition. Nanofibers for drug release systems mainly come from biodegradable polymers, such as PLA, PCL, PDLA, PLLA, PLGA, and hydrophilic polymers such as PVA, PEG, and PEO. Nonbiodegradable polymers, such as PEU, were likewise found out.[6] Nanofiber systems for the release of drugs are needed to fill diverse roles. The mattress should be capable to protect the compound from decomposition and should allow for controlled release in the targeted tissue, over a needed period of time at a constant release rate.[70] Drug release and tissue engineering are closely related regions. Sometimes release of therapeutic causes can increase the efficiency of tissue engineering. Various nanostructured materials are applicable in tissue engineering. Electrospun fiber mats provide the advantage of increased drug release compared to roll-films because of the increased surface area.[17,22]

1.4.2.1.9 *Drug Delivery and Pharmaceutical Composition*

Delivery of drug or pharmaceuticals to patients in the most physiologically acceptable manner has always been an important concern in medicine. In general, the smaller the dimensions of the drug and the coating material wanted to encapsulate the drug, the better the drug to be assimilated by human being. Drug delivery with polymer nanofibers is based on the rule that the dissolution rate of a particulate drug increases with increasing surface area of both the drug and the similar carrier if needed. As the drug and carrier materials can be mixed for electrospinning of nanofibers, the likely modes of the drug in the resulting nanostructure products are as follows:

- Drug as particles attached to the surface of the carrier which is in the form of nanofibers.
- Both drug and carrier are nanofiber form; therefore, the product will be the two kinds of nanofibers interlaced together.

- The blend of drug and carrier materials integrated into one fiber containing both sections.
- The carrier material is electrospun into a tubular frame in which the drug particles are encapsulated.

However, as the drug delivery in the form of nanofibers is still in the early stage exploration, a real delivery mode after production and efficiency has yet to be determined in the future.[42] Drug delivery with electrospun nanofibers is based along the principle that drug-releasing rate increases with increasing surface area of both the drug and the similar carrier used. The increased surface area of drug improved the bioavailability of the poor water-soluble drug. Various drugs such as avandia, eprosartan, carvedilol, hydrochloridethiazide, aspirin, naproxen, nifedipine, indomethacin, and ketoprofen were entrapped into PVP to form pharmaceutical compositions which provided controllable releasing. Not only synthetic polymers, but also natural polymers can be applied for modeling drug delivery system.[7] Controlled drug release over a definite period of time is possible with biocompatible delivery matrices of polymers and biodegradable polymers. They are mostly used as drug delivery systems to deliver therapeutic agents because they can be well designed for programed distribution in a controlled fashion. Nanofiber mats applied as drug carriers in drug delivery system because of their high functional characteristics. The drug delivery system relies on the rule that the dissolution rate of a particulate drug increases with increasing surface area of both the drug and the similar carrier. Importantly, the large surface area associated with nanospun fabrics allows for quick and efficient solvent evaporation, which provides the incorporated drug limited time to recrystallize which favors forming amorphous dispersions or solid solutions. Depending on the polymer carrier used, the release of pharmaceutical dosage can be designed as rapid, immediate, delayed, or varied dissolution. Many researchers successfully encapsulate drugs within electrospun fibers by mixing the drugs in the polymer solution to be electrospun. Various solutions containing low molecular weight drugs have been electrospun, including lipophilic drugs such as ibuprofen, cefazolin, rifampin, paclitaxel, and itraconazole and hydrophilic drugs such as mefoxin and tetracycline hydrochloride. However, they have encapsulated proteins in electrospun polymer fibers. Besides the normal electrospinning process, another path to develop drug-loaded polymer nanofibers for controlling drug release is to use coaxial electrospinning and research has successfully encapsulated two kinds of medicinal pure drugs through this process.[37] Electrospinning affords great flexibility in selecting materials for drug delivery applications. Either biodegradable or nondegradable materials can be utilized to control whether drug release occurs by diffusion alone or diffusion and scaffold degradation. Also, because of the flexibility in material selection, many drugs can be delivered including:

- antibiotics;
- anticancer drugs;
- proteins; and
- DNA

Using the various electrospinning techniques, many different drug-loading methods can also be applied:

- coatings;
- embedded drug; and
- encapsulated drug (coaxial and emulsion electrospinning).

However, as the drug delivery in the form of nanofibers is still in the early stage exploration, a real delivery mode after production and efficiency has yet to be found in the future.[16]

1.4.3 COSMETICS

The current skin masks applied as topical creams, lotions, or ointments. They may include dusts or liquid sprays and more likely than fibrous materials to migrate into sensitive areas of the body, such as the nose and eyes where the skin mask is being utilized to the face. Electrospun-polymer nanofibers have been tried as a cosmetic skin care mask for treating skin healing, skin cleaning, or other therapeutic or medical properties with or without various additives. This nanofibrous skin mask with small interstices and high surface area can make easy far greater utilization and speed up the rate of transfer of the additives to the skin for the fullest potential of the additive. The cosmetic skin mask from the electrospun nanofibers can be applied gently and painlessly as well as directly to the three-dimensional topography of the skin to provide healing or cure treatment to the skin.[42] Electrospun nanofibers have been aimed for use in cosmetic cares such as treating skin healing and skin cleaning with or without various additives in recent years. Despite the growth in the number of electrospun polymer nanofiber publications in the recent years, there is a rare work, including scientific papers and patents, in the cosmetic field about the use of electrospun nanofibers. Developing nanofibers in this field have been focused on skin treatment applications, such as care mask, skin healing, and skin cleaning, with active agents (cosmetics) with controlled release from time to time. The cosmetic application be included in the biomedicine application, which admits the drug delivery system employing the active agents used in cosmetics, body care supplements. Therefore, the cosmetic and drug delivery are closely interrelated areas. The electrospun nanofibers provide the advantage of the increasing drug release when compared to cast-films because of the increased surface area. Besides, it's agreeable to processing different polymers such as natural, synthetic, and blends, according to their solubility or melting point. It is significant that although most of the research in polymeric nanofiber by electrospinning consider the technique simple, cost-effective, and easily scalable from laboratory to commercial production, only a limited number of companies have commercially performed electrospun fibers.[7]

1.4.4   ENERGY AND ELECTRONICS AND CHEMICAL ENGINEERING

The demand for energy use of goods and services has been increasing every year throughout the world. However, it was reported the estimated reserve amounts of petroleum and natural gas in the world are only 41 and 67 years, respectively. To solve this problem, new, clean, renewable, and sustainable energies have to be ground and used to replace the current nonsustainable energies. Wind generator, solar power generator, hydrogen battery, and polymer battery are among the most popular alternatives to produce new energies. In recent years, electrospun nanofibers have presented their potential in these applications (Fig. 1.15):

- super capacitors;
- lithium cells;
- fuel cells;
- solar cells; and
- transistors

Further, electrospun nanofibers with electrical and electro-optical have also got much interest recently because of their potential applications in creating nanoscale electronic and optoelectronic devices.[7]

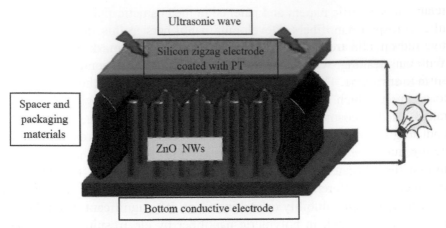

FIGURE 1.15   ZnO nanofibers in energy application.

1.4.5   FILTRATION APPLICATION

Filtration is necessary in many engineering fields. It was estimated that future filtration market would be up to US 700 billion US dollars by the year 2020.[1,62]

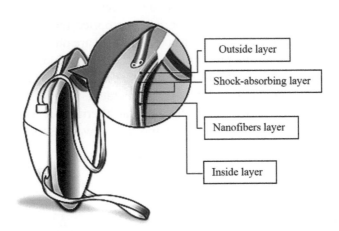

FIGURE 1.16   Applications of nanofibers in filtration.

Applications of nanofibers in filtration were shown in Figure 1.16. Fibrous materials used for filter media provide advantages of high filtration efficiency and low air resistance. Filtration efficiency, which is closely related with the fiber fineness, is one of the most important concerns for the filter performance. One direct way of developing high efficient and effective filter media are by using nanometer sized fibers in the filter structure. With outstanding of polymeric nanofibers properties, such as high specific surface area, high porosity, and excellent surface adhesion, they are suited to be made into filtering media for filtering out particles in the submicron range. Also, this filter system could be used for processing waste-water containing active sludge.[1,21] Electrospun fibers are being widely studied for aerosol filtration, air cleaning applications in industry and for particle collection in clean rooms. The advantage of using electrospun fibers in the filtration media is the fiber diameters can be easily controlled and can produce an impact in high-efficiency particulate air filtrations.[21] The filtration efficiency is commonly influenced by these parameters[6]:

- The filter physical structure like
- fiber fineness;
- matrix structure;
- thickness;
- pore size;
- fiber surface electronic properties;
- its surface chemical characteristic; and
- surface free energy.

The particle collecting capability is also associated with the size range of particles being collected. Besides the filtration efficiency, other properties such as pressure drop and flux resistance are also important factors to be assessed for a filter media.[6] Filter efficiency increases linearly with the decrease of thickness of filter membrane and applied pressure increase.[37]

1.4.6   REINFORCED COMPOSITES/REINFORCEMENT

Electrospun fiber reinforced composites have significant potential for development of high intensity or high toughness materials and materials with good thermal and electrical conductivity. Figure 1.17 is an example of these composites. Few studies have found out the use of electrospun fibers in composites. Traditional reinforcements in polymer matrices can create stress concentration sites because of their irregular shapes and cracks spread by burning through the fillers or traveling up, down, and around the particles. However, electrospun fibers have various advantages over

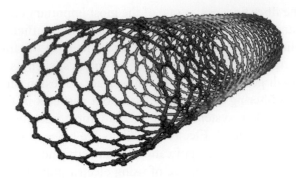

FIGURE 1.17   Carbon nanotubes composite.

traditional fillers. The reinforcing effects of fibers are influenced essentially by fiber size. Smaller size fibers give more efficient support. Fibers with finer diameters have a preferential orientation of polymer chains along the fiber axis. The orientation of macromolecules in the fibers improves with decreasing in diameter, making finer diameter fibers strong. Therefore, the use of nanometer sized fibers can significantly raise the mechanical integrity of polymer matrix compared to micron-sized fibers. However, the high percentage of porosity and irregular pores between fibers can contribute to an interpenetrated structure when spread in the matrix, which also improves the mechanical strength because of the interlocking mechanism. These characteristic features of nanofibers enable the transfer of applied stress to the fiber–matrix in a more serious fashion than most of the commonly used filler materials. Current issues related to the use of electrospun nanofibers as reinforcement materials are the control of dispersion and orientation of the fibers in the polymer matrix. To achieve better reinforcement, electrospun nanofibers may require to be collected as an aligned yarn instead of a randomly distributed felt so the post-electrospinning stretching process could be applied to further improve the mechanical properties. Further, if crack growth is transverse to the fiber orientation, the crack toughness of the composite can be optimized. So, the interfacial adhesion between fibers and matrix material needs to be controlled such the fibers are deflecting the cracks by fiber–matrix interface debonding and fiber pullout. The interfacial adhesion should not be excessively strong or too weak. Ideal control can only be obtained by careful selective fiber-surface treatment. Spreading electrospun mats in the matrix can be improved by cutting down the fibers to shorter fragments. This can be accomplished, if the electrospun fibers are collected as aligned bundles (instead of nonwoven network), which can then be optically or mechanically trimmed to get fiber fragments of several 100

nm in length.[21] Early studies on electrospun nanofibers also included reinforcement of polymers. As electrospun nanofiber mats have a large specific
surface area and an irregular pore structure, mechanical interlocking among
the nanofibers should occur.[6] One of the most significant applications of
traditional fibers, especially engineering fibers such as carbon, glass, and
Kevlar fibers, is to be used as reinforcements in composite developments.
With these reinforcements, the composite materials can provide superior
structural properties such as high modulus and strength to weight ratios,
which cannot be attained by other engineered monolithic materials alone.
Nanofibers will also eventually find important applications in making nanocomposites. This is because nanofibers can have even better mechanical
properties than microfibers of the same materials, and therefore the superior
structural properties of nanocomposites can be anticipated. However, nanofiber reinforced composites may control some extra merits which cannot
be shared by traditional (microfiber) composites. For instance, if there is
a difference in refractive indices between fiber and ground substance, the
resulting composite becomes opaque or nontransparent because of light
scattering. This limit, however, can be avoided when the fiber diameters
become significantly smaller than the wavelength of visible illumination.[62]

1.4.7 DEFENSE AND SECURITY

Military, firefighter, law enforcement, and medical personal need high-level
protection in many environments ranging from combat to urban, agricultural,
and industrial, when dealing with chemical and biological threats like[8,71]:

- nerve agents;
- mustard gas;
- blood agents such as cyanides; and
- biological toxins such as bacterial spores, viruses, and rickettsiae.

Nanostructures with their minuscule size, large surface area, and light
weight will improve, by orders of magnitude, our capability to[71]:

- detect chemical and biological warfare agents with sensitivity and
  selectivity;
- protect through filtration and destructive decomposition of harmful
  toxins; and
- provide site-specific naturally prophylaxis.

Polymer nanofibers are considered as excellent membrane materials owing to their lightweight, high surface area, and breathable (porous) nature. The high sensitivity of nanofibers toward warfare agents makes them excellent candidates as sensing of chemical and biological toxins in concentration levels of parts per billion (Figs. 1.18 and 1.19). Governments across the globe are investing in strengthening the protection levels offered to soldiers in the battlefield. Various methods of varying nanofiber surfaces to improve their capture and decontamination capacity of warfare agents are under investigation. Nanofiber membranes may be employed to replace the activated charcoal in adsorbing toxins from the atmosphere. Active reagents can be planted in the nanofiber membrane by chemical functionalization, postspinning variation, or through using nanoparticle polymer composites. There are many avenues for future research in nanofibers from the defense perspective. As well as serving protection and decontamination roles, nanofiber membranes will also suffer to provide the durability, washability, resistance to intrusion of all liquids, and tear strength needed of battledress fabrics.[8]

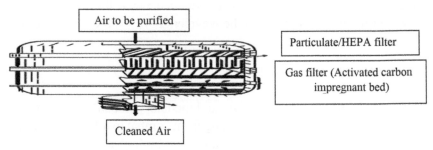

FIGURE 1.18 Cross-sectioning of a facemask canister used for protection from chemical and biological warfare agents.

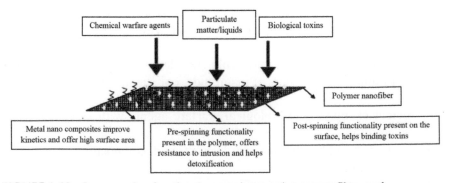

FIGURE 1.19 Incorporating functional groups into a polymer nanofiber mesh.

1.4.8 PROTECTIVE CLOTHING

The protective clothing in the military is largely expected to help to increase the suitability, sustainability, and combat effectiveness of the individual soldier system against extreme climate, ballistics. In peace ages, breathing apparatus and protective clothing with the particular role of against chemical warfare agents such as sarin, soman, tabun become a special concern for combatants in conflicts and civilian populations in terrorist attacks. Current protective clothing containing charcoal absorbents has its terminal points for water permeability, extra-weight imposed to the article of clothing. A lightweight and breathable fabric, which is permeable to both air and water vapor, insoluble in all solvents and reactive with nerve gases and other deadly chemical agents, is worthy. Because of their large surface area, nanofiber fabrics are neutralizing chemical agents and without impedance of the air and water vapor permeability to the clothing. Some examples of protective clothing applications were shown in Figure 1.20. Electrospinning results in nanofibers lay down in a layer that has high porosity but small pore size, offering good resistance to penetrating chemical harms agents in aerosol form. Preliminary investigations indicate that compared to conventional

FIGURE 1.20 Protective clothing applications.

textiles the electrospun nanofibers present both small impedance to moisture vapor diffusion and efficiency in trapping aerosol particles, as well as show strong promises as ideal protective clothing. Conductive nanofibers are expected to be utilized in fabricating tiny electronic or machines such as Schottky junctions, sensors, and actuators. Conduct (of electrical, ionic, and photoelectric) membranes also have potential for applications including electrostatic dissipation, corrosion protection, electromagnetic interference shielding, photovoltaic device, etc.[42]

Electrospun nanofibers can play an important part in textile applications as protective clothing and other functional fabric materials. The electrospun nanofibrous membranes are capable of neutralizing chemical agents without impedance of the air and water vapor permeability to the clothing because of their high specific surface area and high porosity but small pore size. Preliminary investigations suggest the electrospun nanofibers control both minimal impedance to moisture vapor diffusion and efficiency in trapping aerosol particles compared with conventional textiles. Smith prepared a fabric comprising electrospun PEI nanofibers as lightweight protective clothing which was captured and neutralizing chemical warfare agents. This formed fabric also could be used in protective breathing apparatuses because PEI provides multiple amine sites for the nucleophilic decomposition of mustard gases and fluorophosphates nerve gases. A protective mask was constructed by attaching PC/PS electrospun nanofibrous layer to one side of a moist fabric composed of cellulose and wool. The diameter of the nanofibers in the protective layer was in the range of 100–10,000 nm. A nonwoven fabric composed of a submicrosized fiber, which receives a PC shell and a polyurethane core, was made by co-axial electrospinning. The resultant fabric combines the filtration efficiency of the PC and the mechanical effectiveness of polyurethane, which is useful in exposure suits and aviation clothing. A water-resistant and air-permeable laminated fabric was manufactured by utilizing hot-melt polyester as dots onto the surface of the electrospun nylon nonwoven fabric.[7,72] Ideally, protective clothing should have close to essential properties, such as lightweight, breathable fabric, air and water vapor permeability, insoluble in all solvents, and improved toxic chemical resistance. Electrospun nanofiber membranes recognized as potential candidates for protective clothing applications for these causes:

- They're lightweight
- Large surface area
- High porosity (breathable nature)
- Great filtration efficiency

- Resistant to penetration of harmful chemical agents in aerosol form
- Their ability to neutralize the chemical agents without impedance of the air
- Water vapor permeability to the clothing.

Various methods for variation of nanofiber surfaces have been examined to improve protection against toxins. One protection method that has been used includes chemical surface variation and attachment of reactive groups such as axioms, cyclodextrins, and chloramines that bind and detoxify warfare agents.[37]

1.5 CONCLUDING REMARKS

Producing nanofibers by electrospinning is a simple and widely utilized for varied applications. In this chapter, different parameters, which are important in electrospinning process, were studied at first. Then, some commonly used techniques to produce aligned nanofibers were discussed. Finally, various applications of electrospun nanofibers by some samples were reviewed.

KEYWORDS

- **applications of electrospun nanofibers**
- **process parameters**
- **controlling process**
- **morphology**
- **porosity**
- **diameter**

REFERENCES

1. Lu, P.; Ding, B. Applications of Electrospun Fibers. *Rec. Pat. Nanotechnol.* **2008,** *2* (3), 169–182.
2. Reneker, D. H.; Yarin, A. L.; Zussman, E.; Xu, H. Electrospinning of Nanofibers from Polymer Solutions and Melts. *Adv. Appl. Mech.* **2007,** *41*, 43–346.
3. Vonch, J.; Yarin, A.; Megaridis, C. M. Electrospinning: A Study in the Formation of Nanofibers. *J. Undergrad. Res.* **2007,** *1*, 1–6.

4. Nouailhat, A. *An Introduction to Nanosciences and Nanotechnology*. John Wiley & Sons: Hoboken, NJ, 2010; Vol. 10, p 206.

5. Sawhney, A. P. S.; Condon, B.; Singh, K. V.; Pang, S. S.; Li, G.; Hui, D. Modern Applications of Nanotechnology in Textiles. *Text. Res. J.* **2008,** *78* (8), 731–739.

6. Fang, J.; Niu, H. T.; Lin, T.; Wang, X. G. Applications of Electrospun Nanofibers. *Chin. Sci. Bull.* **2008,** *53* (15), 2265–2286.

7. Rafiei, S.; Maghsoodloo, S.; Noroozi, B.; Mottaghitalab, V.; Haghi, A. K. Mathematical Modeling in Electrospinning Process of Nanofibers: A Detailed Review. *Cellulose Chem. Technol.* **2013,** 47 (5–6), 323–338.

8. Ramakrishna, S.; Fujihara, K.; Teo, W. E.; Yong, T.; Ma, Z.; Ramaseshan, R. Electrospun Nanofibers: Solving Global Issues. *Mater. Today* **2006,** *9* (3), 40–50.

9. Hasanzadeh, M.; Mottaghitalab, V.; Ansari, R.; Moghadam, B. H.; Haghi, A. K. Issues in Production of Carbon Nanotubes and Related Nanocomposites: A Comprehensive Review. *Cellulose Chem. Technol.* **2015,** 49 (3–4), 237–257.

10. Brown, P. J.; Stevens, K. *Nanofibers and Nanotechnology in Textiles*; CRC Press: Boca Raton, FL, 2007; p 544.

11. Brooks, H.; Tucker, N. Electrospinning Predictions Using Artificial Neural Networks. *Polymer* **2015,** 58, 22–29.

12. Wang, H. S.; Fu, G. D.; Li, X. S. Functional Polymeric Nanofibers from Electrospinning. *Rec. Pat. Nanotechnol.* **2009,** *3* (1), 21–31.

13. Wang, C.; Cheng, Y. W.; Hsu, C. H.; Chien, H. S.; Tsou, S. Y. How to Manipulate the Electrospinning Jet with Controlled Properties to Obtain Uniform Fibers with the Smallest Diameter?—A Brief Discussion of Solution Electrospinning Process. *J. Polym. Res.* **2011,** *18* (1), 111–123.

14. Lukáš, D.; Sarkar, A.; Martinová, L.; Vodsed'álková, K.; Lubasova, D.; Chaloupek, J.; Pokorný, P.; Mikeš, P.; Chvojka, J.; Komarek, M. Physical Principles of Electrospinning (Electrospinning as a Nano-scale Technology of the Twenty-first Century). *Text. Progr.* **2009,** *41* (2), 59–140.

15. Reneker, D. H.; Yarin, A. L. Electrospinning Jets and Polymer Nanofibers. *Polymer* **2008,** *49* (10), 2387–2425.

16. Sill, T. J.; Recum, H. A. Electrospinning: Applications in Drug Delivery and Tissue Engineering. *Biomaterials* **2008,** *29* (13), 1989–2006.

17. Agarwal, S.; Wendorff, J. H.; Greiner, A. Use of Electrospinning Technique for Biomedical Applications. *Polymer* **2008,** *49* (26), 5603–5621.

18. Tao, J.; Shivkumar, S.; Molecular Weight Dependent Structural Regimes during The Electrospinning of PVA. *Mater. Lett.* **2007,** *61* (11), 2325–2328.

19. Zeng, Y.; Pei, Z.; Wang, X.; Chen, S. Numerical Simulation of Whipping Process in Electrospinning. In *WSEAS International Conference. Proceedings. Mathematics and Computers in Science and Engineering*. World Scientific and Engineering Academy and Society, 2009.

20. Kowalewski, T. A.; NSKI, S. B. Ł. O.; Barral, S. Experiments and Modelling of Electrospinning Process. *Tech. Sci.* **2005,** *53* (4), 1123–1134.

21. Baji, A.; Mai, Y. W.; Wong, S. C.; Abtahi, M.; Chen, P. Electrospinning of Polymer Nanofibers: Effects on Oriented Morphology, Structures and Tensile Properties. *Compos. Sci. Technol.* **2010,** *70* (5), 703–718.

22. Zong, X.; Kim, K.; Fang, D.; Ran, S.; Hsiao, B. S.; Chu, B. Structure and Process Relationship of Electrospun Bioabsorbable Nanofiber Membranes. *Polymer* **2002,** *43* (16), 4403–4412.

23. Haghi, A. K. Electrospun Nanofiber Process Control. *Cellulose Chem. Technol.* **2010,** *44* (9), 343–352.
24. Zanin, M. H. A.; Cerize, N. N. P.; de Oliveira, A. M. Production of Nanofibers by Electrospinning Technology: Overview and Application in Cosmetics. *Nanocosmetics and Nanomedicines*; Springer: Berlin-Heidelberg, 2011; p. 311–332.
25. Pham, Q. P.; Sharma, U.; Mikos, A. G. Electrospinning of Polymeric Nanofibers for Tissue Engineering Applications: A Review. *Tissue Eng.* **2006,** *12* (5), 1197–1211.
26. Ghochaghi, N. Experimental Development of Advanced Air Filtration Media Based on Electrospun Polymer Fibers. *Mechanical and Nuclear Engineering.* Virginia Commonwealth: Richmond, VA, 2014; pp 1–165.
27. Ziabari, M.; Mottaghitalab, V.; Haghi, A. K. Evaluation of Electrospun Nanofiber Pore Structure Parameters. *Kor. J. Chem. Eng.* **2008,** *25* (4), 923–932.
28. Sawicka, K. M.; Gouma, P. Electrospun Composite Nanofibers for Functional Applications. *J. Nanopart. Res.* **2006,** *8* (6), 769–781.
29. Li, W. J.; Laurencin, C. T.; Caterson, E. J.; Tuan, R. S.; Ko, F. K. Electrospun Nanofibrous Structure: A novel Scaffold for Tissue Engineering. *J. Biomed. Mater. Res.* **2002,** *60* (4), 613–621.
30. Sill, T. J.; Recum, H. A. Electrospinning: Applications in Drug Delivery and Tissue Engineering. *Biomaterials* **2008,** *29* (13), 1989–2006.
31. Yousefzadeh, M.; Latifi, M.; Amani. T, M.; Teo, W. E.; Ramakrishna, S. A Note on The 3D Structural Design of Electrospun Nanofibers. *J. Eng. Fabrics Fibers* **2012,** *7* (2), 17–23.
32. Li, Z.; Wang, C. Effects of Working Parameters on Electrospinning. *One-Dimensional Nanostructures*; Springer: Berlin-Heidelberg, 2013; pp 15–28.
33. Bognitzki, M.; Czado, W.; Frese, T.; Schaper, A.; Hellwig, M.; Steinhart, M.; Greiner, A.; Wendorff, J. Nanostructured Fibers via Electrospinning. *Adv. Mater.* **2001,** *13* (1), 70–72.
34. De Vrieze, S.; Van Camp, T.; Nelvig, A.; Hagström, B.; Westbroek, P.; De Clerck, K. The Effect of Temperature and Humidity on Electrospinning. *J. Mater. Sci.* **2009,** *44* (5), 1357–1362.
35. Yarin, A. L. Koombhongse, S.; Reneker, D. H. Bending Instability in Electrospinning of Nanofibers. *J. Appl. Phys.* **2001,** *89* (5), 3018–3026.
36. Huang, Z. M.; Zhang, Y. Z.; Kotaki, M.; Ramakrishna, S. A review on Polymer Nanofibers by Electrospinning and their applications in Nanocomposites. *Compos. Sci. Technol.* **2003,** *63*, 2223–2253.
37. Angammana, C. J. A Study of the Effects of Solution and Process Parameters on the Electrospinning Process and Nanofibre Morphology. 2011, University of Waterloo.
38. Bhardwaj, N.; Kundu, S. C. Electrospinning: A Fascinating Fiber Fabrication Technique. *Biotechnol. Adv.* **2010,** *28* (3), 325–347.
39. Rafiei, S.; Maghsoodloo, S.; Saberi, M.; Lotfi, S.; Motaghitalab, V.; Noroozi, B.; Haghi, A. K. New Horizons in Modeling and Simulation of Electrospun Nanofibers: A Detailed Review. *Cellulose Chem. Technol.* **2014,** *48* (5–6), 401–424.
40. Tan, S. H.; Inai, R.; Kotaki, M.; Ramakrishna, S. Systematic Parameter Study for Ultra-Fine Fiber Fabrication via Electrospinning Process. *Polymer* **2005,** *46* (16), 6128–6134.
41. Frenot, A.; Chronakis, I. S. Polymer Nanofibers Assembled by Electrospinning. Curr. Opin. Colloid Interface Sci. **2003,** *8* (1), 64–75.
42. Deitzel, J. M.; Kleinmeyer, J.; Harris, D. E. A.; Tan, N. C. B. The Effect of Processing Variables on the Morphology of Electrospun Nanofibers and Textiles. *Polymer* **2001,** *42* (1), 261–272.

43. Luo, C. J.; Nangrejo, M.; Edirisinghe, M. A Novel Method of Selecting Solvents for Polymer Electrospinning. *Polymer* **2010,** *51* (7), 1654–1662.
44. Rutledge, G. C.; Fridrikh, S. V. Formation of Fibers by Electrospinning. *Adv. Drug Deliv. Rev.* **2007,** *59* (14), 1384–1391.
45. Thompson, C. J.; Chase, G. G.; Yarin, A. L.; Reneker, D. H. Effects of Parameters on Nanofiber Diameter Determined from Electrospinning Model. *Polymer* **2007,** *48* (23), 6913–6922.
46. Reneker, D. H. Chun, I. Nanometre Diameter Fibres of Polymer, Produced by Electrospinning. *Nanotechnology* **1996,** *7* (3), 216–223.
47. Zhang, S. *Mechanical and Physical Properties of Electrospun Nanofibers,* 2009; pp 1–83.
48. Vaquette, C.; C. White, J. J. Increasing Electrospun Scaffold Pore Size with Tailored Collectors for Improved Cell Penetration. *Acta Biomater.* **2011,** 7 (6), 2544–2557.
49. Katta, P.; Alessandro, M.; Ramsier, R. D.; Chase, G. G. Continuous Electrospinning of Aligned Polymer Nanofibers onto a Wire Drum Collector. *Nano Lett.* **2004,** *4* (11), 2215–2218.
50. Park, H. S.; Park, Y. O. Filtration Properties of Electrospun Ultrafine Fiber Webs. *Kor. J. Chem. Eng.* **2005,** *22* (1), 165–172.
51. 51. Dalton, P. D.; Klee, D.; Möller, M. Electrospinning with Dual Collection Rings. *Polymer* **2005,** *46* (3), 611–614.
52. Kim, G. H. Electrospinning Process using Field-Controllable Electrodes. *J. Polym. Sci., B: Polym. Phys.* **2006,** *44* (10), 1426–1433.
53. Chuangchote, S.; Supaphol, P. Fabrication of Aligned Poly(Vinyl Alcohol) Nanofibers by Electrospinning. *J. Nanosci. Nanotechnol.* **2006,** *6* (1), 125–129.
54. Zheng, Y. F.; Gong, R. H.; Zeng, Y. Multijet Motion and Deviation in Electrospinning. *RSC Adv.* **2015,** *5* (60), 48533–48540.
55. Yang, D.; Lu, B.; Zhao, Y.; Jiang, X. Fabrication of Aligned Fibrous Arrays by Magnetic Electrospinning. *Adv. Mater.* **2007,** 19 (21), 3702–3706.
56. Persano, L.; Camposeo, A.; Tekmen, C.; Pisignano, D. Industrial Upscaling of Electrospinning and Applications of Polymer Nanofibers: A Review. *Macromol. Mater. Eng.* **2013,** *298* (5), 504–520.
57. Keun, S, W.; Ho, Y, J.; Seung, L, T.; Park, W. H. Effect of PH on Electrospinning of Poly(Vinyl Alcohol). *Mater. Lett.* **2005,** *59* (12), 1571–1575.
58. Zhou, H. *Electrospun Fibers from Both Solution and Melt: Processing, Structure and Property,* Cornell University: Ithaca, NY, 2007.
59. Feng, J. J. The Stretching of an Electrified Non-Newtonian Jet: A Model for Electrospinning. *Physics of Fluids (1994–present)* **2002,** *14* (11), 3912–3926.
60. Maleki, M.; Latifi, M.; Amani, T. M. Optimizing Electrospinning Parameters for Finest Diameter of Nano Fibers. *World Acad. Sci., Eng. Technol.* **2010,** *40,* 389–392.
61. Thompson, C. J. *An Analysis of Variable Effects on a Theoretical Model of the Electrospin Process for Making Nanofibers.* University of Akron: Akron, 2006.
62. Patan, A. K.; Sasikanth, K.; Sreekanth, N.; Suresh, P.; Brahmaiah, B. Nanofibers—A New Trend in Nano Drug Delivery Systems. *Int. J. Pharm. Res. Anal.* **2013,** *3,* 47–55.
63. Venugopal, J.; Ramakrishna, S. Applications of Polymer Nanofibers in Biomedicine and Biotechnology. *Appl. Biochem. Biotechnol.* **2005,** *125* (3), 147–157.
64. Thavasi, V.; Singh, G.; Ramakrishna, S. Electrospun Nanofibers in Energy and Environmental Applications. *Energy Environ. Sci.* **2008,** *1* (2), 205–221.

65. Hasan, A.; Memic, A.; Annabi, N.; Hossain, M. D.; Paul, A.; Dokmeci, M. R.; Dehghani, F.; Khademhosseini, A. Electrospun Scaffolds for Tissue Engineering of Vascular Grafts. *Acta Biomater.* **2014,** *10* (1), 11–25.

66. Fang, J.; Wang, X.; Lin, T. Functional Applications of Electrospun Nanofibers. *Nanofibers—Production, Properties and Functional Applications*; InTech: Rijeka, 2011; pp 287–326.

67. Loh, Q. L.; Choong, C. Three-Dimensional Scaffolds for Tissue Engineering Applications: Role of Porosity and Pore Size. *Tissue Eng., B: Rev.* **2013,** *19* (6), 485–502.

68. Bölgen, N.; Vaseashta. A. Nanofibers for Tissue Engineering and Regenerative Medicine. In *Third International Conference on Nanotechnologies and Biomedical Engineering.* Springer: Berlin-Heidelberg, 2016.

69. Zhang, Y.; Lim, C. T.; Ramakrishna, S.; Huang, Z. M. Recent Development of Polymer Nanofibers for Biomedical and Biotechnological Applications. *J. Mater. Sci.: Mater. Med.* **2005,** *16* (10), 933–946.

70. Garg, K.; Bowlin, G. L. Electrospinning Jets and Nanofibrous Structures. *Biomicrofluidics* **2011,** *5* (1), 013403-1–013403-19.

71. Tan, S.; Huang, X.; Wu, B. Some Fascinating Phenomena in Electrospinning Processes and Applications of Electrospun Nanofibers. *Polym. Int.* **2007,** *56* (11), 1330–1339.

72. Lee, S. C.; Obendorf, S. K. Use of Electrospun Nanofiber Web for Protective Textile Materials as Barriers to Liquid Penetration. *Text. Res. J.* **2007,** *77* (9), 696–702.

UPDATES ON ELECTROSPINNING PROCESS MODELS (PART I)

SHIMA MAGHSOODLOU* and S. PORESKANDAR

Textile Engineering, University of Guilan, Rasht, Iran

Corresponding author. E-mail: sh.maghsoodlou@gmail.com

CONTENTS

ABSTRACT

Understanding of electrospinning flows behavior can be complicated because of the variation of different forces. Theoretical and numerical comprehension of these phenomena can assist in overcoming the existing restrictions throughout the process. For achieving higher mass production and nanofibers orientation for special application, it is necessary to comprehend and control the dynamic and mechanic behavior of the electrospinning jet. Modeling, simulations, and design of experiments will offer a better understanding of the electrospinning jet mechanics and dynamics. It would be sufficient to describe some of these methods. A review of these methods will be discussed in this chapter.

2.1　INTRODUCTION

Systematical understanding of flows behavior in processes such as electrospinning can be extremely difficult because of the variation of different forces that may be involved, including capillarity, viscosity, inertia, gravity, as well as the additional stresses resulting from the extensional deformation of the microstructure within the fluid. Theoretical and numerical comprehension of these phenomena can assist in overcoming the existing restrictions throughout the process. For example, as regards this approach, in spite of individual applications of the electrospinning process, its mass production is still a challenge. For achieving higher mass production and nanofibers orientation for special application, like tissue engineering and microelectronics, it is necessary to comprehend and control the dynamic and mechanic behavior of the electrospinning jet. These numerical approaches have been investigated widely by means of modeling and subsequent simulation.[1] The modeling, simulation, data mining, genetic algorithm, design of experiment, and Taguchi design are the best methods for achieving this purpose. Overviews of these methods will be investigated in this chapter.

2.2　DESIGN OF EXPERIMENT

When the aim is well defined, the problem should be analyzed with the help of the following questions:

What is known? What is unknown? What do we need to investigate?

Experimental design and optimization are tools that are used to systematically examine different types of problems that arise within, for example, research, development, and production. It is obvious that if experiments are performed randomly, the result obtained will also be random. Therefore, it is a necessity to plan the experiments in such a way that the interesting information will be obtained.[2]

Two important questions in the final designs are:

Which experimental variables can be investigated? Which responses can be measured?

Thus, when the experimental variables and the responses have been defined the experiments can be planned and performed in such a way that a maximum of information is gained from a minimum of experiments. To simplify the communication, a few different terms are introduced and defined (Table 2.1).[3]

TABLE 2.1   Important Terms in Experimental Design.

Experimental domain	The experimental area defined by the variation of the experimental variables
Factors	Experimental variables that can be changed independently of each other
Independent variables	Same as factors
Continuous variables	Independent variables that can be changed continuously
Discrete variables	Independent variables that are changed step-wise, e.g., type of solvent
Responses	The measured value of the results from experiments
Residual	The difference between the calculated and the experimental result

In any experimental procedure, several experimental variables or factors may influence the result. A screening experiment is performed to determine the experimental variables and interactions that have significant influence on the result, measured in one or several responses. When the variables to be investigated are selected, it is also decided which variables that should *not* be investigated. These variables have to be kept at a fixed level in all experiments included in the experimental design. However,

remember that it is always more economical to include a few extra variables in the first screening, than adding one variable later. Think about *how*
the different variables should be defined. It is sometimes possible to lower
the number of experiments needed, to achieve the important information, just by redefining the original variables. When all aspects have been
penetrated, and variables, responses as well as experimental domain are
selected, then it is time for the next step in the planning procedure. When
a list of variables to be investigated has been completed, an experimental
design is chosen to estimate the influence of the different variables on the
result. Different designs such as full factorial or fractional factorial design,
Taguchi design, and mixture design can be used.[4] Taguchi designs were
investigated deeply with some examples of different designs and different
in this chapter.

2.3   TAGUCHI DESIGN OF EXPERIMENT

The most significant challenge in the electrospinning process is to achieve
uniform nanofibers consistently and reproducibly.[5-8] Controlling the physical
characteristics of electrospun nanofibers such as fiber diameter depend on
various parameters, which are principally divided into three categories: solution properties (solution viscosity, solution concentration, polymer molecular weight, and surface tension), processing conditions (applied voltage,
volume flow rate, spinning distance, and needle diameter), and ambient
conditions (temperature, humidity, and atmospheric pressure).[1,5,9-11] As
mentioned before, the aim of the experimental design is to provide reasonable and scientific answers to such questions. In other words, experimental
design comprises sequential steps to ensure efficient data gathering process
and leading to valid statistical inferences.[12] Thus, Taguchi experimental
design is a simple technique which can optimize process parameters by less
number of experiments.[2]

The Taguchi method was introduced to reduce the period and cost of
product development. In the Taguchi method, it is possible to control the
alternations made by uncontrollable parameters, which are not considered
in the classical DoE design. It works by converting the amount of target
features to a signal-to-noise (S/N) ratio for measuring the performance of
the level of controlling parameters in contradiction to these parameters.
The S/N ratio is the favorite signal ratio for the nonfavorite random noise
and displays the quality of the experimental values. Three dissimilar functions were utilized as the following target based on the S/N ratio: "larger

is better," "nominal is best," and "smaller is better." In addition, analysis of variance (ANOVA) was utilized to conclude the statistical significance of the electrospinning factors. By using the ANOVA and S/N ratios, the optimum condition of electrospinning can be achieved. Finally, the confirmation test of experiments was performed utilizing the optimum electrospinning conditions, which were determined by the Taguchi optimization method and, by this means, validation of the optimization was tested. Also, optimization of the process factors has been done accurately. The performance characteristic function that the smaller is better was utilized for obtaining the optimal process factors for the final aim. The S/N ratio η formulate is expressed as[13,14]

$$\frac{S}{N(\eta)} = -10 \cdot \log\left(\frac{1}{n}\sum_{i=1}^{n} Y_i^2\right) \tag{2.1}$$

where Y_i is the achieved value in the experimental test and n is the number of tests. Equation 2.1 is used to calculate the S/N ratios of selected factors and their levels for final aim. The significance level of the variables for final aim was determined using the certain confidence level of the ANOVA. Software Minitab 14 was utilized to optimize the process according to the Taguchi approach. Minitab is strong software that is acknowledged to accurately solve numerous statistical issues and improve quality in the areas of engineering, statistics and mathematics.[13,14]

The experimental design using Taguchi method involves arranging an orthogonal array to organize the parameters affecting the process and the levels they should be varied. It determines the factors affecting the product quality with the least number of experiments, thus saving time and resources.

The most significant, influential factors (i.e., solution concentration which changes viscosity and surface tension, spinning distance which determined the enough space for elongation and charge accumulation on the nanofiber, applied voltage which determines the elongation force and the strength of electric field, and volume flow rate, which changes of the mass transfer rate), were chosen for creating different polymers nanofibers with no beads in this research. Each of these factors is considered to be varied at three or four levels for better observation (summarized in Tables 2.2–2.4) for investigating similar influencing of these parameters in all experimental conditions. The final design experiments were summarized in Table 2.5–2.7.

TABLE 2.2 L_9 Orthogonal Array for Selected Factors and Levels for PVA Nanofibers.

Factors	Symbol	Level		
		1	2	3
Concentration (%)	A	8	10	12
Voltage (V)	B	15	20	25
Flow rate (mL/h)	C	0.2	0.3	0.4
Spinning distance (cm)	D	10	15	20

TABLE 2.3 L_{18} Orthogonal Array for Selected Factors and Levels for PAN Nanofibers.

Factors	Symbol	Level					
		1	2	3	4	5	6
Concentration (%)	A	8	10	12	14	16	18
Voltage (V)	B	10	14	18	–	–	–
Flow rate (mL/h)	C	0.5	0.75	1	–	–	–
Spinning distance (cm)	D	10	13	15	–	–	–

TABLE 2.4 L_{18} orthogonal array for selected factors and levels for PAN/CNT nanofibers.

Factors	Symbol	Level					
		1	2	3	4	5	6
Concentration (%)	A	8	10	12	14	16	18
Voltage (V)	B	10	14	18	–	–	–
Flow rate (mL/h)	C	0.5	0.75	1	–	–	–
Spinning distance (cm)	D	10	13	15	–	–	–

TABLE 2.5 Summary of Final Design Experiment with L_9 Orthogonal Array for Selected Factors and their Corresponding Average Nanofiber Diameter for PVA.

Experiment number	Factor				Average fiber diameter (nm)
	Concentration (%)	Spinning distance (cm)	Applied voltage (V)	Flow rate (mL/h)	
1	8	10	15,000	0.2	241
2	8	15	20,000	0.3	206
3	8	20	25,000	0.4	231
4	10	20	15,000	0.3	232
5	10	10	20,000	0.4	299
6	10	15	25,000	0.2	249
7	12	15	15,000	0.4	315
8	12	20	20,000	0.2	297
9	12	10	25,000	0.3	283

TABLE 2.6 Summary of Final Design Experiment with L_{18} Orthogonal Array for Selected Factors and their Corresponding Average Nanofiber Diameter for PAN.

Experiment number	Factor				Average fiber diameter (nm)
	Concentration (%)	Spinning distance (cm)	Applied voltage (V)	Flow rate (mL/h)	
1	8	10	10,000	0.5	299
2	8	13	14,000	0.75	274
3	8	15	18,000	1	292
4	10	13	10,000	0.5	431
5	10	15	14,000	0.75	341
6	10	10	18,000	1	369
7	12	10	10,000	0.75	655
8	12	13	14,000	1	676
9	12	15	18,000	0.5	682
10	14	15	10,000	1	1205
11	14	10	14,000	0.5	1042
12	14	13	18,000	0.75	787
13	16	15	10,000	0.75	892
14	16	10	14,000	1	825
15	16	13	18,000	0.5	682
16	18	13	10,000	1	921
17	18	15	14,000	0.5	1143
18	18	10	18,000	0.75	1458

TABLE 2.7 Summary of Final Design Experiment with L_{18} Orthogonal Array for Selected Factors and their Corresponding Average Nanofiber Diameter for PAN/CNT.

Experiment number	Factor				Average fiber diameter (nm)
	Concentration (%)	Spinning distance (cm)	Applied voltage (V)	Flow rate (mL/h)	
1	8	10	10,000	0.5	169
2	8	13	14,000	0.75	154
3	8	15	18,000	1	223
4	10	13	10,000	0.5	408
5	10	15	14,000	0.75	381
6	10	10	18,000	1	371
7	12	10	10,000	0.75	652

TABLE 2.7   *(Continued)*

Experiment number	Factor				Average fiber diameter (nm)
	Concentration (%)	Spinning distance (cm)	Applied voltage (V)	Flow rate (mL/h)	
8	12	13	14,000	1	580
9	12	15	18,000	0.5	548
10	14	15	10,000	1	848
11	14	10	14,000	0.5	723
12	14	13	18,000	0.75	789
13	16	15	10,000	0.75	1062
14	16	10	14,000	1	981
15	16	13	18,000	0.5	1191
16	18	13	10,000	1	1747
17	18	15	14,000	0.5	1650
18	18	10	18,000	0.75	1877

2.4   SIMULATION OF THE PROCESS

Simulation is transition from a mathematical or computational model to the description of the system behavior based on sets of input parameters. It is often the only means for accurately predicting the performance of the modeled system. The investigation of simulation techniques is fairly a new area and various research in different fields are talking about it. Process simulation is a model-based representation that can be used for the design, development, analysis, and optimization of technical processes. Knowing about chemical and physical properties is a basic requisite to have an appropriate mathematical model (Fig. 2.1).[15,16]

System simulation is a powerful tool. Using simulation is generally cheaper and safer than conducting experiments with a prototype of the final product. In addition, in experimental situations when the possibility of error is high, simulation can be even more realistic than experiments, as they allow the free configuration of environmental and operational parameters and is able to be run faster than in real time.[17] Simulation yield appropriate information about how something will act without actual testing in real. The real value of systems representation, and more particularly systems simulation, is that it provides a succinct and concise encapsulation of knowledge. It is an active, usable, and testable symbolic representation of how a process works. Researchers can explicate and exemplify their understanding and test it without the risk of running a real prototype. In a situation with different alternatives analysis, simulation can improve the efficiency, particularly when the necessary data for initializing can easily be obtained from operational data. Finally, applying simulation adds decision support systems to the toolbox of traditional decision support systems.[18]

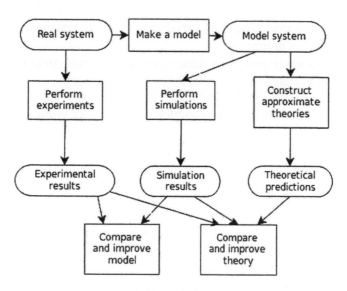

FIGURE 2.1   Simulation process schematic.

2.5   SYSTEMS AND SYSTEM ENVIRONMENT

To model a system, it is necessary to understand the concept of a system and the system constraints. A system is defined as a group of objects that are joined together in some interaction or interdependence toward the accomplishment of some purpose. A system is often affected by changes occurring outside the system. Such changes are said to occur in the system environment. In modeling systems, it is necessary to decide on the boundary between the system and its environment.[19]

2.5.1   COMPONENTS OF A SYSTEM

To understand and analyze a system, a number of terms should be defined[19]:

- *Entity*: This is an object of interest in the system. An attribute is a property of an entity.
- *Activity*: Represents a time period of specified length.
- State of a system: A collection of variables necessary to describe a system at a particular time, relative to the objectives of a study.

Systems are categorized to be one of two types, discrete and continuous.

- *Discrete system*: The state variables change instantaneously at separated points in time.
- *Continuous system*: The state variables change continuously with respect to time.

Few systems in practice are completely discrete or completely continuous, but since one type of change predominates for most systems, it will usually be possible to classify a system as being either discrete or continuous.

2.5.2 MODEL OF A SYSTEM

A model is defined as a representation of a system for the purpose of studying the system. In practice, what is meant by "the system" depends on the objectives of a particular study. For most studies, it is not necessary to consider all the details of a system; thus, a model is not only a substitute for a system, it is also a simplification of the system. However, there should be sufficient detail in the model to permit valid conclusions to be drawn about the real system.[18]

Different models of the same system may be required as the purpose of investigation can change.

Just as the components of a system are entities, attributes, and activities, models are represented similarly. However, the model contains only those components that are considered to be relevant to the study. Simulation concepts can be described as[19]

System: A collection of entities that interact together over time to accomplish one or more goals.

Model: An abstract representation of a system, usually containing logical and/or mathematical relationships, which describe a system in terms of state, entities, and their attributes, sets, events, activities, and delays.

System state: A collection of variables that contain all the information necessary to describe the system at any time.

Entity: Any component in the system which requires explicit representation in the model and that can change the state of the system.

Item: Any component in the system which requires explicit representation in the model and that cannot change the state of the system.

Attributes: The properties of a given entity or item.

Event: An instantaneous occurrence that changes the state of a system.

Activity: Duration of time of specified length, which length is known when it begins (although it may be defined in terms of a statistical distribution).

Action: An action is a series of changes to the state; every individual change is called an action.

Transform: An action that changes the attributes, but not the number of components (entities or items).

Split/Join: An action that splits a number of new components, or a number of components, are joined together to one new component.

Process: A series of state changes within a component during a particular time-span.

Process: The system is modeled as a combination of processes and interactions (relations) interaction.

2.5.3 EXPERIMENTATION AND SIMULATION

At some point in the lives of most systems, there is a need to study them to try to gain some insight into the relationship among various components, or to predict performance under some new conditions being considered.[18] Figure 2.2 maps out different ways in which a system might be studied.

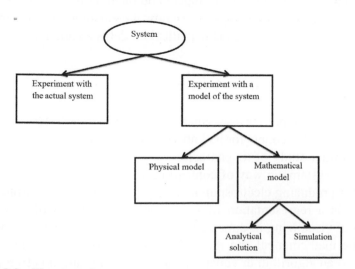

FIGURE 2.2   Ways to study a system.

2.5.4 EXPERIMENT WITH THE ACTUAL SYSTEM VERSUS EXPERIMENT WITH A MODEL OF THE SYSTEM

If it is possible (and cost-effective) to alter the system physically and then let it operate under the new conditions, it is probably desirable to do so. In this case, there is no question about whether what we study is relevant. However, it is rarely feasible to do this, because such an experiment would often be too costly or too disruptive to the system. In more abstract terms, "system" might not even exist, but we nevertheless want to study it in its various proposed alternative configurations to see how it should be built in the first place. For these reasons, it is usually necessary to build a model as a representation of the system and study it as a surrogate for the actual system. When using a model, there is always the question of whether it accurately reflects the system for the purposes of the decisions to be made.[20]

2.5.5 PHYSICAL MODEL VERSUS MATHEMATICAL MODEL

The physical models (also called iconic models) are not typical of the kinds of models that are usually of interest in systems analysis and computer simulations. In some circumstances, it has been found useful to build physical models to study engineering or management systems.

The vast majority of models built for such purposes are mathematical, representing a system in terms of logical and quantitative relationships that are then manipulated and changed to see how the model reacts, and thus how the system would react, if the mathematical model is a valid one.[20]

2.5.6 MATHEMATICAL MODELING

As a definition, a model is a schematic description of a system, theory, or phenomenon that accounts for its known or inferred properties and may be used for further study of its characteristics.[17] Mathematical modeling, as an easy and systematic way of controlling the influence of variables, can be used for producing electrospun nanofiber properties and morphology.[21] Modeling is a representation of a real object or system of objects for purposes of visualizing its appearance or analyzing its behavior. The main aim of the mathematical and theoretical analysis of engineering process is access to an algorithm or equation for creating a relation between input and output variables. In addition, for observation of the physical behavior of a sample, we should respect this behavior in many samples.[22,23] The basic principle used in modeling of the chemical engineering process is the concept of balance of momentum, mass, and energy, which can be expressed in a general form as[1]

$$A = I + G - O - C \tag{2.2}$$

where A is the accumulation built up within the system; I is the input entering through the system surface; G is the generation produced in system volume; O is the output leaving through system boundary; C is the consumption used in system volume.

The form of the expression depends on the level of the process phenomenon description. Various developed models can be employed for the analysis of jet deposition and alignment mechanisms on different collecting devices.[24] All models start with some assumptions and have shortcomings that need to be addressed. Different stages of electrospun jets have been investigated by different mathematical models during the last decade by one or three-dimensional techniques.[25] A comprehend review of these models will be discussed in the future chapters.

A suitable theoretical model of the electrospinning process is one that can show a strong–moderate–minor rating effects of these parameters on the fiber diameter.[26] In addition, they can be useful for investigating the effects of factors perception that cannot be measured experimentally.[27] So, it is necessary to develop theoretical/numerical models of electrospinning.[28] The analysis and comparison of the model with experiments identify the critical role of the spinning fluid's parameters. In each model, the researcher tried to improve the existing models are changing the tools in electrospinning by using another survey. Therefore, it is attempting to accept a whole view on important models after an investigation about basic objects. Results from modeling also explain how processing parameters and fluid behavior lead to nanofibers with appropriate properties.[24,29]

2.5.7 ANALYTICAL SOLUTION VERSUS SIMULATION

Once we have built a mathematical model, it must then be examined to see how it can be used to answer the questions of interest about the system it is supposed to represent. If the model is simple enough, it may be possible to work with its relationships and quantities to get an exact, analytical solution. Some analytical solutions can become extraordinarily complex, requiring vast computing resources, inverting a large nonsparse matrix is a well-known example of a situation in which there is an analytical formula known in principle, but obtaining it numerically in a given instance is far from trivial. If an analytical solution to a mathematical model is available and is computationally efficient, it is usually desirable to study the model in this way rather than via a simulation. However, many systems are highly complex, so that valid mathematical models of them are themselves complex, precluding any possibility of an analytical solution. In this case, the model must be studied by means of simulation, that is, numerically exercising the model for the inputs in question to see how they affect the output measures of performance. Due to the sheer complexity of the systems of interest and of the models, it is necessary to represent them in a valid way. Given, then, that we have a mathematical model to be studied by means of simulation (henceforth, referred to as a simulation model), we must then look for particular tools to execute this model (i.e., actual simulation). It is useful for this purpose to classify simulation models along three different dimensions[30]:

2.5.8 STATIC VERSUS DYNAMIC SIMULATION MODELS

A static simulation model is a representation of a system at a particular time, or one that may be used to represent a system in which time simply plays no role; examples of static simulations are Monte Carlo models. On the other hand, a dynamic simulation model represents a system as it evolves over time, such as a conveyor system in a factory.[30]

2.5.9 DETERMINISTIC VERSUS STOCHASTIC SIMULATION MODELS

If a simulation model does not contain any probabilistic (i.e., random) components, it is called deterministic. A complicated (and analytically intractable) system of differential equations describing a fluid flow is an example of such a model. In deterministic models, the output is "determined" once the set of input quantities and relationships in the model have been specified; even though it might take a lot of computer time to evaluate what it is. Many systems, however, must be modeled as having at least some random input components, and these give rise to stochastic simulation models. Most queuing and inventory systems are modeled stochastic. Stochastic simulation models produce output that is itself random and must therefore be treated as only an estimate of the true characteristics of the model; this is one of the main disadvantages of simulation.[30]

2.5.10   *CONTINUOUS VERSUS DISCRETE SIMULATION MODELS*

Loosely speaking, we define discrete and continuous simulation models analogously to the way discrete and continuous systems were defined above. It should be mentioned that a discrete model is not always used to model a discrete system and vice versa. The decision whether to use a discrete or a continuous model for a particular system depends on the specific objectives of the study.[18]

2.5.11   *A CLOSER LOOK AT SYSTEM MODELS*

Although many attempts have been made throughout the years to categorize systems and models, no consensus has been arrived at. However, it is convenient to make the following distinction between the different models[18]:

2.5.11.1   *CONTINUOUS-TIME MODELS*

Here, the state of a system changes continuously over time. These types of models are usually represented by sets of differential equations. A further subdivision would be (Fig. 2.3)

- Lumped parameter models expressed in ordinary differential equations:

$$\frac{dx}{dt} = f(x, u, t) \tag{2.3}$$

- Distributed parameter models expressed in partial differential equations:

$$\frac{\partial u}{\partial t} = \frac{\alpha \partial^2 u}{\partial t^2} \tag{2.4}$$

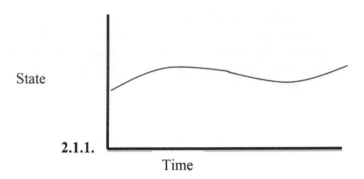

2.1.1.

FIGURE 2.3   Trajectory of continuous-time model.

2.5.11.2   DISCRETE-TIME MODELS

With discrete-time models, the time axis is discretized. The system state changes are commonly represented by difference equations. These types of models are typical to engineering systems and computer-controlled systems. They can also arise from discrete versions of continuous-time models.

The time-step used in the discrete-time model is constant (Fig. 2.4).

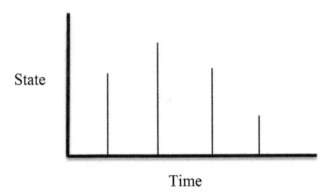

FIGURE 2.4   Trajectory of a discrete-time model.

2.5.11.3   DISCRETE-EVENT MODELS

In discrete-event models, the state is discretized and "jumps" in time. Events can happen any time but only every now and then at (stochastic) time intervals (Fig. 2.5).

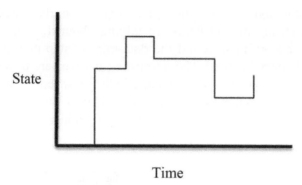

Time

FIGURE 2.5   Trajectory of discrete-event model.

2.6   DATA MINING

Over the past few decades, rapid advancements in computer technologies have allowed organizations to gather vast quantities of data for various purposes. Database technology provides efficient access to a myriad of information and allows the creation and maintenance of massive databases. However, it is far easier to collect data than to analyze it and extract information from it. Traditionally, data were analyzed manually but as it grows in size, many hidden patterns and potentially useful relationships may not be recognized by the researcher. As data-manipulation technologies rapidly advance, people rely increasingly on computers to accumulate data, process data, and make use of data. Knowledge discovery in databases consists of intelligent tools that handle massive data sets and useful patterns that help people make use of the data. Also, accurate classification is important for making effective decisions. Accuracy of the results produced by a classification system greatly depends on how well a problem is represented using a set of features. This resulted in the growing interest in the field of Knowledge Discovery in Databases, and a particular process of it, data mining. Data mining applies data analysis and discovery algorithms to identify patterns in data.[15]

As a simple definition, data mining refers to extracting or mining knowledge from large amounts of data. Data mining, a major process of knowledge discovery, is concerned with uncovering patterns hidden in the data. Data classification is a common data-mining task that deals with methods for assigning a set of input objects to a set of decision classes. The input objects are usually described by a set of features (such as numbers or a string of symbols). Each feature represents a characteristic of the given object.[31]

The investigation of data-mining technique is fairly a new area and various research in different fields are talking about it. Data mining is a knowledge discovery process and the data-mining step may interact with the user or a knowledge base. The interesting patterns are presented to the user and may be stored as new knowledge in the knowledge base (Fig. 2.6). Note that according to this view, data mining is only one step in the entire process, albeit an essential one, since it uncovers hidden patterns for evaluation (Fig. 2.2).[15,16]

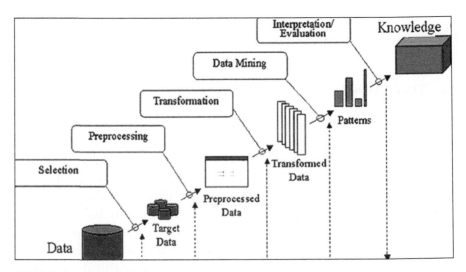

FIGURE 2.6   Data mining steps.

The major reason that data mining has attracted a great deal of attention in information industry in recent years is due to the wide availability of huge amounts of data and the imminent need for turning such data into useful information and knowledge. System data mining is a powerful tool. Using it separately will help any process make better and more informed decisions. However, by utilizing simulation and data mining together, they form a far more powerful combination for driving value through justified knowledge and insight than when used separately.[31]

The real value of systems representation, and more particularly systems simulation, is that it provides a succinct and concise encapsulation of knowledge. It is an active, usable, and testable symbolic representation of how a process works. Researchers can explicate and exemplify their understanding and test it without the risk of running a real prototype. A system representation can provide a useful and valuable representation of the problem or opportunity domain. When grounded in reality the models are highly valuable—otherwise, they can lead decision makers astray. But how can the necessary grounding best be achieved? When data are available, data mining provides the answer.[15]

A powerful tool for knowledge discovery in its own right, data mining is the ideal companion for generating systems representations and simulations. Data mining is an analytical tool that enables to discover—in data of course—relevant objects and the actual relationships that exist between those objects. Systems simulation provides a vehicle to represent those objects and relationships and then to add user insight from experience and prior knowledge that is not represented in available data.

Data mining involves six common classes of tasks[31]:

- *Anomaly detection (outlier/change/deviation detection)*: The identification of unusual data records that might be interesting or data errors that require further investigation.
- *Association rule learning (dependency modeling)*: Searches for relationships between variables. For example, a supermarket might gather data on customer purchasing habits. Using association rule learning, the supermarket can determine which products are frequently bought together and use this information for marketing purposes. This is sometimes referred to as market-basket analysis.
- *Clustering*: It is the task of discovering groups and structures in the data that are in some way or another "similar," without using known structures in the data.
- *Classification*: It is the task of generalizing known structure to apply to new data. For example, an e-mail program might attempt to classify an e-mail as "legitimate" or as "spam."
- *Regression*: It attempts to find a function which models the data with the least error.
- *Summarization*: It provides a more compact representation of the data set, including visualization and report generation.

In conjunction, data mining will be used to extract from a large volume of data knowledge or hidden information, previously unknown, potentially useful and interesting information. The application of these techniques of data mining can uncover behavioral patterns and therefore nontrivial knowledge from the simulation data. The analysis of these results allows to correct and to improve the quality of the simulation study.[32]

2.7   GENETIC PROGRAMMING

A general assumption inmost computer science unquestioningly is that any effective problem-solving process must be logically sound and deterministic. The consecution of this hypothesis is that almost all conventional approaches to artificial intelligence and machine learning possess these characteristics. Genetic programing is a systematic method for getting computers to solve a problem. Genetic programing can automatically create, in a single run, a parameterized solution to a problem in the form of a graphical structure whose nodes or edges represent components and where the parameter values of the components are specified by mathematical expressions containing free variables.[33,34]

Genetic programing starts from a high-level statement of what needs to be done to solve the problem. The five major preparatory steps for the basic version of genetic programing require the human user to specify[35,36]

1. the set of terminals (e.g., the independent variables of the problem, zero-argument functions, and random constants) for each branch of the to-be-evolved computer program,
2. the set of primitive functions for each branch of the to-be-evolved computer program,
3. the fitness measure (for explicitly or implicitly measuring the fitness of candidate individuals in the population),
4. certain parameters for controlling the run, and
5. a termination criterion and method for designating the result of the run.

Genetic programing requires a set of primitive ingredients to get started. All of basic steps which are indicated in Figure 2.7 will be explained in following. The first two preparatory steps specify the primitive ingredients that are to be used to create the to-be-evolved programs. The universe of allowable compositions of these ingredients defines the search space for a run of genetic programing.

The identification of the function set and terminal set for a particular problem (or category of problems) is often a mundane and straightforward process that requires only de minimus knowledge and platitudinous information about the problem domain.[34,37]

The fundamental elements of an individual are its genes, which come together to form code. An individual's program is a tree-like structure and as such there are two types of genes: functions and terminals. Terminals, in tree terminology, are leaves (nodes without branches), while functions are nodes with children. The function's children provide the arguments for the function.[38]

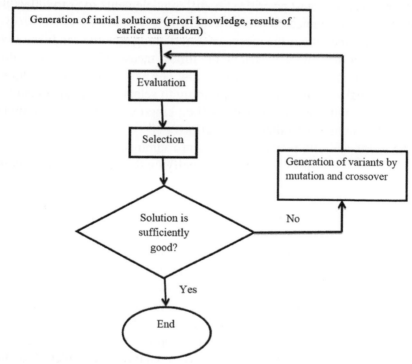

FIGURE 2.7 Basic evolutionary algorithm.

The third preparatory step concerns the fitness measure for the problem. The individuals need to be assessed for their fitness. The fitness measure specifies what needs to be done. The result that is produced by genetic programing specifies "how to do it."

The fitness measure is the primary mechanism for communicating the high-level statement of the problem's requirements to the genetic programing system. The fitness measure typically assigns a single numeric value reflecting the extent to which a candidate individual satisfies the problem's high-level requirements.[34]

If one views the first two preparatory steps as defining the search space for the problem, one can then view the third preparatory step (the fitness measure) as specifying the search's desired direction. Having applied the fitness test to all the individuals in the initial random population, the evolutionary process starts. The fitness measure for a real-world problem is typically multi-objective, that is, there may be more than one element that is considered in ascertaining fitness. In practice, the elements of a multi-objective fitness measure usually conflict with one another. Thus, a multi-objective fitness measure must prioritize the different elements so as to reflect the tradeoffs that the engineer is willing to accept.[39]

The fitness measure is the means of ascertaining that one candidate individual is better than another. That is, the fitness measure is used to establish a partial order among candidate individuals. The partial order is used during the executional steps of genetic programing to select individuals to participate in the various genetic operations (i.e., crossover, reproduction, mutation, and the architecture-altering operations).[40,41]

The fourth and fifth preparatory steps are administrative.

The fourth preparatory step entails specifying the control parameters for the run.

The major control parameters are the population size and the number of generations to be run. Some analytic methods are available for suggesting optimal population sizes for runs of the genetic algorithm on particular problems. However, the practical reality is that we generally do not use any such analytic method to choose the population size.[42]

Instead, we determine the population size such that genetic programing can execute a reasonably large number of generations within the amount of computer time we are willing to devote to the problem.

Individuals in the new population are formed by two main methods: reproduction and crossover. Once the new population is complete (i.e., the same size as the old) the old population is destroyed.[43,44]

An asexual method, reproduction is where a selected individual copies itself into the new population. It is effectively the same as one individual surviving into the next generation.

Organisms' sexual reproduction is the analogy for crossover. Crossover requires two individuals and produces two different individuals for the new population. In this technique, genetic material from two individuals is mixed to form off-spring. The process can be used crossover on 90% of the population—it is the more important of the two methods because it provides the source of new (and eventually better) individuals.[44]

Mutation operates on only one individual. Normally, after crossover has occurred, each child produced by the crossover undergoes mutation with a low probability. The probability of mutation is a parameter of the run. A separate application of crossover and mutation, however, is also possible and provides another reasonable procedure.[45]

When an individual has been selected for mutation, one type of muta-tion operator in tree GP selects a point in the tree randomly and replaces the existing subtree at that point with a new randomly generated subtree. The new randomly generated subtree is created in the same way and subject to the same limitations (on depth or size) as programs in the initial random population. The altered individual is then placed back into the population.

The fifth preparatory step consists of specifying the termination criterion and the method of designating the result of the run.[45,46]

2.8   GENETIC PROGRAMMING TREES FOR NANOFIBERS AND COMPOSITE NANOFIBERS

By modeling and simulation, the relationship between electrospinning parameters cannot obtain exactly. Also, for composite nanofibers, there is not an appropriate model for simulation. By using data-mining methods like genetic programing, it is possible to relate all selecting parameters by gener-ating trees. In this method, we need initial data for estimation. By using Taguchi design, results for PAN nanofiber and PAN/CNT nanofibers, which were mentioned in first section of this chapter, were used for generation. Two hundred individuals of solution have been generated. The best results are showed in Figures 2.8 and 2.9.

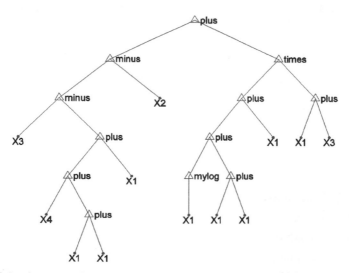

FIGURE 2.8   An appropriate genetic programing tree for PAN nanofiber: X_1 = concentration, X_2 = voltage, X_3 = feeding rate, X_4 = spinning distance.

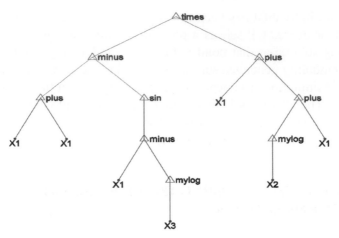

FIGURE 2.9 An appropriate genetic programing tree for PAN/CNT nanofiber: X_1 = concentration, X_2 = voltage, X_3 = feeding rate, X_4 = spinning distance.

2.9 CONCLUDING REMARKS

Producing nanofibers by electrospinning is a simple and widely utilized for varied applications. During the recent years, more attentions have focused on optimization of this method to solve the problems which make electrospinning uncontrollable. In addition, the electrospinning process is a fluid-dynamics-related problem. For achieving this goal, using different simple methods (i.e., modeling, simulation, data mining, genetic algorithm, and design of experiment) were useful. They were utilized for predicting ideal electrospinning process or electrospun fiber properties. In this chapter, we put forward a group of analysis methods for investigating and optimizing the problem of process.

KEYWORDS

- analysis methods
- fluid behavior
- process analysis
- electrospinning
- nanofibers orientation

REFERENCES

1. Rafiei, S.; Maghsoodloo, S.; Saberi, M.; Lotfi, S.; Motaghitalab, V.; Noroozi, B.; Haghi, A. K. New Horizons in Modeling and Simulation of Electrospun Nanofibers: A Detailed Review. *Cell. Chem. Technol.* **2014,** *48* (5–6), 401–424.

2. Kalita, G.; Adhikari, S.; Aryal, H. R.; Somani, P. R.; Somani, S. P.; Sharon, M.; Umeno, M. Taguchi Optimization of Device Parameters for Fullerene and Poly(3-Octylthiophene) based Heterojunction Photovoltaic Devices. *Diamond Relat. Mater.* **2008,** *17* (4), 799–803.

3. Lundstedt, T.; Seifert, E.; Abramo, L.; Thelin, B.; Nyström, Å.; Pettersen, J.; Bergman, R. Experimental Design and Optimization. *Chemometr. Intell. Lab. Syst.* **1998,** *42* (1), 3–40.

4. Kirk, R. E. *Experimental Design.* Wiley Online Library, 1982.

5. Lu, P.; Ding, B. Applications of Electrospun Fibers. *Rec. Pat. Nanotechnol.* **2008,** *2* (3), 169–182.

6. Li, Z.; Wang, C. Effects of Working Parameters on Electrospinning. In *One-Dimensional nanostructures*; Springer: Berlin, 2013; pp 15–28.

7. Bognitzki, M.; Czado, W.; Frese, T.; Schaper, A.; Hellwig, M.; Steinhart, M.; Greiner, A.; Wendorff, J. Nanostructured Fibers via Electrospinning. *Adv. Mater.* **2001,** *13* (1), 70–72.

8. De Vrieze, S.; Van Camp, T.; Nelvig, A.; Hagström, B.; Westbroek, P.; De Clerck, K. The Effect of Temperature and Humidity on Electrospinning. *J. Mater. Sci.* **2009,** *44* (5), 1357–1362.

9. Tan, S. H.; Inai, R.; Kotaki, M.; Ramakrishna, S. Systematic Parameter Study for Ultra-Fine Fiber Fabrication via Electrospinning Process. *Polymer* **2005,** *46* (16), 6128–6134.

10. Angammana, C. J. *A Study of the Effects of Solution and Process Parameters on the Electrospinning Process and Nanofibre Morphology,* University of Waterloo: Waterloo, 2011.

11. Bhardwaj, N.; Kundu, S. C. Electrospinning: A Fascinating Fiber Fabrication Technique. *Biotechnol. Adv.* **2010,** *28* (3), 325–347.

12. Dean, A.; Voss, D. Response Surface Methodology. *Des. Anal. Experim.* **1999,** *1,* 483–529.

13. Khanlou, H. M.; Ang, B. C.; Talebian, S.; Afifi, A. M.; Andriyana, A. Electrospinning of Polymethyl Methacrylate Nanofibers: Optimization of Processing Parameters using the Taguchi Design of Experiments. *Text. Res. J.* **2014,** *7,* 0040517514547208 1–13.

14. Albetran, H.; Dong, Y. U.; Low, I. M. Characterization and Optimization of Electrospun TiO_2/PVP Nanofibers using Taguchi Design of Experiment Method. *J. Asian Ceram. Soc.* **2015,** *3* (3), 292–300.

15. Han, J.; Kamber, M.; Pei, J. *Data Mining: Concepts and Techniques,* Elsevier, 2011.

16. Law, A. M. *Simulation Modeling and Analysis* McGraw Hill: Boston, 2015.

17. Rafiei, S.; Maghsoodloo, S.; Noroozi, B.; Mottaghitalab, V.; Haghi, A. K.; Mathematical Modeling in Electrospinning Process of Nanofibers: A Detailed Review. *Cell. Chem. Technol.* **2013,** *47* (5–6), 323–338.

18. Kelton, W. D.; Law, A. M. *Simulation Modeling and Analysis*; McGraw Hill: Boston, 2000.

19. Ford, F. A. *Modeling the Environment: An Introduction to System Dynamics Models of Environmental Systems.* Island Press, 1999.

20. Jeruchim, M. C.; Balaban, P.; Shanmugan, K. S. *Simulation of Communication Systems: Modeling, Methodology and Techniques*. Springer Science & Business Media: Berlin, 2006.
21. Thompson, C. J. *An Analysis of Variable Effects on a Theoretical Model of the Electrospin Process for Making Nanofibers*, University of Akron, 2006.
22. Denn, M. M. Issues in Viscoelastic Fluid Mechanics. *Annu. Rev. Fluid Mech.* **1990,** *22* (1), 13–32.
23. Kröger, M. Simple Models for Complex Nonequilibrium Fluids. *Phys. Rep.* **2004,** *390* (6), 453–551.
24. Greenfeld, I.; Arinstein, A.; Fezzaa, K.; Rafailovich, M. H.; Zussman, E. Polymer Dynamics in Semidilute Solution during Electrospinning: A Simple Model and Experimental Observations. *Phys. Rev. E* **2011,** *84* (4), 041806.
25. Haghi, A. K.; Zalkov, G. E. Mathematical Models on the Transport Properties of Electrospun Nanofibers. *Nanopolymers and Modern Materials: Preparation, Properties, and Applications*. 2013, CRC Press: Boca Raton, FL; pp 195–218.
26. Fridrikh, S. V.; Yu, J. H.; Brenner, M. P.; Rutledge, G. C. Controlling the Fiber Diameter during Electrospinning. *Physical Review Letters*. American Physical Society, 2003; pp 144502–144505.
27. Patanaik, A.; Jacobs, V.; Anandjiwala, R. D. *Experimental Study and Modeling of the Electrospinning Process*, 2008; pp 1160–1168.
28. Kowalewski, T. A.; Barral, S.; Kowalczyk, T. Modeling Electrospinning of Nanofibers. *IUTAM Symposium on Modelling Nanomaterials and Nanosystems*. Springer: Berlin, 2009.
29. Liu, L.; Dzenis, Y. Simulation of Electrospun Nanofibre Deposition on Stationary and Moving Substrates. *Micro Nano Lett.* **2011,** *6* (6), 408–411.
30. Bellomo, N.; Pulvirenti, M. *Modeling in Applied Sciences: A Kinetic Theory Approach*. Springer Science & Business Media: Berlin, 2013.
31. Hand, D. J.; Mannila, H.; Smyth, P. *Principles of Data Mining*. MIT Press: Cambridge, MA, 2001.
32. Saoud, M. S.; Boubetra, A.; Attia, S. How Data Mining Techniques Can Improve Simulation Studies. *Int. J. Comput. Theory Eng.* **2014,** *6* (1), 15–19.
33. Koza, J. R.; Keane, M. A.; Streeter, M. J.; Mydlowec, W.; Yu, J.; Lanza, G. *Genetic Programming IV: Routine Human-Competitive Machine Intelligence*; Springer Science & Business Media: Berlin, vol. 5. 2006; p 590.
34. Koza, J. R. *Genetic Programming: On the Programming of Computers by Means of Natural Selection*; MIT Press: Cambridge, MA, 1992; vol. 1, p 680.
35. Koza, J. R.; Poli, R. Genetic Programming. *Search Methodologies*; Springer: Berlin, 2005; pp 127–164.
36. Khan, M. W.; Alam, M. A Survey of Application: Genomics and Genetic Programming, a New Frontier. *Genomics* **2012,** *100* (2), 65–71.
37. Sivanandam, S. N.; Deepa, S. N. Genetic Programming. *Introduction to Genetic Algorithms*; Springer: Berlin, 2008; pp 131–163.
38. Tsakonas, A. A Comparison of Classification Accuracy of Four Genetic Programming-Evolved Intelligent Structures. *Informat. Sci.* **2006,** *176* (6), 691–724.
39. R. Vazquez, K.; Fonseca, C. M.; Fleming, P. J. Identifying the Structure of Nonlinear Dynamic Systems Using Multiobjective Genetic Programming. *IEEE Trans. Syst., Man, and Cybern., A: Syst. Hum.* **2004,** *34* (4), 531–545.

40. Tomassini, M.; Vanneschi, L.; Collard, P.; Clergue, M. A Study of Fitness Distance Correlation as a Difficulty Measure in Genetic Programming. *Evol. Comput.* **2005,** *13* (2), 213–239.
41. Burke, E. K.; Gustafson, S.; Kendall, G. Diversity in Genetic Programming: An Analysis of Measures and Correlation with Fitness. *IEEE Trans. Evol. Comput.* **2004,** *8* (1), 47–62.
42. Eiben, Á. E.; Hinterding, R.; Michalewicz, Z. Parameter Control in Evolutionary Algorithms. *IEEE Trans. Evol. Comput.* **1999,** *3* (2), 124–141.
43. Whitley, D.; Rana, S.; Heckendorn, R. B. The Island Model Genetic Algorithm: On Separability, Population Size and Convergence. *J. Comput. Informat. Technol.* **1999,** *7,* 33–48.
44. Koza, J. R. Genetic Programming as a Means for Programming Computers by Natural Selection. *Stat. Comput.* **1994,** *4* (2), 87–112.
45. Weimer, W.; Nguyen, T.; Le Goues, C.; Forrest, S. Automatically Finding Patches Using Genetic Programming. In *Proceedings of the 31st International Conference on Software Engineering.* IEEE Computer Society, 2009.
46. Yang, J.; Honavar, V. Feature Subset Selection Using a Genetic Algorithm. *Feature Extraction, Construction and Selection*; Springer: Berlin, 1998; pp 117–136.

CHAPTER 3

UPDATES ON ELECTROSPINNING PROCESS MODELS (PART II)

SHIMA MAGHSOODLOU* and S. PORESKANDAR

Textile Engineering, University of Guilan, Rasht, Iran

Corresponding author. E-mail: sh.maghsoodlou@gmail.com

CONTENTS

ABSTRACT

Electrospinning has appeared as a widespread and novel technology to create synthetic nanofibrous. The most vital challenge in the electrospinning proces is to achieve uniform nanofibers consistently. In addition, the jet shows different behaviors during the process. The most suitable way for understanding and controlling electrospun jet movement will perform by utilizing modeling and simulation method. Modeling and simulating procedure will permit to offer an in-depth insight into the physical apprehension of complex phenomena during electrospinning. The main aim of this chapter is to investigate a brief review of the electrospun nanofiber process parts by utilizing modeling methods.

3.1   INTRODUCTION

As mentioned in Chapter 1, this process is an efficient and simplest technique for producing nanofibers with different structures[1–8] which can be distinguished by four main sections.[9] In addition, controlling the physical characteristics of electrospun nanofibers such as fiber diameter depend on various parameters that are principally divided into three categories: solution properties (solution viscosity, solution concentration, polymer molecular weight, and surface tension), processing conditions (applied voltage, volume flow rate, spinning distance, and needle diameter), and ambient conditions (temperature, humidity, and atmospheric pressure).[6,10–13] The most important parameters are summarized in Figure 3.1.

Different part of electrospinning process can be shown as following in Figures 3.2 and 3.3. Descriptions of these stages are briefly discussed in the following sections.

3.2   A BRIEF DESCRIPTION STAGES OF THE ELECTROSPINNING PROCESS

3.2.1   FIRST STAGE: FORMATION OF TAYLOR CONE

The electrospinning solution is usually an ionic solution that contains charged ions. The amounts of positive and negative charged particles are equal; therefore, the solution is electrically neutral. When an electrical

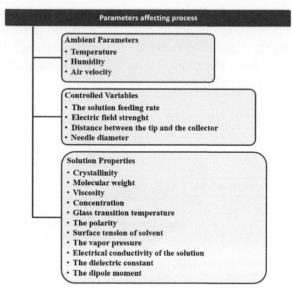

FIGURE 3.1   Parameters affecting the morphology and size of electrospun nanofibers.

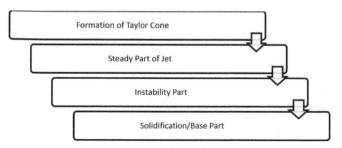

FIGURE 3.2   Different parts of electrospinning process.

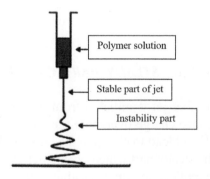

FIGURE 3.3   Schematic of different parts of electrospinning process.

potential difference is given between needle and collector, a hemispherical surface of the polymeric droplet at the orifice of the needle is gradually expanded. When the potential comes into a critical value eq 3.1 a flow of jet starts formation to drop. Therefore, Taylor's cone is formed.[14,15] A schematic of these steps is shown in Figure 3.4.

$$V^2C = 4\frac{H^2}{L^2}\left(\ln\frac{2L}{R}-\frac{3}{2}\right)(0.117\,\pi\gamma R).$$ (3.1)

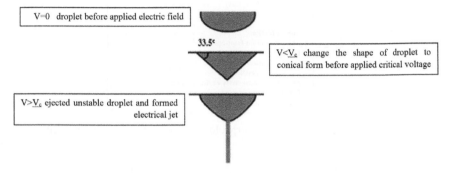

FIGURE 3.4   Changes in the polymer droplet with applied potential.

It can be assumed that the driving is purely sinusoidal, so that the speed at the nozzle is[16]

$$\upsilon_{nozzle} = \upsilon_0 + A\left(\frac{\gamma}{\rho a_0}\right)^{1/2}\sin(2\pi ft).$$ (3.2)

As an aside, a_0 is in general not equal to the nozzle radius, as the jet contracts for high jet speeds and expands for low speeds. This effect depends on the fluid parameters as well as on the velocity profile in the nozzle.

3.2.2   SECOND STAGE: STEADY PART OF JET

The electrospun nanofiber jet is initiated from the droplet when the repelling forces of the surface charge overcome the surface tension and viscous forces of the droplet[17] and lead to elongate the jet straight in the direction of its axis.[14,18] During the steady part which occurs near the Taylor cone, the jet has only asymmetric motion, so remains stable in time.[19]

A stable formation jet, shown in Figure 3.3, travels to a collector. During the elongation of the electrified liquid jet, the jet surface area increases dramatically.[20]

Therefore, for a thin, stable jet, it is significant to look for a balance of viscosity charge density and surface tension.[9]

As mentioned before, parameters such as solution concentration, field strength, flow rate, jet velocity, and shear rate have a strong effect on the electrospinning process. Now, assume that if the molecular weight of fluid is low, the flow rate of the polymer solution will be determinative for the rate at which the radius of an electrostatically driven jet decays. So by decreasing the flow rate, the radius of the jet will be decayed more rapidly. On the other hand, increasing the flow rate will be terminated to slowly decrease of the jet radius.[15]

This phenomenon can be explained by assuming a simple cylindrical geometry to represent the volume element of the electrospinning jet which consummated to macroscopic models for the process. In this simple way, this assumption can only illustrate a rough layout of the interaction between the solution feed rate, electrospinning voltage, and the charge-to-mass ratio.

$$\frac{A}{V} = \frac{2}{R}.$$ (3.3)

From eq 3.3, the ratio of surface area to volume (specific surface area) is inversely proportional to the jet radius. Therefore, an increase in the jet radius results in a corresponding decrease in the specific surface area associated with that specific volume element. If the density of the polymer solution and the surface charge density are assumed to be constant, then it follows that the charge-to-mass ratio will decrease with increasing jet radius. From eq 3.4, the acceleration is directly proportional to the ratio of charge to mass; hence, an increase in jet radius results in a decrease in the acceleration of the fluid.[21]

$$a = E\left(\frac{q}{m}\right).$$ (3.4)

As the viscosity of the polymer solvent decreased, the spinning drop changed from hemispherical to conical. By using equipotential line approximation calculations, the radius r_0 of a spherical drop can be calculated as follows[22]:

$$r_0^3 = \frac{4\varepsilon m_0}{k\pi\sigma\rho}. \tag{3.5}$$

3.2.3 THIRD STAGE: INSTABILITY PART

After a small distance of stable traveling the jet, it will start unstable behavior because of axisymmetric and nonaxisymmetric instabilities and sprays or separates into many fibers.[14,17,18]

As the jet spirals toward the collector, higher order instabilities have resulted in spinning distance. These instabilities are separated into three sections (summarized in Fig. 3.5)[23]:

- Rayleigh instability (Fig. 3.5a)
- Bending instability (Fig. 3.5b)
- Whipping instability (Fig. 3.5c)

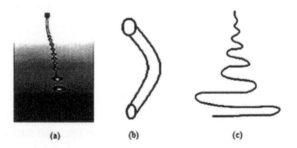

FIGURE 3.5   (a) Rayleigh instability, (b) bending instability, and (c) whipping instability.

The polymer jet is influenced by these instabilities. Recently, instability in electrospinning has received much attention.[14] Each instability grows at different rates.[9,24]

These instabilities arise owing to the charge–charge repulsion between the excess charges present in the jet, which encourages the thinning and elongation of the jet.[25] This happens when changes occur in the shape and charge per unit area of the jet due to its elongation and the evaporation of the solvent. This shifts the balance between the surface tension and the electrical forces, and the jet becomes unstable. To reduce its local charge per unit surface area, the unstable jet ejects a smaller jet from the surface of the primary jet.[26] At high electric forces, the jet is dominated by bending (axisymmetric) and whipping instability (nonaxisymmetric), causing the jet to move around and produces wave in the jet. At higher electric fields and at enough charge density in the jet, the axisymmetric (i.e., Rayleigh and bending) instabilities are suppressed and the nonaxisymmetric instability is increased.[25]

Bending or stretching of the jet is due to a nonaxisymmetric or whipping instability. However, the key role in reducing the jet diameter from micrometer to nanometer is played by nonaxisymmetric or whipping instability, which causes bending or stretching. When the polymer jet becomes very long and thin, the time required for the excess charge to redistribute itself along the full length of the jet becomes longer. The location of the excess charge then tends to change with the elongation. The repulsive Coulomb forces between the charges carried with the jet elongate the jet in the direction of its axis until the jet solidifies. This leads to an incredibly high velocity at the thin leading end of the straight jet. As a result, the jet bends and develops a series of lateral excursions that grow into spiraling loops (Figure 3.6). Each of these loops grows larger in diameter as the jet grows longer and thinner.[27]

FIGURE 3.6 Developments of bending instabilities by considering jet axis in z coordinate.

These instabilities vary and increase with distance, electrical field, and fiber diameter at different rates depending on the fluid parameters and performing conditions. Also, they influence the size and geometry of the deposited fibers.[25]

At high field, a second axisymmetric instability and a third, nonaxisymmetric instability, which is due to fluctuations in the dipolar component of the charge distribution, dominate. These conducting modes are electrically driven and essentially independent of surface tension. The axisymmetric conducting mode arises as a consequence of the finite, nonzero conductivity of the fluid, which produces an extra root to the growth rate equations. The whipping instability can occur through either: (a) small lateral fluctuations in the centerline of the jet result in the induction of a dipolar charge distribution, as the free-charge adjusts to screen out the field inside the jet. The dipoles interact with the external electric field, producing a torque that further bends the jet; and (b) mutual repulsion of surface charges carried by jet causes the centerline to bend. Of crucial importance to understanding the electrospinning process is the competition between modes of instability. Whether whipping or axisymmetric breakup of the jet dominates for a given set of conditions depends on the jet radius and surface charge experienced by a fluid element as it travels downstream. In general, the whipping mode dominates at high charge density, whereas the axisymmetric mode dominates at low charge density. As a jet thins away from the nozzle, both the charge density and the radius change.[28]

Suitable methods for analyzing these behaviors were important. In future parts, methods for better analyzing of these behaviors will be discussed.

3.2.4 FOURTH STAGE: SOLIDIFICATION

For polymers dissolved in nonvolatile solvents, water or other suitable liquids can be used to collect the jet, remove the solvent, and develop the polymer fiber.[18]

Solvent selection for polymers is another issue worth mentioning. Different solvents carry different properties in the aspect of conductivity, viscosity, volatility, and surface tension. In some cases, a solvent with two components is favored because of the balanced viscosity and volatility. Higher conductivity of solvent can be more favored for the process of electrospinning.[29,30]

As the jet moves toward the collector, it continues to expand by going past through the loops. Jet solidification is based on the traveling distance of the fibers. The elongation and thinning of the charged jet continue until solidification takes place. The distance between the collector and the capillary tip has a direct effect on the jet solidification and fiber diameter. Evaporation of the solvent changed the viscoelastic properties of the polymer solutions and stopped the elongation. If the nozzle-to-collector distance is long enough or the whipping instability is high, there is more time for fibers to dry before being picked up.[14,31]

The decreased evaporation rate of solvent from the jet allowed the charged jet to remain fluid, to continue to elongate, and to become thinner. On the other hand, during the deposition, the fibers will bend as it lies on top of one another given their high aspect ratio and flexibility due to the presence of residual solvents within. However, if the jet solidification can be accelerated, the fiber will have a greater stiffness and this may give it a greater resistance to bending which makes it less likely to be compacted. Increased stiffness before the fibers are deposited will result in less compaction and greater volume, especially for polymers with higher intrinsic stiffness.[31]

3.3   THE IMPORTANCE OF THE USAGE SIMULATION AND DATA MINING IN ELECTROSPINNING PROCESS ANALYSIS

Electrospinning can produce nanofibers with different structures and functionalities. However, the key part of electrospinning is how to control the process. Thus, understanding the influence of variable parameters is very important. These parameters include fluid viscosity, elasticity, conductivity, solvent volatility, spinning voltage and distance as well as ambient parameters such as humidity, temperature.[30]

The most significant challenge in this process is to attain uniform nanofibers consistently and reproducibly.[3,4,12,32]

Depending on several solution parameters, different results can be obtained using the same polymer and electrospinning setup.[7]

A successful electrospinning has involved an understanding of the complex interaction of electrostatic fields, properties of polymer solutions and component design and system geometry.[33]

Studying the dynamical behavior of the jet is of interest for the development of a possible control system.[34]

The process is complex with the resulting jet (fiber) diameter being influenced by numerous material, design, and operating parameters. A significant part of our information of the electrospinning process comes from empirical observations, but the complexity of the process makes an empirical determination of parameter effects very difficult if not impractical.[35]

The mechanics of this process deserving a specific attention and necessary to predictive tools or way for better understanding and optimization and controlling process.[36] In addition, studying the dynamical behavior of the jet is of interest for the maturation of a possible control system.[34]

For prediction of electrospun fiber properties and morphology, modeling/ simulating or data mining can be used.[37]

Results from modeling also explain how processing parameters and fluid behavior lead to nanofibers with appropriate properties.[38,39]

In addition, mathematical and theoretical modeling and simulating procedure will assist in offering an in-depth insight into the physical understanding of complex phenomena during electrospinning and might be very applicable to manage contributing factors toward increasing the production rate.[13]

A brief review on the most important of models will be investigated in following.

3.4   SIMULATION VERSUS DATA MINING

The investigation of simulation and data mining techniques is fairly a new area and various research in different fields are talking about it. Process simulation is a model-based representation that can be used for the design, development, analysis, and optimization of technical processes. Knowing about chemical and physical properties is a basic requisite to have an appropriate mathematical model (Fig. 3.7). Data mining is a knowledge discovery process and the data mining step may interact with the user or a knowledge base. The interesting patterns are presented to the user and may be stored as new knowledge in the knowledge base. Note that according to this view, data mining is only one step in the entire process, albeit an essential one since it uncovers hidden patterns for evaluation (Fig. 3.8).[40,41]

System simulation and data mining are both powerful tools. Using them separately will help any process make better and more informed decisions. However, by utilizing together, they form a far more powerful combination for driving value through justified knowledge and insight than when used separately.

The real value of systems representation, and more particularly systems simulation, is that it provides a succinct and concise encapsulation of knowledge. It is an active, usable, and testable symbolic representation of how a process works. Researchers can explicate and exemplify their understanding and test it without the risk of running a real prototype. A system representation can provide a useful and valuable representation of the problem or opportunity domain. When grounded in reality, the models are highly valuable—otherwise, they can lead decision makers astray. But how can the necessary grounding best be achieved? When data are available, data mining provides the answer.

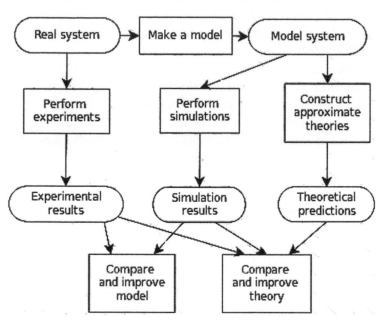

FIGURE 3.7   Simulation process schematic.

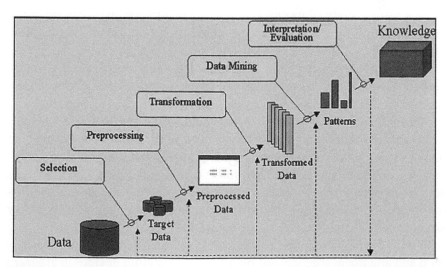

FIGURE 3.8   Data mining steps.

A powerful tool for knowledge discovery in its own right, data mining is the ideal companion for generating systems representations and simulations. Data mining is an analytical tool that enables to discover—in data of course—relevant objects and the actual relationships that exist between those objects. Systems simulation provides a vehicle to represent those objects and relationships and then to add user insight from experience and prior knowledge that is not represented in available data.

In conjunction, data mining will be used to extract from a large volume of data knowledge or hidden information, previously unknown, potentially useful and interesting information. The application of these techniques of data mining can uncover behavioral patterns and therefore on-trivial knowledge from the simulation data. The analysis of these results allows to correct and to improve the quality of the simulation study.[42]

3.5   MATHEMATICAL MODELING

As mentioned before, for prediction electrospun nanofiber properties and morphology, modeling/simulating can be used. In fact, an easy and systematic way of controlling the influence of variables is created by utilizing models.[37] So, it is necessary to develop theoretical/numerical models of electrospinning.[43]

As a definition, a model is a schematic description of a system, theory, or phenomenon that accounts for its known or inferred properties and may be used for further study of its characteristics.[44] In addition, they can be useful for investigating the effects of factors perception that cannot be measured experimentally.[45]

A real mathematical model might initiate a revolution in the understanding of dynamic and quantum-like phenomena in the electrospinning process.[46,47]

The analysis and comparison of the model with experiments identify the critical role of the spinning fluid's parameters. In each model, the researcher tried to improve the existing models are changing the tools in electrospinning by using another survey. Therefore, it is attempting to accept a whole view on important models after an investigation about basic objects. Results from modeling also explain how processing parameters and fluid behavior lead to nanofibers with appropriate properties.[38,39]

Various developed models can be employed for the analysis of jet deposition and alignment mechanisms on different collecting devices in arbitrary electric fields.[39]

A suitable theoretical model of the electrospinning process is one that can show a strong–moderate–minor rating effects of these parameters on the fiber diameter. Some disadvantages of this method are low production rate, nonoriented nanofiber production, difficulty in diameter prediction and controlling nanofiber morphology, the absence of enough information on the rheological behavior of polymer solution, and difficulty in precise process control that emphasis necessity of modeling.[46] A summary of investigated models is shown in Table 3.1.

TABLE 3.1 A Summary of Investigated Models Used in Electrospinning Process.

Models name	Name of researcher	Year	References
The electro hydrodynamic of electrospinning (leaky dielectric model)	Taylor	1969	[48]
Equations	**Suppositions**		
(1) Navier–Stokes	(1) Dielectric fluid		
(2) Maxwell stress	(2) Bulk charge = 0		
(3) Coulomb force equation	(3) Axial motion		
	(4) Steady part		
	(5) Newtonian		
	(6) Incompressible		
Models name	**Name of researcher**	**Year**	**References**
One-dimensional modes of steady, inviscid, annular liquid jets	Ramos	1996	[49]
Equations	**Suppositions**		
(1) Laplace equation	(1) Steady jet		
(2) Bernulli'sequation	(2) Inviscid liquid jet		
(3) Euler equation	(3) Annular liquid jet		
	(4) Slender jet		
Models name	**Name of researcher**	**Year**	**References**
Asymptotic decay of radius of a weakly conductive viscous jet in an external electric field	Spivak, Dzenis	1998	[50]
Equations	**Suppositions**		
(1) Momentum balance	(1) Newtonian		
(2) Mass balance	(2) Infinite viscose jet		
(3) Electric charge balance	(3) Weakly conductive jet		
	(4) No magnetic effects		
	(5) Long slender jet		

TABLE 3.1 *(Continued)*

Models name	Name of researcher	Year	References
Bending instability of electrically charged liquid jets of polymer solutions in electrospinning	Reneker	2000	[51]
Effects of parameters on nanofiber diameter determined from electrospinning model	Thompson	2007	[35]
Mathematical models of bead-spring jets during electrospinning for fabrication of nanofibers	Darsi	2013	[52]

Equations	Suppositions
(1) Momentum balance	(1) Viscoelastic jet
(2) Viscoelastic force equation	(2) Entire jet

Models name	Name of researcher	Year	References
Experimental characterization of electrospinning: the electrically forced jet and instabilities	Shin, Hohman	2001	[53]

Equations	Suppositions
(1) Momentum balance	(1) Newtonian
(2) Mass balance	(2) Incompressible
(3) Electric charge balance	(3) Long slender jet
(4) Instabilities analysis	

Models name	Name of researcher	Year	References
The Stretching of an electrified non-newtonian jet: a model for electrospinning	Feng	2002	[54]

Equations	Suppositions
(1) Momentum balance	(1) Non-Newtonian
(2) Mass balance	(2) Steady jet
(3) Electric charge balance	(3) Slender jet

Models name	Name of researcher	Year	References
Thermo-electro-hydrodynamic model for electrospinning process	Wan, Gue, Pan	2004	[55]
A mathematical model for preparation by AC-electrospinning process	He, Wu, Pang	2005	[56]
A thermo-electro-hydrodynamic model for vibration-electrospinning process	Xu, Wang	2011	[57]

Equations	Suppositions
(1) Momentum balance	(1) Thermo-electro-hydrodynamic
(2) Mass balance	(2) Steady part
(3) Electric charge balance	(3) Thermal effects
(4) Bratueq	

All models start with some assumptions and have shortcomings that need to be addressed. Different stages of electrospun jets have been investigated by different mathematical models during the last decade by one or three-dimensional techniques.[58]

The main focus of this chapter is investigated models to give a better understanding of electrospinning process. In the future chapters of this book, each model with summary basic equations were investigated.

3.6   CONCLUDING REMARKS

Producing nanofibers by electrospinning is a simple and widely utilized for varied applications. During the recent years, more attentions have focused on optimization of this method to solve the problems which make electrospinning uncontrollable. In addition, the electrospinning process is a fluid dynamics related problem. For achieving this goal, using simple models can be useful. The analysis and comparability of the model with experiments identify the critical role of the spinning fluid's parameters. Different models were utilized for predicting ideal electrospinning process or electrospun fiber properties. In each model, the researchers tried to improve the existing models or changing the tools in electrospinning by using another view. In this chapter, a brief review was applied for investigating electrospun nanofiber process and also, the importance of utilizing mathematical models for developing the electrospinning process. All mathematical models and governing equations were investigated in future chapters.

KEYWORDS

- **electrospinning process formulation**
- **electrospun nanofiber models**
- **uniform nanofibers**
- **modeling**
- **simulation**

REFERENCES

1. Lyons, J.; Li, C.; Ko, F. Melt-Electrospinning. Part I: Processing Parameters and Geometric Properties. *Polymer* **2004,** *45* (22), 7597–7603.
2. Reneker, D. H.; Yarin, A. L. Electrospinning Jets and Polymer Nanofibers. *Polymer* **2008,** *49* (10), 2387–2425.
3. De, V, S.; Van, C, T.; Nelvig, A.; Hagström, B.; Westbroek, P.; De, C. K. The Effect of Temperature and Humidity on Electrospinning. *J. Mater. Sci.* **2009,** *44* (5), 1357–1362.
4. Bognitzki, M.; Czado, W.; Frese, T.; Schaper, A.; Hellwig, M.; Steinhart, M.; Greiner, A.; Wendorff, J. Nanostructured Fibers via Electrospinning. *Adv. Mater.* **2001,** *13* (1), 70–72.
5. Deitzel, J. M.; Kleinmeyer, J.; Harris, D. E. A.; Tan, N. C. B. The Effect of Processing Variables on the Morphology of Electrospun Nanofibers and Textiles. *Polymer* **2001,** *42* (1), 261–272.
6. Bhardwaj, N.; Kundu, S. C. Electrospinning: A Fascinating Fiber Fabrication Technique. *Biotechnol. Adv.* **2010,** *28* (3), 325–347.
7. Sill, T. J.; Recum, H. A. Electrospinning: Applications in Drug Delivery and Tissue Engineering. *Biomaterials* **2008,** *29* (13), 1989–2006.
8. Zhou, H. *Electrospun Fibers from Both Solution and Melt: Processing, Structure and Property.* Cornell University, 2007.
9. Zhang, S. *Mechanical and Physical Properties of Electrospun Nanofibers*; 2009; pp 1–83.
10. Tan, S. H.; Inai, R.; Kotaki, M.; Ramakrishna, S. Systematic Parameter Study for Ultra-Fine Fiber Fabrication via Electrospinning Process. *Polymer* **2005,** *46* (16), 6128–6134.
11. Angammana, C. J. *A Study of the Effects of Solution and Process Parameters on the Electrospinning Process and Nanofibre Morphology*, University of Waterloo, 2011.
12. Lu, P.; Ding, B. Applications of Electrospun Fibers. *Rec. Pat. Nanotechnol.* **2008,** *2* (3), 169–182.
13. Rafiei, S.; Maghsoodloo, S.; Saberi, M.; Lotfi, S.; Motaghitalab, V.; Noroozi, B.; Haghi, A. K. New Horizons in Modeling and Simulation of Electrospun Nanofibers: A Detailed Review. *Cellulose Chem. Technol.* **2014,** *48* (5–6), 401–424.
14. Ghochaghi, N. Experimental Development of Advanced Air Filtration Media Based on Electrospun Polymer Fibers. In *Mechnical and Nuclear Engineering*; Virginia Commonwealth: Richmond, VA, 2014; pp 1–165.
15. Yarin, A. L.; Koombhongse, S.; Reneker, D. H. Taylor Cone and Jetting from Liquid Droplets in Electrospinning of Nanofibers. *J. Appl. Phys.* **2001,** *90* (9), 4836–4846.
16. Eggers, J.; Villermaux, E. Physics of Liquid Jets. *Rep. Progr. Phys.* **2008,** *71* (3), 036601–036679.
17. Brooks, H.; Tucker, N. Electrospinning Predictions Using Artificial Neural Networks. *Polymer* **2015,** *58*, 22–29.
18. Reneker, D. H.; Chun, I. Nanometre Diameter Fibres of Polymer, Produced by Electrospinning. *Nanotechnology* **1996,** *7* (3), 216–223.
19. Zaikov, G. E. *Chemical Process in Liquid and Solid Phase Properties, Performance and Applications.* Apple Academic Press: Oakville, ON, 2013; p 526.
20. Šimko, M.; Erhart, J.; Lukáš, D. A Mathematical Model of External Electrostatic Field of a Special Collector for Electrospinning of Nanofibers. *J. Electrostat.* **2014,** *72* (2), 161–165.

21. Deitzel, J. M.; Krauthauser, C.; Harris, D.; Perganits, C.; Kleinmeyer, J. Key Parameters Influencing the Onset and Maintenance of the Electrospinning Jet. In *Polymeric Nanofibers*; Reneker, D. H., Fong, H., Eds.; 2006; pp 56–73.

22. Baumgarten, P. K. Electrostatic Spinning of Acrylic Microfibers. *J. Colloid Interface Sci.* **1971**, *36* (1), 71–79.

23. He, J. H.; Wu, Y.; Zuo, W. W. Critical Length of Straight Jet in Electrospinning. *Polymer* **2005**, *46* (26), 12637–12640.

24. Wu, Y.; Yu, J. Y.; He, J. H.; Wan, Y. Q. Controlling Stability of the Electrospun Fiber by Magnetic Field. *Chaos, Solit. Fract.* **2007**, *32* (1), 5–7.

25. Baji, A.; Mai, Y. W.; Wong, S. C.; Abtahi, M.; Chen, P. Electrospinning of Polymer Nanofibers: Effects on Oriented Morphology, Structures and Tensile Properties. *Compos. Sci. Technol.* **2010**, *70* (5), 703–718.

26. Koombhongse, S.; Liu, W.; Reneker, D. H. Flat Polymer Ribbons and Other Shapes by Electrospinning. *J. Polym. Sci., B: Polym. Phys.* **2001**, *39* (21), 2598–2606.

27. Reneker, D. H.; Yarin, A. L.; Fong, H.; Koombhongse, S. Bending Instability of Electrically Charged Liquid Jets of Polymer Solutions in Electrospinning. *J. Appl. Phys.* **2000**, *87*, 4531–4547.

28. Shin, Y. M.; Hohman, M. M.; Brenner, M. P.; Rutledge, G. C. Electrospinning: A Whipping Fluid Jet Generates Submicron Polymer Fibers. *Appl. Phys. Lett.* **2001**, *78* (8), 1149–1151.

29. Jarusuwannapoom, T.; Hongrojjanawiwat, W.; Jitjaicham, S.; Wannatong, L.; Nithitanakul, M.; Pattamaprom, C.; Koombhongse, P.; Rangkupan, R.; Supaphol, P. Effect of Solvents on Electro-Spinnability of Polystyrene Solutions and Morphological Appearance of Resulting Electrospun Polystyrene Fibers. *Eur. Polym. J.* **2005**, *41* (3), 409–421.

30. Reneker, D. H.; Yarin, A. L.; Zussman, E.; Xu, H. Electrospinning of Nanofibers from Polymer Solutions and Melts. *Adv. Appl. Mech.* **2007**, *41*, 343–346.

31. Tripatanasuwan, S.; Zhong, Z.; Reneker, D. H. Effect of Evaporation and Solidification of the Charged Jet in Electrospinning of Poly(Ethylene Oxide) Aqueous Solution. *Polymer* **2007**, *48* (19), 5742–5746.

32. Li, Z.; Wang, C. *Effects of Working Parameters on Electrospinning, in One-Dimensional Nanostructures*; Springer: Berlin, 2013; pp 15–28.

33. Lukáš, D.; Sarkar, A.; Martinová, L.; Vodsed'álková, K.; Lubasova, D.; Chaloupek, J.; Pokorný, P.; Mikeš, P.; Chvojka, J.; Komarek, M. Physical Principles of Electrospinning (Electrospinning as a Nano-scale Technology of the Twenty-first Century). *Text. Progr.* **2009**, *41* (2), 59–140.

34. Feng, J. J. Stretching of a Straight Electrically Charged Viscoelastic Jet. *J. Non-Newtonian Fluid Mech.* **2003**, *116* (1), 55–70.

35. Thompson, C. J.; Chase, G. G.; Yarin, A. L.; Reneker, D. H. Effects of Parameters on Nanofiber Diameter Determined from Electrospinning Model. *Polymer* **2007**, *48* (23), 6913–6922.

36. Yarin, A. L.; Koombhongse, S.; Reneker, D. H. Bending Instability in Electrospinning of Nanofibers. *J. Appl. Phys.* **2001**, *89* (5), 3018–3026.

37. Thompson, C. J. *An Analysis of Variable Effects on a Theoretical Model of the Electrospin Process for Making Nanofibers*, University of Akron: Akron, 2006.

38. Liu, L.; Dzenis, Y. Simulation of Electrospun Nanofibre Deposition on Stationary and Moving Substrates. *Micro Nano Lett.* **2011**, *6* (6), 408–411.

39. Greenfeld, I.; Arinstein, A.; Fezzaa, K.; Rafailovich, M. H.; Zussman, E. Polymer Dynamics in Semidilute Solution during Electrospinning: A Simple Model and Experimental Observations. *Phys. Rev. E*, **2011,** *84* (4), 041806.

40. Han, J.; Kamber, M.; Pei, J. *Data Mining: Concepts and Techniques*, third ed. Elsevier: Amsterdam, 2011.

41. Law, A. M. *Simulation Modeling and Analysis.* McGraw Hill: Boston, MA, 2015.

42. Saoud, M. S.; Boubetra, A.; Attia, S. How Data Mining Techniques Can Improve Simulation Studies. *Int. J. Comput. Theory Eng.* **2014,** *6* (1), 15–19.

43. Kowalewski, T. A.; Barral, S.; Kowalczyk, T. Modeling Electrospinning of Nanofibers. In *IUTAM Symposium on Modelling Nanomaterials and Nanosystems.* Springer: Berlin, 2009.

44. Rafiei, S.; Maghsoodloo, S.; Noroozi, B.; Mottaghitalab, V.; Haghi, A. K. Mathematical Modeling in Electrospinning Process of Nanofibers: A Detailed review. *Cell. Chem. Technol.* **2013,** *47* (5–6), 323–338.

45. Patanaik, A.; Jacobs, V.; Anandjiwala, R. D. *Experimental Study and Modeling of the Electrospinning Process*; 2008; pp 1160–1168.

46. Fridrikh, S. V.; Yu, J. H.; Brenner, M. P.; Rutledge, G. C. Controlling the Fiber Diameter during Electrospinning. In *Physical Review Letters*; American Physical Society, 2003; pp 144502–144505.

47. Ciechańska, D. Multifunctional Bacterial Cellulose/Chitosan Composite Materials for Medical Applications. *Fibres Text. East. Eur.* **2004,** *12* (4), 69–72.

48. Melcher, J. R.; Taylor, G. I. Electrohydrodynamics: A Review of the Role of Interfacial Shear Stresses. *Annu. Rev. Fluid Mech.* **1969,** *1* (1), 111–146.

49. Ramos, J. I. One-Dimensional Models of Steady, Inviscid, Annular Liquid Jets. *Appl. Math. Model.* **1996,** *20* (8), 593–607.

50. Spivak, A. F.; Dzenis, Y. A. Asymptotic Decay of Radius of a Weakly Conductive Viscous Jet in an External Electric Field. *Appl. Phys. Lett.* **1998,** *73* (21), 3067–3069.

51. Reneker, D. H.; Yarin, A. L.; Fong, H.; Koombhongse, S. Bending Instability of Electrically Charged Liquid Jets of Polymer Solutions in Electrospinning. *J. Appl. Phys.* **2000,** *87* (9), 4531–4547.

52. Dasri, T. Mathematical Models of Bead-Spring Jets during Electrospinning for Fabrication of Nanofibers. *Walailak J. Sci. Technol.* **2012,** *9* (4), 287–296.

53. Shin, Y. M.; Hohman, M. M.; Brenner, M. P.; Rutledge, G. C. Experimental Characterization of Electrospinning: The Electrically Forced Jet and Instabilities. *Polymer* **2001,** *42* (25), 09955–09967.

54. Feng, J. J. The Stretching of An Electrified Non-Newtonian Jet: A Model for Electrospinning. *Phys. Fluids (1994–present)* **2002,** *14* (11), 3912–3926.

55. Wan, Y. Q.; Guo, Q.; Pan, N. Thermo-electro-hydrodynamic Model for Electrospinning Process. *Int. J. Nonlin. Sci. Numer. Simul.* **2004,** *5* (1), 5–8.

56. He, J. H.; Wu, Y.; Pang, N. A Mathematical Model for Preparation by AC-Electrospinning Process. *Int. J. Nonlin. Sci. Numer. Simul.* **2005,** *6* (3), 243–248.

57. Xu, L.; Wang, L.; Faraz, N. A Thermo-Electro-Hydrodynamic Model for Vibration-Electrospinning Process. *Therm. Sci.* **2011,** *15*, S131–S135.

58. Haghi, A. K.; Zalkov, G. E. Mathematical Models on the Transport Properties of Electrospun Nanofibers. In *Nanopolymers and Modern Materials: Preparation, Properties, and Applications*; CRC Press: Boca Raton, FL, 2013; pp 195–218.

CHAPTER 4

A CONDUCTIVE VISCOUS JET AND ELECTROHYDRODYNAMICS: UPDATE ON MODELS

SHIMA MAGHSOODLOU* and S. PORESKANDAR

Textile Engineering, University of Guilan, Rasht, Iran

Corresponding author. E-mail: sh.maghsoodlou@gmail.com

CONTENTS

ABSTRACT

In this chapter, motion of a weakly conductive viscous jet accelerated by an external electric field will be considered. Nonlinear rheological constitutive equation applicable for polymer fluids (Ostwald–de Waele law) is utilized. A differential equation for the variation of jet radius with axial coordinate is derived. Asymptotic variation of the jet radius at large distances from the jet origin is analyzed. Allometric relationships will be extracted from allometric scaling, which are different depend on polymer properties.

4.1   INTRODUCTION

The flow of liquid jets deformed and accelerated by an external electric field is a coupled electromechanics problem that has attracted considerable interest.[1,2] Electrostatically driven jets are involved in a variety of applications, including electrostatic atomization of liquids.[3] The behavior of electrostatically driven jets at large distances from their origin is hard to study due to the fact that low viscosity and low molecular weight fluids are utilized in most electrostatic jet applications break up into droplets long before the jet reaches its asymptotic length. This breakup is due to the longitudinal Rayleigh instability, caused by surface capillary waves.[4] The Rayleigh instability is not typically observed in electrospinning of polymer fluids.[5,6] However, a transverse instability or splaying of the jet into two or more smaller jets is sometimes observed. As a rule, the transverse jet splaying occurs further away from the jet origin. Therefore, peculiarities of jet flow at large distances are important for jet splaying analysis. When it's close to the jet breakup, surface-charge advection is dominant, and the velocity in z-direction keeps unchanged.

The process just likes parachute jump. Initially, it accelerates due to its velocity becomes higher and higher; while the velocity increases, the air resistance increases while its acceleration decelerates until it become zero. In addition, asymptotic results can be used to evaluate diameters of polymer fibers electrospun in a single-jet flow regime. A simple model of an electrostatically driven Newtonian jet will be developed.[7] Jet evaporation is shown in Figure 4.1.

Many experiment shows scaling relationship between R and z, which can be expressed as an allometric equation of the form (Fig. 4.2)

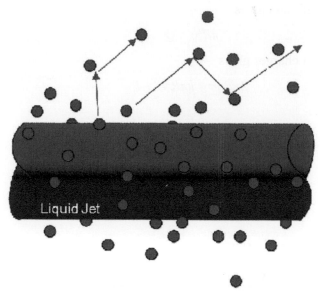

FIGURE 4.1 Evaporating molecules surrounding the liquid jet.

$$R \sim z^b. \tag{4.1}$$

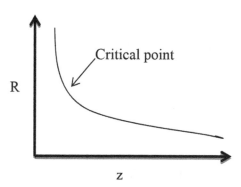

FIGURE 4.2 Relationship between R and z.

Assume that the volume flow rate and the current keep unchanged during the electrospinning procedure

$$Q \approx R^0 \tag{4.2}$$

and

$$I \approx R^0 \tag{4.3}$$

Conservation of mass gives

$$\pi R^2 v = Q \tag{4.4}$$

Conservation of charges gives

$$2\pi R v \sigma + k\pi R^2 E = I \tag{4.5}$$

So

$$v \approx R^{-2} \tag{4.6}$$

$$\sigma \approx R \tag{4.7}$$

$$E \approx R^{-2} \tag{4.8}$$

4.2   STEADY PART OF JET

At the initial stage of the electrospinning, electrical force is dominant over other forces acting on the jet[7]:

$$\frac{d}{dz}\left(\frac{v^2}{2}\right) = \frac{2\sigma E}{\rho R}. \tag{4.9}$$

From the previous equations

$$\frac{d}{dz}\left(R^{-4}\right) \approx R^{-2} \tag{4.10}$$

that leads to the following scaling:

$$R \approx z^{-1/2}. \tag{4.11}$$

4.3   INSTABILITY PART OF JET

Instability occurs when combining force of the electric and viscous forces approximately vanishes[7–9]

$$\frac{2\sigma E}{R} + \frac{d\tau}{dz} \approx 0. \tag{4.12}$$

The gradient of pressure was assumed to be kept unchanged during this stage

$$\frac{dp}{dz} \approx \text{Constant.} \tag{4.13a}$$

Under the instability condition, reduces to

$$\frac{d}{dz}\left(\frac{v^2}{2}\right) = -\frac{1}{\rho}\frac{dp}{dz} = \text{Constant.} \tag{4.13b}$$

In view of the scaling relation

$$\frac{d}{dz}\left(R^{-4}\right) \approx R^0. \tag{4.14}$$

Asymptotic behavior of the model is evaluated under the assumption that the effects of the viscous forces are negligible (ideal liquid approximation). The power-law asymptote with the exponent 1/4 is obtained for the jet radius[7]:

$$R \approx z^{-1/4}. \tag{4.15}$$

However, the polymer fluids are highly viscous and the effects of the viscous forces cannot be neglected. In addition, the polymer fluids often exhibit nonlinear rheological behavior.

4.4 REACHING TO COLLECTOR

When $z \to \infty$, acceleration in z-direction vanishes completely

$$\frac{d}{dz}\left(\frac{v^2}{2}\right) = 0 \tag{4.16}$$

that leads to the scaling law

$$R \approx z^0. \tag{4.17}$$

In this chapter, a broader class of fluids will be described by the nonlinear power-law rheological constitutive equation.

4.5 A SIMPLE MATHEMATICAL MODEL

A model of the jet motion will be formulated taking into account inertial, hydrostatic, viscous, electric, and surface-tension forces and the asymptotic behavior of the jet at large distances will be analyzed. For this purpose, consider an infinite viscous jet pulled from the capillary orifice and accelerated by a constant external electric field. In Figure 4.3, the effect of electric field on jet can be seen.[8,9]

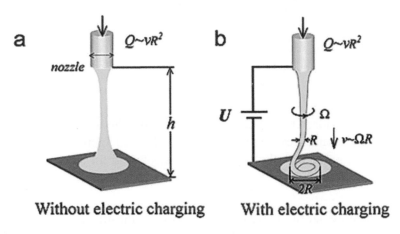

Without electric charging   With electric charging

FIGURE 4.3 Electric field (a) forms a straight jet without electric charging and (b) is induced to coil under an applied voltage.

Neglecting the magnetic effects, the general three-dimensional linear momentum balance equation for the jet element is

$$\rho(\vec{v}\cdot\nabla)\vec{v}+\nabla\rho=\nabla\vec{\tau}+\nabla\vec{\tau}. \tag{4.18}$$

Rheological behavior of many fluids, including the polymer fluids (Fig. 4.4), can be described by the power-law constitutive equation, known as the Ostwald–de Waele law[4]:

$$\vec{\tau} = \mu\left[tr-\left(\vec{\gamma}\right)\right]^{(m-1)/2}\vec{\gamma}. \tag{4.19}$$

Viscous Newtonian fluids are described by a special case of eq 4.19 with the flow index $m = 1$. Pseudoplastic (shear thinning) fluids are described by the flow indices $0 \leq m \leq 1$. Dilatant (shear thickening) fluids are described by the flow indices $m > 1$.[8]

The differential equation of momentum balance, eq 4.18, is complemented by the equations of mass balance, electric charge balance, and the electrostatic field equation.[8,9]

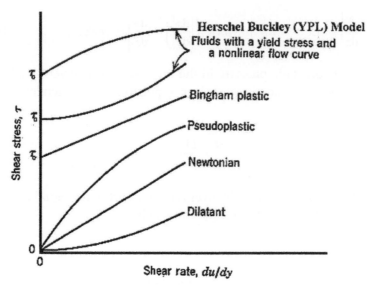

FIGURE 4.4 Flow curves for various types of fluids.

Assume that the jet flow is an extensional axisymmetric flow in the direction of the external electric field. Denote z as the axial coordinate. Consider a weakly conductive jet. In such a jet, electrical current due to electronic or ionic conductivity of the fluid is small compared with the current provided by the convective charge transfer with moving jet particles. However, the conductivity is sufficient for the electric charges to migrate the short distance to the jet surface. The bulk electric charge can then be assumed zero in the asymptotic limit. The surface charge will interact with the external electric field creating the pulling force responsible for jet acceleration. In addition, the surface charge will cause transverse electric repulsion that will lower the hydrostatic pressure. The overall electric potential, φ, can be obtained as a sum of the potential of the surface charge, φ_s, and the potential of the external field, $\varphi_{ext} = -E_0 z$, where E_0 is the electric field.

$$\varphi = \varphi_s + \varphi_{ext} - E \cdot z. \tag{4.20}$$

Assume that the slope of the jet surface in the direction of the flow is small, $dR/dz \ll 1$, where R is the jet radius. Further, assume that the effect of the surface charge on the axial component of the electric field is negligible. The linear momentum balance, eq 4.18, can then be averaged over the jet cross section.[4] The resulting equation in the axial direction is

$$\frac{d}{dz}\left[\frac{\rho}{2}\pi R^2 v^2 + \pi R^2 p - \mu\pi R^2 \left(\left|\frac{dv}{dz}\right|\right)^{m-1}\frac{dv}{dz}\right] = 2\pi R\Omega E_0. \tag{4.21}$$

The average hydrostatic pressure in the cross section is determined by the surface tension and transverse electric repulsion. For a slender jet, it is approximated by

$$p = \frac{\sigma_s}{R} - \frac{\Omega^2}{2\varepsilon_0}. \tag{4.22}$$

Equation 4.22 can be obtained by averaging eq 4.18 in the radial direction. Averaging the mass-balance equation yields

$$\pi R^2 v = Q. \tag{4.23}$$

Similarly, the electric charge balance equation reduces to

$$\frac{2Q\Omega}{R} - \pi\sigma R^2\frac{d\Phi}{dz} = I. \tag{4.24}$$

In the weakly conductive jet, the electric current is defined primarily by the convective charge transfer. Therefore, in the asymptotic limit, the electric current can be approximated by

$$I = \frac{2Q\Omega}{R} \tag{4.25}$$

and the ratio of the surface charge density and jet radius is constant

$$\frac{\Omega}{R} = \frac{I}{2Q}. \tag{4.26}$$

The flow rate and the electric current are considered external parameters of the problem. Let us introduce the dimensionless jet radius $\tilde{R} = R/R_0$ and axial coordinate $\tilde{z} = z/z_0$, where

$$z_0 = \frac{\rho Q^3}{2\pi^2 R_0^4 E_0 I}. \tag{4.27}$$

The characteristic jet radius, R_0, is sometimes taken equal to the radius of the capillary orifice.[2] Equations 4.21 and 4.22 reduce to the following dimensionless equation for the jet radius:

The most significant parameter to jet decay is the reduced wave number

$$\chi = \frac{2\pi h_0}{\lambda}.$$ (4.29)

At a resonance or "Rayleigh" wave number $x = x_R$ perturbations grow fastest, and the distance between the nozzle and the first drop "breakup length," is the shortest. For $x > 1$ or without driving the Rayleigh mode is selected from a spectrum of tiny initial perturbations by virtue of its dominant growth, but breakup becomes much more irregular.

The Weber number, $We = 2\pi^2 R_0^3 \sigma_s / \rho Q^2$, describes the ratio of the surface-tension forces to the inertia forces.

It measures the ratio of the kinetic energy of a drop issuing from the jet relative to its surface energy. The temporal perturbation on a jet is translated into space by convection with velocity. As perturbations grow along the jet on a timescale measures, how much a disturbance can grow from one swell to the next.

The parameter $\gamma = \pi^2 I^2 R_0^6 / 4\varepsilon_0 \rho Q^4$ describes the ratio of the electric forces to the inertia forces.

The effective Reynolds number for the fluid characterized by the power-law constitutive eq 4.20, $Re = \left(Q^2 \rho \right) / \left(2\pi^2 R_0^4 \mu \right) \left[\left(4\pi E_0 IR_0^2 \right) / Q^2 \rho \right]^{-m}$, describes the ratio of the inertia forces to the viscous forces.

The Bernoulli integral obtained from eq 4.28 is

$$\tilde{R}^{-4} + We\tilde{R}^{-1} - \gamma R^2 - \frac{1}{Re} \left(\frac{1}{2} \frac{d}{d\tilde{z}} \tilde{R}^{-2} \right)^m = \tilde{z} + c.$$ (4.30)

A general closed-form solution of eq 4.30 is not available. Then, the power-law asymptotic approximation of the jet radius were considered as below

$$\tilde{R} \sim \tilde{z}^{-\alpha}$$ (4.31)

where the exponent, α, is a positive constant. Substituting eq 4.31 into eq 4.30 gives

$$\tilde{z}^{4\alpha} + We\tilde{z}^{-\alpha} - \gamma \tilde{z}^{-2\alpha} - \frac{\alpha^m}{Re} \tilde{z}^{(2\alpha-1)m} - \tilde{z} = O(1).$$ (4.32)

The power balance at $\tilde{z} \to +\infty$ yields

$$4\alpha = \max\left[1, (2\alpha-1)m \right].$$ (4.33)

The analysis presented here extends the limit of applicability of the power-law asymptote,[7] eq 4.15, to pseudoplastic and dilatant fluids with the flow index m between 0 and 2. More complicated asymptotic behavior of the dilatant fluids with the flow index $m > 2$ is discovered. The obtained results can be used as a basis for stability analysis of viscous polymer jets.

4.6   CONCLUDING REMARKS

In this chapter, a mathematical model was utilized for understanding the rheological behavior of fluids. Thus, nonlinear rheological constitutive equation applicable for polymer fluids (Ostwald–de Waele law) was utilized. Asymptotic variation of the jet radius at large distances from the jet origin was analyzed. Different part of jet was studied by this method. It was found that the well-known power-law asymptote for Newtonian fluids with the exponent 1/4 holds for more general class of fluids like pseudoplastic and dilatant fluids with the flow index between 0 and 2. Dilatant fluids with the flow index greater than 2 exhibit power-law asymptotes with the exponents depending on the flow index. The obtained results can be used as a basis for stability analysis of viscous polymer jets.

KEYWORDS

- viscous jet
- rheological behavior of fluids
- mathematical model
- allometric scaling
- Ostwald–de Waele equation

REFERENCES

1. Saville, D. A. Electrohydrodynamics: The Taylor–Melcher Leaky Dielectric Model. *Annu. Rev. Fluid Mech.* **1997,** *29* (1), 27–64.
2. Calvo, G.; Davila, A. M.; Barrero, J. A. Current and Droplet Size in the Electrospraying of Liquids. Scaling Laws. *J. Aerosol Sci.* **1997,** *28* (2), 249–275.
3. Bailey, A. G. *Electrostatic Spraying of Liquids.* Wiley: New York, 1988.

4. Yarin, A. L. *Free Liquid Jets and Films: Hydrodynamics and Rheology*; Longman. New York, 1993.
5. Doshi, J.; Reneker, D. H. Electrospinning Process and Applications of Electrospun Fibers. *J. Electrostat.* **1995,** *35* (2), 151–160.
6. Reneker, D. H.; Chun, I. Nanometre Diameter Fibres of Polymer, Produced by Electro-spinning. *Nanotechnology* **1996,** *7* (3), 216–223.
7. Kirichenko, V. N.; Petryanov, S. I.; Suprun, N. N.; Shutov, A. A. *Asymptotic Radius of a Slightly Conducting Liquid Jet in an Electric Field.* Soviet Physics Doklady, 1986.
8. Melcher, J. R.; Taylor, G. I. Electrohydrodynamics: A Review of the Role of Interfacial Shear Stresses. *Annu. Rev. Fluid Mech.* **1969,** *1* (1), 111–146.
9. Groot, S. R. D.; Suttorp, L. G. *Foundations of Electrodynamics*, 1972.

CHAPTER 5

MACROSCOPIC MODELS FOR ELECTROSPINNING

SHIMA MAGHSOODLOU* and S. PORESKANDAR

Textile Engineering, University of Guilan, Rasht, Iran

Corresponding author. E-mail: sh.maghsoodlou@gmail.com

CONTENTS

ABSTRACT

A charged polymer jet uses accelerate and stretch by an external electric field for producing ultrafine nanofibers. Electrohydrodynamic models for electrospinning Newtonian jets are proposed in this chapter. A problem arises, however, with the boundary condition at the nozzle. Unless the initial surface charge density is very small near zero, the jet bulges out upon exiting the nozzle in a ballooning instability, which never occurs in reality. The stretching of an electrified jet is governed by the interplay among electrostatics, fluid mechanics and rheology, and the role of viscoelasticity. This chapter presents a slender-body theory for the stretching of a straight charged jet of Giesekus fluid. Results show strain-hardening as the most influential rheological property. It causes the tensile force to rise at the start, which enhances stretching of the jet. Further downstream, however, the higher elongational viscosity tends to suppress jet stretching. In the end, strain-hardening leads to thicker fibers. This confirms the main result of a previous study using empirical rheological models. The behavior of the electrically driven jet forms an interesting contrast to that in conventional fiber spinning.

5.1   INTRODUCTION

As mentioned before in previous chapters, electrospinning typically involves two stages. In the first, a polymer jet issues from the nozzle and is accelerated and stretched smoothly by electrostatic forces. In the second stage, a bending instability occurs farther downstream when the jet gets sufficiently thin, and the fiber spirals violently. The enormously increased contour length produces a very large stretch ratio and a nanoscale diameter. For the steady stretching in stage one, Spivak and Dzenis[1] published a simple model that assumes the electric field to be uniform and constant, unaffected by the charges carried by the jet. Hohman et al.[2,3] developed a slender-body theory for electrospinning that couples jet stretching, charge transport, and the electric field. The model encounters difficulties, however, with the boundary condition at the nozzle. Steady solutions may be obtained only if the surface charge density at the nozzle is set to zero or a very low value. Even after this drastic assumption, no steady solution was possible for fluids with higher conductivities. For stage two, the bending instability has been carefully documented by two group;[4-6] each has proposed a theory for the instability. Reneker et al. modeled the polymer jet by a linear Maxwell equation. Like-charge repulsion generates a bending force that destabilizes the jet.

Electrospinning is an example of an electrohydrodynamic phenomenon. In electrohydrodynamics (EHD), charges induce fluid motion within an electric field. During the process, the transport and distribution of these charges generate stresses that result in the movement of the fluid. The leaky dielectric EHD model is an appropriate model to use because the model of the fluid's electrical properties as a poorly conducting liquid is comparable to the behavior of most polymer solutions, the most commonly used type of fluid in electrospinning.[7,8] Hohman et al.[2] built an electrohydrodynamic instability theory and predicted that under favorable conditions, a nonaxisymmetric instability prevails over the familiar Rayleigh instability and a varicose instability due to electric charges. In theoretical work to date, the rheology of the polymer jet has been represented by a Newtonian viscosity,[2,3] a power-law viscosity,[1] and the linear Maxwell equation.[4,5]

On the other hand, for a polymer with high molten temperature, thermal factor is critical for the process. So a rigorous thermo-electrohydrodynamics description of electrospinning is needed for better understanding of the process.[9] This study establishes a mathematical model to explore the physics behind electrospinning.

This chapter concerns the first stage only. The steady stretching process is important in that it not only contributes to the thinning directly but also sets up the conditions for the onset of the bending instability. Therefore, three main macroscopic models will be reviewed in detail: leaky dielectric model, EHD model, and thermo-electrohydrodynamic model.

5.2 SLENDER-BODY THEORY

Jet dynamics probes a wide range of physical properties, such as liquid surface tension, viscosity, or non-Newtonian rheology and density contrast with its environment. Jets are also sensitive, on very small scales (typically nanometers), to thermal fluctuations. On very large scales, on the other hand, gravitational interactions are important. The basic flow state can be both laminar and turbulent. The carrying fluid can be electrically charged, or magnetic.

Experimentally, various laboratory techniques, and notably high-speed digital cinematography in recent years, have revived the subject. Minute details of the breakup process can now be documented in real time, as well as the structure of the resulting spray, thus providing us with a rich source of information for comparison with theory.

On the analytical side, the most basic tool is linear stability analysis around the cylindrical base state. However, there are many important features of the breakup process for which nonlinear effects are dominant. Numerically, this remains an extremely hard problem.[10]

In electrospinning, the jet is elongated by electrostatic forces and gravity, while surface tension, viscosity, and inertia also play a part. As the jet thins, the surface charge density is varied, which in turn affects the electric field and the pulling force. The flow field and electric field are thus intimately coupled. It is assumed that the liquid is weakly conducting so the leaky dielectric model applies.[7] Thus, the jet carries electric charges only on its surface; any charges in the interior are quickly conducted to the surface. Meanwhile, the fluid is sufficiently dielectric so as to sustain an electric field tangential to the jet surface. The slender-body approximation is a familiar concept in dealing with jets and drops. It has been widely used in fiber spinning of viscoelastic liquids and in electrospraying. The standard assumptions for slender jets are

a) the jet radius R decreases lowly along the axial direction $z : \left| dR(z)/dz \right| \ll 1$,
b) the axial velocity is uniform in the cross section of the jet.

Thus, the flow to a nonuniform elongation can be simplified, with all quantities depending only on the axial position z.

5.3   GOVERNING EQUATIONS

The jet is governed by four steady-state equations representing the conservation of mass and electric charges, the linear momentum balance, and Coulomb's law. Mass conservation requires that

$$\pi R^2 v = Q \tag{5.1}$$

Charge conservation may be expressed by

$$\pi R^2 kE + 2\pi R v\sigma = I \tag{5.2}$$

The momentum equation is formulated by considering the forces on a short segment of the jet (Fig. 5.1):

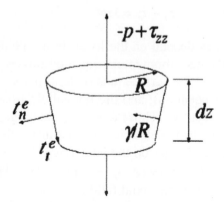

FIGURE 5.1 Momentum balance on a short section of the jet.

$$\frac{d}{dz}\left(\pi R^2 \rho v^2\right) = \pi R^2 \rho g + \frac{d}{dz}\left[\pi R^2\left(-p+\tau_{zz}\right)\right] + \frac{\gamma}{R}\cdot 2\pi RR' + 2\pi R\left(t_t^e - t_n^e R'\right) \quad (5.3)$$

The prime indicates derivative with respect to z, and R is the slope of the jet surface. The ambient pressure has been set to zero. The electrostatic tractions are determined by the surface charge density and the electric field[7]:

$$t_n^e = \left\|\frac{\varepsilon}{2}\left(E_n^2 - E_t^2\right)\right\| \approx \frac{\sigma^2}{2\overline{\varepsilon}} - \frac{\overline{\varepsilon}-\varepsilon}{2}E^2 \quad (5.4)$$

$$t_t^e = \sigma E_t \approx \sigma E. \quad (5.5)$$

We have used the jump conditions for E_n and E_t:
$\left\|\varepsilon E_n\right\| = \overline{\varepsilon}\overline{E}_n - \varepsilon E_n = \sigma, \left\|E_t\right\| = \overline{E}_t - E_t = 0$ and assumed that $\varepsilon E_n \lll \overline{\varepsilon}\overline{E}_n$ and $E_t \approx E$. The over bar indicates quantities in the surrounding air. The pressure $p(z)$ is determined by the radial momentum balance, and applying the normal force balance at the jet surface leads to

$$-p+\tau_{zz} = t_t^e - \frac{\gamma}{R}. \quad (5.6)$$

Inserting eqs 5.4–5.6 into eq 3.3 yields

$$\rho v v' = \rho g + \frac{3}{R^2}\frac{d}{dz}\left(\eta R^2 v'\right) + \frac{vR'}{R^2} + \frac{\sigma\sigma'}{\overline{\varepsilon}} + \left(\varepsilon - \overline{\varepsilon}\right)EE' + \frac{2\sigma E}{R} \quad (5.7)$$

where a generalized Newtonian constitutive relation has been used for the viscous normal stress difference:

$$\tau_{zz} - \tau_{rr} = 3\eta\dot{v} \tag{5.8}$$

and the viscosity η may depend on the local strain rate or the accumulated strain. Spivak and Dzenis'[1] momentum equation misses several terms: the viscous normal stress t_{rr}; the surface tension term in the z equation (though it is included in the r equation); and the normal electrostatic traction t_n^e in both axial and radial balances. Those terms are not, in general, smaller than the terms retained. Finally, the electric potential inside the jet is determined by the free and induced charges on the jet surface. The induced charges are determined by E_n and $\bar{E}_n$: $\sigma_{ind} = (\varepsilon - \varepsilon_0)E_n - (\bar{\varepsilon} - \varepsilon_0)\bar{E}_n$. The normal field E_n is related, via Gauss' law, to the axial field E:

$$2\pi R E_n + \frac{d}{dz}(\pi R^2 E) = 0. \tag{5.9}$$

Since $\bar{\varepsilon}\bar{E}_n - \varepsilon E_n = \sigma$, E_n $\bar{E}_n$ can also be expressed in terms of E. Now, the potential along the centerline of the jet, due to the total surface charges $\sigma + \sigma_{ind}$ may be obtained by Coulomb's law:

$$\Phi(z) = \Phi_\infty(z) + \frac{1}{2\bar{\varepsilon}}\int\frac{R d\xi}{\sqrt{(z-\xi)^2 + R^2}} - \frac{\beta}{4}\int\frac{d(ER^2)/d\xi}{\sqrt{(z-\xi)^2 + R^2}}d\xi \tag{5.10}$$

where Φ_∞ is the potential due to the *external* field in the absence of the jet, $\beta = (\varepsilon/\bar{\varepsilon}) - 1$, and the integration is over the entire length of the fiber L. To avoid solving an integer differential equation, one realizes that the dominant contribute onto the integrals comes from the two regions satisfying $R \lll |z - \xi| \lll L$. An asymptotic estimation leads to

$$\Phi(z) \approx \Phi_\infty(z) + \ln\chi\left(\frac{1}{\bar{\varepsilon}}\sigma R - \frac{\beta}{2}\frac{d(ER^2)}{dz}\right) \tag{5.11}$$

The axial field is, therefore,

$$E(z) = E_\infty(z) - \ln\chi\left(\frac{1}{\bar{\varepsilon}}\frac{d(\sigma R)}{dz} - \frac{\beta}{2}\frac{d^2(ER^2)}{dz^2}\right). \tag{5.12}$$

From this point on, we will take the external field E_∞ to be spatially uniform. Equations 5.9 and 5.10 have been previously derived by Hohman et al.[2] by using the idea of an effective line charge along the axis of the jet. Spivak and Dzenis neglected the axial field due to surface charges and assumed a constant $E = E_\infty$ everywhere.[1] This will be seen to be a poor approximation since $E(z)$ typically varies greatly. Now, we have eqs 5.1, 5.2, 5.7, and 5.10 for the four unknown functions R, v, E, and σ. To achieve better agreement with experimental data, however, Hohman et al.[3] used a more sophisticated method to compute E. First, eq 5.9 is not reduced to eq 5.10. To account for the capacitor plates in the experiment and the protruding nozzle, image charges are added to the integro-differential equation, as is a fringe field due to the charged nozzle. Interestingly, the more complex model suffers from a difficulty with the boundary condition at the origin of the jet, while the simpler model does not.

5.4 DIMENSIONLESS PARAMETERS

Parameters in electrospinning fall into three categories: process parameters (Q, I, and E_∞), geometric parameters (R_0 and L), and material parameters (ρ, η_0, ε, $\bar{\varepsilon}$ K, and γ). Among those, R_0 is the radius at the origin of the jet just outside the nozzle, and η_0 is the zero-shear-rate viscosity. We adopt the following characteristic quantities:

Length: R_0

Velocity: $v_0 = \dfrac{Q}{\pi R_0^2}$

Electric field: $E_0 = \dfrac{I}{\pi R_0^2 k}$

Surface charge density: $\sigma_0 = \bar{\varepsilon} E_0$

Scaling all quantities using these characteristic values, and denoting the dimensionless quantities using the same symbols, we arrive at the following dimensionless governing equations:

$$R^2 v = 1 \tag{5.13}$$

$$ER^2 + PeRv\sigma = 1 \tag{5.14}$$

$$vv' = \frac{1}{Fr} + \frac{3}{Re}\frac{1}{R^2}\frac{d\left(\eta R^2 v'\right)}{dz} + \frac{1}{We}\frac{R'}{R^2} + \varepsilon\left(\sigma\sigma' + \beta EE' + \frac{2E\sigma}{R}\right) \tag{5.15}$$

where

$$\varepsilon = \frac{\overline{\varepsilon} E_0^2}{\rho v_0^2}$$

$$E = E_\infty - \ln \chi \left(\frac{d(\sigma R)}{dz} - \frac{\beta}{2} \frac{d^2 (ER^2)}{dz^2} \right). \qquad (5.16)$$

5.5 ELECTROHYDRODYNAMIC MODEL FOR ELECTROSPINNING PROCESS

In 1969, Melcher and Taylor[8] investigated on a branch of fluid mechanics named electrohydrodynamic which involves the effects of moving media on electric fields for the first time after William Gilbert's experiments in the seventeenth century. Next, in 1997, Savile[7] studied this phenomena. Applications of EHD abound: spraying, the dispersion of one liquid in another, coalescence, ink-jet printing, boiling, augmentation of heat and mass transfer, fluidized bed stabilization, pumping, and polymer dispersion are but a few.

The leaky dielectric model consists of the Stokes equations to describe fluid motion and an expression for the conservation of current employing an Ohmic conductivity. Electromechanical coupling occurs only at fluid–fluid boundaries where the charge, carried to the interface by conduction, produces electric stresses different from those present in perfect dielectrics or perfect conductors. With perfect conductors or dielectrics the electrical stress is perpendicular to the interface, and alterations of interface shape combined with interfacial tension serve to balance the electric stress. Leaky dielectrics are different because free charge accumulated on the interface modifies the field. Viscous flow develops to provide stresses to balance the action of the tangential components of the field acting on interface charge.

5.6 BALANCE LAWS

The differential equations describing EHD arise from equations describing the conservation of mass and momentum, coupled with Maxwell's equations. To establish a context for the approximations inherent in the leaky dielectric model, it is necessary to look on a deeper level.[7]

The hydrodynamic model consists of the Stokes equations without any electrical forces which are coupled to the electric field occurs at boundaries, so forces from the bulk-free charge must be negligible.

5.7   LEAKY DIELECTRIC MODEL FOR ELECTROHYDRODYNAMIC

To summarize, the leaky dielectric electrohydrodynamic model consists of the following five equations. The derivation is given here identifies the approximations in the leaky dielectric model. Except for the electrical body force terms, it is essentially the model proposed by Melcher and Taylor.[8]

a)   Assumptions:
b)   Dielectric fluid (poorly conductivity liquid)
c)   Bulk charge $= 0$
d)   Axial motion
e)   Steady part.

Fluid movement and a formation of a current balance equation are written by these considerations[7,11]:

a)   Electromechanical coupling: fluid-to-fluid interface.
b)   Normal stress balances by changing the shape of the interface and interfacial tension.
c)   If electric field $\neq 0$, tangential stress balances by viscous force and causes the fluid motion.
d)   Liquid bulk is considered as quasi-neutral and free charges confined by a very thin layer under the liquid–gas interface.

Maxwell stress depends on electrostatic phenomena and hydrodynamic behavior. So the electrical phenomena are described by[7,8]

$$\nabla \cdot \varepsilon\varepsilon_0 E = \rho^e \qquad (5.17)$$

and

$$\nabla \times E = 0 \qquad (5.18)$$

The relationship between Maxwell stresses and the electrical body force is to suppose that electrical forces exerted on free charge and charge dipoles are transferred directly to the fluid.[8,12]

$$P = NQd \qquad (5.19)$$

Polarization depends on the volumetric charge density (ρ^P) and surface charge density.

$$P = N\alpha\varepsilon\varepsilon_0 E \quad \text{and} \quad \nabla \cdot P = -\rho^P \tag{5.20}$$

The Coulomb force (due to free charge) is

$$F_{\text{Coulomb}} = \rho^F E \tag{5.21}$$

Total electrical force per unit volume is

$$\rho^F E + P \cdot \nabla E = \nabla \cdot \left(\varepsilon\varepsilon_0 EE - \frac{1}{2}\varepsilon\varepsilon_0 E \cdot E\delta \right) \tag{5.22}$$

which balance by pressure gradient can be shown as

$$-\nabla P^* + \rho^F E + P \cdot \nabla E = 0 \tag{5.23}$$

and the isotropic effect of E is

$$P^* = P + \frac{1}{2}\varepsilon_0 \left[\varepsilon - 1 - \rho\left(\frac{\partial\varepsilon}{\partial\rho}\right)_T \right] E \cdot E. \tag{5.24}$$

Therefore,

$$-\nabla P + \underbrace{\nabla \cdot \left(\varepsilon\varepsilon_0 EE - \frac{1}{2}\varepsilon\varepsilon_0 \left[1 - \frac{\rho}{\varepsilon}\left(\frac{\partial\varepsilon}{\partial\rho}\right)_T \right] E \cdot E\delta \right)}_{\text{Maxwell stress tensor} = \sigma^M} = 0. \tag{5.25}$$

The equation of motion of an incompressible Newtonian fluid of uniform viscosity[7]:

$$\rho\frac{Du}{Dt} = -\nabla P + \nabla \cdot \sigma^M + \mu\nabla^2 u \tag{5.26}$$

The electrical stresses emerge as body forces due to a nonhomogeneous dielectric permeability and free charge:

$$\rho\frac{Du}{Dt} = -\nabla\left[p - \frac{1}{2}\varepsilon_0\rho\left|\frac{\partial\varepsilon}{\partial\rho}\right|_T E \cdot E \right] - \frac{1}{2}\varepsilon_0 E \cdot E\nabla\varepsilon + \mu\nabla^2 u. \tag{5.27}$$

EHD motions are driven by the electrical forces on boundaries or in the bulk. The net Maxwell stress at a sharp boundary has the normal and tangential components.[8]

$$\left[\sigma^M \cdot n\right] \cdot n = \frac{1}{2}\left\| \varepsilon\varepsilon_0 \left(E \cdot n\right)^2 - \varepsilon\varepsilon_0 \left(E \cdot t_1\right)^2 - \varepsilon\varepsilon_0 \left(E \cdot t_2\right)^2 \right\|$$

$$\left[\sigma^M \cdot n\right] \cdot t_i = qE \cdot t_i \qquad (5.28)$$

Finally, the equations are written in dimensionless variables using the scales defined. The equation of motion is for nonhomogeneous fluids with electrical body forces. The hydrodynamic boundary conditions, continuity of velocity and stress, including the viscous and Maxwell stress are assumed.[7]

$$\frac{\tau_\mu}{\tau_p}\frac{\partial u}{\partial t} + Reu \cdot \nabla u = -\nabla p - \frac{1}{2}E \cdot E\nabla\varepsilon + \nabla \cdot (\varepsilon E)E + \nabla^2 u \quad \text{and} \quad \nabla \cdot u = 0 \quad (5.29)$$

$$\nabla \cdot \sigma E = 0 \qquad (5.30)$$

$$\frac{\tau_c}{\tau_p}\frac{\partial q}{\partial t} + \frac{\tau_c}{\tau_F}\left[u \cdot \nabla_s q - qn \cdot (n \cdot \nabla)u\right] = \left\|-\sigma E\right\| \cdot n \qquad (5.31)$$

$$\left\|\varepsilon E\right\| \cdot n = q \qquad (5.32)$$

$$\left[\sigma^M \cdot n\right] \cdot n = \frac{1}{2}\left\| \varepsilon\left(E \cdot n\right)^2 - \varepsilon\left(E \cdot t_1\right)^2 - \varepsilon\left(E \cdot t_2\right)^2 \right\|$$

$$\left[\sigma^M \cdot n\right] \cdot t_i = qE \cdot t_i. \qquad (5.33)$$

5.8 THERMO-ELECTROHYDRODYNAMIC MODEL

Electrohydrodynamic model is a simple model without considering the thermal effect. In 2004, the researchers[9] consider the coupling effects of thermal, electrical, and hydrodynamics. A complete set of balance laws governing the general thermo-electrohydrodynamics flows has been derived before. It consists of modified Maxwell's equations governing electrical field in a moving fluid, the modified Navier–Stokes equations governing heat and fluid flow under the influence of electric field, and constitutive equations describing behavior of the fluid (Fig. 5.2).

Governing equations without thermal effects are[1,13]

1. Mass balance:

$$\nabla \cdot u = 0 \qquad (5.34)$$

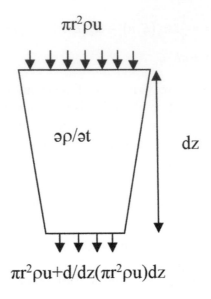

$$\pi r^2 \rho u$$

$$\partial \rho / \partial t$$

$$dz$$

$$\pi r^2 \rho u + d/dz(\pi r^2 \rho u)dz$$

FIGURE 5.2   Schematic of a control volume.

2.   Linear momentum balance:

$$\rho(u \cdot \nabla)u = \underbrace{\nabla T^m + \nabla T^e}_{\text{viscous and electric forces}} \tag{5.35}$$

3.   Electric charges balance:

$$\nabla \cdot J = 0 \tag{5.36}$$

Modified Navier–Stokes equations are[9]

$$\frac{\partial q_e}{\partial t} + \nabla \cdot J = 0 \tag{5.37}$$

$$\rho \frac{Du}{Dt} = \nabla \cdot t + \rho f + q_e E + (\nabla E) \cdot P \tag{5.38}$$

$$\rho c_P \frac{Du}{Dt} = Q_h + \nabla \cdot q + J \cdot E + E \cdot \frac{DP}{Dt} \tag{5.39}$$

This set of conservation laws can constitute a closed system when it is supplemented by appropriate constitutive equations for the field variables such as polarization. The most general theory of constitutive equations determining the polarization, electric conduction current, heat flux, and Cauchy stress tensor.[9,14]

$$P = \varepsilon_p E \tag{5.40}$$

$$J = kE + \sigma u + \sigma_T \nabla T \tag{5.41}$$

$$q = K\nabla T + K_E E \tag{5.42}$$

$$t = -\tilde{P}\underline{\underline{I}} + \eta \left[\nabla \underline{v} + (\nabla \underline{v})^t \right]. \tag{5.43}$$

Here, coefficients ε_p, μ_m, k, σ, σ_T, K, K_E, η are material properties and depend only on temperature in the case of an incompressible fluid.

5.9 MATHEMATICAL MODEL

An unsteady flow of an infinite viscous jet pulled from a capillary orifice and accelerated by a constant external electric field (Table 5.1).[9]

1. Mass conservation:

$$\frac{\partial}{\partial t}\left(r^2\right) + \frac{\partial}{\partial z}\left(r^2 u\right) = 0 \tag{5.44}$$

2. Charge conservation:

$$\frac{\partial}{\partial t}\left(2\pi r(\sigma + \varepsilon_p E)\right) + \frac{\partial}{\partial z}\left(2\pi r\left(\sigma + \varepsilon_p E\right)u + \pi r^2 kE + \pi r^2 \sigma_T \frac{\partial T}{\partial z}\right) = 0 \tag{5.45}$$

TABLE 5.1 Different Types of Current.

Current	Formula
The Ohmic bulk conduction current	$J_c = \pi r^2 kE$
Surface convection current	$J_s = 2\pi r \sigma u$
Current caused by temperature gradients	$J_T = \pi r^2 \sigma_T (\partial T/\partial z)$

The Navier–Stokes equations becomes[9,13,14]
Momentum equation:

$$\frac{\partial u}{\partial t} + u\frac{\partial u}{\partial z} = -\frac{1}{\rho}\frac{\partial p}{\partial z} + g + \frac{2\sigma E}{\rho r} + \frac{1}{r^2}\frac{\partial \tau}{\partial z} + \frac{1}{r^2}\varepsilon_p E \frac{\partial E}{\partial z} \tag{5.46}$$

Energy equation:

$$\rho c_p \left(\frac{\partial T}{\partial t} + u \frac{\partial T}{\partial z} \right) = Q + \frac{\partial}{\partial z} (k \frac{\partial T}{\partial z} + k_E E + 2\pi r \sigma u + \pi r^2 k E + \pi r^2 \sigma_T \frac{\partial T}{\partial z})$$

$$E + \varepsilon_p E \left(\frac{\partial E}{\partial t} + u \frac{\partial E}{\partial z} \right)$$

(5.47)

Internal pressure of the fluid:

$$p = K\gamma - \frac{\varepsilon - \varepsilon_0}{g\pi} E^2 - \frac{2\pi}{\varepsilon_0} \sigma^2$$

(5.48)

Twice the mean curvature of the interface:

$$K = \frac{1}{R_1} + \frac{1}{R_2}$$

(5.49)

5.10   CONSTITUTIVE EQUATIONS

Rheological behavior of many polymer fluids can be described by power-law constitutive equation in the form[15]

$$\tau = \mu_0 \frac{\partial u}{\partial z} + \sum_{n=1}^{m} a_n \left(\frac{\partial u}{\partial z} \right)^{2n+1}.$$

(5.50)

In dielectrics the charges are not completely free to move, but the positive and negative charges that compose the body may be displaced in relation to one another when a field is applied so the following equations can be used for solving governing equations.[9]

$$q_\rho = -\nabla \cdot P \quad \text{and} \quad P = \varepsilon_p E.$$

(5.51)

5.11   CONCLUDING REMARKS

In this chapter, macroscopic models were investigated. These models can offer in-depth insight into physical understanding of many complex phenomena which cannot be fully explained experimentally. It is a powerful tool to control over physical characters. First, the leaky dielectric model was reviewed. Then, the development of this model formed an important step in the construction of a unified treatment of EHD. The model encompasses a wide range of phenomena in liquids and represents behavior in the former when free charge is induced by the field. The model was investigated in detail. Finally, a complete thermo-electrohydrodynamic model which considers the couple effects of thermal field, electric field was investigated. The disadvantage of this model is that it is too complex for numerical analysis. Therefore, a one-dimensional thermo-electrohydrodynamic model which can be applied to numerical study can be selected.

KEYWORDS

- **polymer solution**
- **electrospinning**
- **electrohydrodynamic model**
- **Newtonian jet**
- **rheological models**

REFERENCES

1. Spivak, A. F.; Dzenis, Y. A. Asymptotic Decay of Radius of a Weakly Conductive Viscous Jet in an External Electric Field. *Appl. Phys. Lett.* **1998,** *73* (21), 3067–3069.
2. Hohman, M. M.; Shin, M.; Rutledge, G.; Brenner, M. P. Electrospinning and Electrically Forced Jets. I. Stability Theory. *Phys. Fluids (1994–present)* **2001,** *13* (8), 2201–2220.
3. Hohman, M. M.; Shin, M.; Rutledge, G.; Brenner, M. P. Electrospinning and Electrically Forced Jets. II. Applications. *Phys. Fluids (1994–present)* **2001,** *13* (8), 2221–2236.
4. Reneker, D. H.; Yarin, A. L.; Fong, H.; Koombhongse, S. Bending Instability of Electrically Charged Liquid Jets of Polymer Solutions in Electrospinning. *J. Appl. Phys.* **2000,** *87* (9), 4531–4547.
5. Yarin, A. L.; Koombhongse, S.; Reneker, D. H. Bending Instability in Electrospinning of Nanofibers. *J. Appl. Phys.* **2001,** *89* (5), 3018–3026.
6. Shin, Y. M.; Hohman, M. M.; Brenner, M. P.; Rutledge, G. C. Electrospinning: A Whipping Fluid Jet Generates Submicron Polymer Fibers. *Appl. Phys. Lett.* **2001,** *78* (8), 1149–1151.
7. Saville, D. A. Electrohydrodynamics: The Taylor–Melcher Leaky Dielectric Model. *Annu. Rev. Fluid Mech.* **1997,** *29* (1), 27–64.
8. Melcher, J. R.; Taylor, G. I. Electrohydrodynamics: A Review of the Role of Interfacial Shear Stresses. *Annu. Rev. Fluid Mech.* **1969,** *1* (1), 111–146.
9. Wan, Y. Q.; Guo, Q.; Pan, N. Thermo-electro-hydrodynamic Model for Electrospinning Process. *Int. J. Nonlin. Sci. Numer. Simul.* **2004,** *5* (1), 5–8.

10. Eggers, J.; Villermaux, E. Physics of Liquid Jets. *Rep. Progr. Phys.* **2008,** *71* (3), 036601–036679.
11. Zhang, J.; Kwok, D. Y. A 2D Lattice Boltzmann Study on Electrohydrodynamic Drop Deformation with the Leaky Dielectric Theory. *J. Comput. Phys.* **2005,** *206* (1), 150–161.
12. Griffiths, D. J.; College, R. *Introduction to Electrodynamics*; Prentice Hall: Upper Saddle River, NJ, 1999; vol. 3, p 576.
13. He, J. H.; Liu, H. M. Variational Approach to Nonlinear Problems and A Review on Mathematical Model of Electrospinning. *Nonlin. Anal.: Theory, Methods Appl.* **2005,** *63* (5), e919–e929.
14. Xu, L.; Wang, L.; Faraz, N. A Thermo-electro-hydrodynamic Model for Vibration-electrospinning Process. *Therm. Sci.* **2011,** *15* (suppl. 1), 131–135.
15. Munir, M. M.; Suryamas, A. B.; Iskandar, F.; Okuyama, K. Scaling Law on Particle-to-fiber Formation during Electrospinning. *Polymer* **2009,** *50* (20), 4935–4943.

CHAPTER 6

UPDATE ON ONE-DIMENSIONAL MODELS OF STEADY, INVISCID, ANNULAR LIQUID JETS

SHIMA MAGHSOODLOU* and S. PORESKANDAR

Textile Engineering, University of Guilan, Rasht, Iran

Corresponding author. E-mail: sh.maghsoodlou@gmail.com

CONTENTS

ABSTRACT

Regular perturbation expansions are used to analyze the fluid dynamics of unsteady, inviscid, slender, thin, incompressible (constant density), axisymmetric, upward and downward, annular liquid jets subjected to nonhomogeneous, conservative body forces. In this chapter, one-dimensional models for inviscid, incompressible, axisymmetric, annular liquid jets falling under gravity will be obtained by means of methods of regular perturbations for slender or long jets, integral formulations, Taylor's series expansions, weighted residuals, and variational principles. Then, the hydraulic model will be asymptotically presented Boussinesq for slender and thin annular liquid jets. These models are based on Taylor's series expansions, method of weighted residuals, and the Kantorovich–Galerkin technique so will be resulted in third-order ordinary differential equations. For these purposes, the control-volume formulation can be utilized for assumption to be uniform throughout the annular jet.

6.1 INTRODUCTION

The instability and atomization problems of liquid jets have attracted the interest of researchers for a long time since they have applications in many industrial processes and machines. Most of these processes are related to the disintegration of a jet into small droplets after being discharged into a stagnant gas. Fuel injection and combustion in internal combustion engines or gas generators, surface coating, painting, and even the flow of ink in inkjet printers can be given as examples of such processes. In each of these applications, it is desired that the droplets have a high surface area/volume ratio, meaning that the droplets must be as small as possible. For example, in a jet engine or an internal combustion engine, the combustion efficiency increases as the fuel droplets become smaller.[1] The disintegration of the jet and its subsequent breakup into droplets have been studied both experimentally and theoretically.[2]

Many experimental and theoretical investigations have been performed on the instability mechanisms of electrified liquid jets.[3] In practical applications, polymer solutions are often utilized. In these processes of material fabrication, liquid viscoelasticity is believed to influence profoundly the instability characteristic and breakup of jets. But up to now, the instability mechanism of electrified viscoelastic liquid jets is complicated. For the viscoelastic model, the theory predicts the viscoelastic jet to be less stable than a Newtonian jet, under identical dynamic conditions.[4]

In the experiments, the atomization of around liquid jet has been investigated by using nozzles with varying inlet geometries and diameters. It has been concluded that the experimental observations can be explained by a mechanism including both liquid–gas interaction and nozzle geometry effects. Theoretically, a dispersion relation has been obtained for an incompressible jet, where the stability of infinitesimal waves on an axially symmetric jet of infinite length is considered. The results show that as the parameter $T = (\rho_2^* / \rho_1^*)\sigma/ \mu_2^* U_0^*$ increases, the wavelengths corresponding to the maximum instability get shorter. Here, ρ_1^* and ρ_2^* denote the densities of the gas and liquid phases, respectively, while μ_2^* is the viscosity of the liquid and U_0^* is the jet speed, according to the authors' definition. Investigations on inviscid instability analysis have showed that a dispersion relation which describes the instability of an inviscid liquid jet discharging into an inviscid gas.[5] The studies have indicated three important results:

- As the velocity profile of the jet approaches uniformity, the instability becomes stronger.
- Increasing the Weber number and gas/liquid density ratio promote instability.
- The electrical charge of the liquid jet should be accounted as an additional parameter.

There are different methods for instability formation study:

- For liquid jet instability and atomization in a coaxial gas stream, various regimes of liquid jet breakup in the parameter space of the liquid Reynolds number, the aerodynamic Weber number, and the ratio of the momentum fluxes between the gas and the liquid streams can be considered.[6] The underlying physical mechanism involved in the primary breakup of the liquid jet has shown to consist of the periodic stripping of liquid sheets, which subsequently break up into smaller drops.[7]
- By extending Rayleigh's theory, a theoretical expression can be derived for the droplet size, radius, and spacing, in terms of the jet parameters and applied frequency. For wavelengths slightly smaller than the circumference of the jet, electrical charges promote instability, while larger wavelengths promote stability.
- A perturbation method[8] can be used to study the instability of the interface between a cylinder formed by the flow of an inviscid liquid dielectric and air in the presence of an external electrostatic field

produced by coaxial electrodes. The outer electrode has a uniform electrical potential, while the inner electrode is grounded. An equation has been obtained for the neutral stability curve, that is, the boundary between stable and unstable conditions, and the threshold values of the electrical potential for instability.

• The theoretical study of electrohydrodynamic stability can also utilize. The interactions between the electrical charges at the interface of a liquid jet and instability phenomena can be studied, with an emphasis on effects related to interfacial charge relaxation. It has been found that charge relaxation causes the growth of harmonic perturbations. It has also been found that viscous effects are dominant only for axisymmetric disturbances.[9]

The instability of an electrostatically sprayed liquid jet has been investigated. Calculations have been shown that a charge with no tangential field stabilizes long waves, while the same effect destabilizes short waves. It can be also concluded that by using fields and charge densities that are achievable in an experiment, a stable jet can be produced.[10]

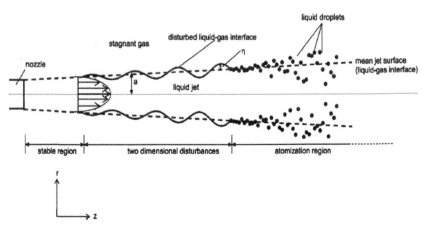

FIGURE 6.1 Flow geometry of electrospinning jet by focusing on stable, 2D dimensional, and atomization regions.

The three part of the electrospinning jet, which is indicated in Figure 6.1, can be analyzed with different methods:

• As it can be seen from the figure, an axisymmetric laminar liquid jet discharges into a stagnant gas through a nozzle. While r^* is the radial direction, z^* is the axial or the stream direction. After leaving the nozzle, the jet flow remains stable until a certain distance from the nozzle exit. However, after a certain distance, two-dimensional disturbances start to occur which, upon amplification, result in either the Rayleigh mode of instability or atomization.

- In an experimental effort, Artana et al. the breakup phenomena of a high-velocity jet when is stressed by an electric field can be investigated. The electric field increases the radial/axial velocity component ratio and promotes the earlier detachment of droplets from the jet. However, no change in the droplet sizes is observed.[11]

- In a theoretical analysis, a temporal linear stability analysis of a circular electrified jet can be developed. The most important finding of the study is that the electrification acts on the stability of a jet in a different way, depending on whether the surface is considered equipotential or not. For the equipotential case, an increase in the velocity or in the electric field or a decrease in the surface tension destabilizes a jet. For the case of frozen charges on the jet surface, the effect of the electric field is opposed to the equipotential case.[12]

- In a theoretical and experimental study, Priol et al. explained the stability of jets, taking the effects of inertia, surface tension, viscous, aerodynamic, and electric forces into account. It is concluded from the theoretical analysis that the atomization of the jet is enhanced with the injection of electrical charges and the experimental results can be confirmed this finding.[13]

The aim of the chapter is to parametrically investigate the stability characteristics of this flow system using the linear stability theory. The densities of the liquid and the gas, surface tension, and the laminar velocity profile shape of the liquid jet are the parameters of the problem. Additionally, an electrically charged liquid jet can also be taken into account. Both the liquid and the gas are considered to be inviscid. The analysis starts with the equations of motion for incompressible, inviscid, axisymmetric flows in cylindrical coordinates. By using the small disturbance theory followed by a normal mode analysis, a dispersion relation can be obtained for the complex wave velocity as a function of the disturbance wave number, where the laminar velocity profile shape parameter, surface tension, density stratification, and electrical charge parameters are the parameters of the problem. The solution of the dispersion relation using a numerical method yields the wavelengths, wave speeds, and the frequencies of the interfacial disturbances.[1]

Investigations on the temporal instability behavior of non-Newtonian liquid jets moving in an inviscid gaseous environment or axisymmetrical disturbances gain the corresponding dispersion relation between the wave growth rate and the wavenumber. The linear stability analysis has shown that a jet of a viscoelastic fluid exhibits a larger growth rate of axisymmetric disturbances than a jet of a Newtonian fluid with the same Ohnesorge number, indicating that non-Newtonian liquid jets are more unstable than their Newtonian counterparts. The breakup of viscoelastic jets has been theoretically analyzed and found that there are major qualitative differences between different kinds of models, even where the elongational viscosities are similar.[14] The nonlinear solution for the non-Newtonian liquid jet breakup obtained by Yarin showed that the onset of this effect may be described by a linear analysis when an unrelaxed axial tension is included in the momentum equation.[15]

The fluid dynamics of annular liquid jets (Fig. 6.2) has been a subject of some interest because of the potential applications of these jets.

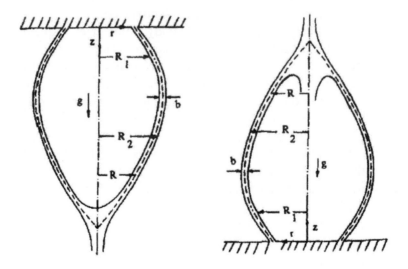

FIGURE 6.2   Schematic diagrams of downward (left) and upward (right) annular liquid jet.

The first study of the fluid dynamics of annular liquid–lithium jets was performed by Hovingh who projected the equations developed by Boussinesq, Lance and Perry, Hopwood, Taylor and Dumbleton along the normal and tangential directions to the mean radius of the annular liquid jet onto a cylindrical polar coordinate system and neglected the friction of the gases that surround the annular liquid jet. The equations derived by Boussinesq assume that the annular jet is steady, incompressible, inviscid, and thin, the gases surrounding and enclosed by the jet are dynamically passive, the pressure is uniform throughout the whole jet, and there are no velocity variations across the jet. Furthermore, these equations may be written as ordinary differential equations for the jet's mean radius and axial and radial velocity components as a function of the axial coordinate. Researchers have improved Hovingh's formulation by accounting for surface tension on both the inner and outer surfaces of the annular liquid jet.[16]

Studying on annular liquid jets to determine the dynamic surface tension of liquids showed when the Weber number exceeds one long-thin jets are obtained, while when the Weber number is less than one the jets take up a rounded form. At the transition point, that is, when the Weber number is one, a discontinuity appears in the liquid sheet. For the fluid dynamics of thick annular jets, convergence lengths were observed 10–30% smaller than the results of theory prediction. These discrepancies can attribute to the meniscus which is formed at the convergence point.

In these studies, the researchers used the hydraulic approximation, assumed steady, thin, incompressible, inviscid fluids, and obtained equations along and normal to the mean radius of the annular liquid jet. In addition, gravitational forces and the curvature of the jet in the vertical plane (compared with that in the horizontal one) were neglected and it was assumed that the jet moves almost vertically and obtained the convergence length, that is, the axial distance at which the annular jet merges onto the symmetry axis to become a solid jet, as a function of the pressure difference between the gases enclosed by and surrounding the annular liquid jet.[16]

Also, a one-dimensional time-dependent model for axisymmetric, incompressible, inviscid, annular liquid jets based on the integration of the Euler equations along cross-sections of the jet and the use of Taylor's series expansions and perturbation techniques for thin annular jets can be appropriate.

In this chapter, perturbation methods, integral formulations, methods of weighted residuals, Taylor's series expansions, and the method of Kantorovich are used to derive one-dimensional models of incompressible, inviscid, axisymmetric, annular liquid jets. Both the inertia- and the surface-tension-dominated flow regimes are considered, and the equations resulting from the different methods are compared among each other and with the hydraulic model developed by Boussinesq. The hydraulic model present here coincides with that derived by Ramo for steady jets even though the approximations employed in the two models are different. Also, the fluid dynamics of slender, thin, annular liquid jets subject to surface tension, gravity and nonhomogeneous, conservative body forces which depend on the jet's thickness is analyzed by means of perturbation methods using the Euler equations.[7,16]

6.2 MODELS BASED ON PERTURBATION METHODS

The fluid dynamics of steady, axisymmetric, incompressible (constant density), inviscid, irrotational, annular liquid jets is governed by the velocity potential, φ^*, which satisfies the Laplace equation, that is,

$$\frac{\partial^2 \varphi^*}{\partial z^{*2}} + \frac{1}{r^*}\frac{\partial}{\partial r^*}\left(r^*\frac{\partial \varphi^*}{\partial r^*}\right) = 0. \tag{6.1}$$

Equation 6.1 is subject to the following boundary conditions at the annular liquid jet's interfaces

$$\frac{\partial \varphi^*\left(Rj^*,z^*\right)}{\partial r^*} = \frac{\partial \varphi^*\left(Rj^*,z^*\right)}{\partial z^*}\frac{dRj^*}{dz^*} \quad (j=1,2,\ldots) \tag{6.2}$$

$$p_1^* - p^*\left[R_1^*\left(z^*\right),z^*\right] = \sigma^* J_1^*$$
$$p^*\left[R_2^*\left(z^*\right),z^*\right] - p_2^* = \sigma^* J_z^* \tag{6.3}$$

In what follows, the gases enclosed by and surrounding the annular liquid jet will be assumed to be dynamically passive, that is, p_1^* and p_2^* will be assumed to be constant. This assumption is justified since the density of gases is, in general, much smaller than that of liquids.

In addition, boundary conditions are to be provided at $z^* = 0$, that is, at the nozzle exit. These boundary conditions must be obtained by matching the potential flow inside the nozzle with that of the free annular jet. Since the flow inside the nozzle must satisfy the no-penetration condition at the solid walls, whereas the boundary conditions for the free annular jet involve free surfaces, a transition from the no-penetration to the free-surface flow is expected. Such a transition is not considered in this chapter where the interest lies in the region below the nozzle exit.

Since the flow is irrotational, the liquid pressure can be determined from Bernoulli's equation as

$$p^* = p_s^* - \frac{1}{2}R^*\left[\left(\frac{\partial \varphi^*}{\partial r^*}\right)^2 + \left(\frac{\partial \varphi^*}{\partial z^*}\right)^2\right] + \rho^* g^* z^*. \tag{6.4}$$

For long or slender annular liquid jets,

$$\varepsilon = \frac{R_0^*}{L^*} \ll 1. \tag{6.5}$$

If the radial and axial coordinates are nondimensionalized with respect to R_0^* and L^*, respectively, the potential with respect to $v_1^* L^*$ and the pressure with respect to $\rho^* v_0^{*2}/2$, where u (is a constant) reference axial velocity component, for example, the cross-section averaged axial velocity component at the nozzle exit, eqs 6.1–6.5 become, after substituting eq 6.5 into eq 6.3.

$$\varepsilon^2 \frac{\partial^2 \phi}{\partial z^2} + \frac{1}{r}\frac{\partial}{\partial r}\left(r\frac{\partial \phi}{\partial r}\right) = 0 \tag{6.6}$$

$$\frac{\partial \varphi(R_i, z)}{\partial r} = \varepsilon^2 \frac{\partial \varphi(R_i, z)}{\partial z}\frac{dR_i}{dz} \quad (i = 1, 2) \tag{6.7}$$

$$\left[\frac{\partial \varphi(R_1, z)}{\partial r}\right]^2 = \varepsilon^2 \left[\alpha_1 - \left(\frac{\partial \varphi(R_1, z)}{\partial z}\right)^2 + \frac{z}{Fr} + \frac{1}{We}\left\{\frac{1}{R_1\left[1 + \varepsilon^2\,(dR_1/dz)^2\right]^{1/2}} - \frac{\varepsilon^2\,(d^2R_1/dz^2)}{\left[1 + \varepsilon^2\,(dR_1/dz)^2\right]^{3/2}}\right\}\right] \tag{6.8}$$

$$\left[\frac{\partial \varphi(R_2, z)}{\partial r}\right]^2 = \varepsilon^2 \left\{\alpha_2 - \left[\frac{\partial \varphi(R_2, z)}{\partial z}\right]^2 + \frac{z}{Fr} - \frac{1}{We}\left(\frac{1}{R_2\left[1 + \varepsilon^2\,(dR_2/dz)^2\right]^{1/2}} - \frac{\varepsilon^2\,(d^2R_2/dz^2)}{\left[1 + \varepsilon^2\,(dR_2/dz)^2\right]^{3/2}}\right)\right\} \tag{6.9}$$

$$p = p_s - \left[\frac{1}{\varepsilon^2}\left(\frac{\partial \varphi^2}{\partial r}\right)^2 + \left(\frac{\partial \varphi}{\partial z}\right)^2\right] + \frac{2}{Fr} \tag{6.10}$$

where $Fr = u_0^{*2}/2g^*L^*$ is the Froude number, $\alpha_i = 2\{p_s^* - p^*(R_i)(z), z)\}/\rho^* u_0^{*2}$ and $We = \rho^* R_0^* u_0^{*2}/2\sigma^*$ is the Weber number. The length used to nondimensionalize the axial coordinate in this section may be replaced by $u_0^{*2}/2g^*$, which corresponds to a Froude number equal to one, and the condition of slenderness implies that $u_0^{*2}/2g^* \geq R_0^*$.

Depending on the magnitude of the Froude and Weber numbers, different flow regimes are possible as indicated in the next sections.

6.3 MODELS BASED ON ELECTRIFIED LINEAR VISCOELASTIC LIQUID JET

Consider an incompressible viscoelastic liquid jet of diameter $2R$, moving through a quiescent, inviscid, incompressible gas medium, for axisymmetric disturbances. In the undisturbed state, the liquid in the sheet has velocity $U = U_z + U_r$, where U_z is the unperturbed axial velocity, U_r is the unperturbed radial velocity, and U_r is assumed to be zero in the present study. The liquid and gas have densities of ρ_l and ρ_g, respectively, and the surface tension of liquid is σ. μ_l denotes the viscosity of the viscoelastic fluid. Figure 6.3 shows the 2D centric section; moreover, the coordinate system is assumed to move with the jet velocity U.

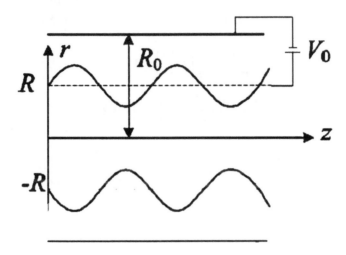

FIGURE 6.3   Schematic of liquid jet.

The viscoelastic properties of the fluid are described using the corotational model with the constitutive equation for the stress tensor τ given by

$$\tau + \lambda_1 \frac{D\tau}{Dt} + \frac{1}{2}\mu_0 (tr\,\tau)\,\dot\gamma - \frac{1}{2}\mu_1 (\tau \cdot \dot\gamma + \dot\gamma \cdot \tau) + \frac{1}{2}\upsilon_1 (\tau : \dot\gamma)\,\delta = -\mu_0$$
$$\left(\dot\gamma + \lambda_2 \frac{D\dot\gamma}{Dt} - \mu_2 (\dot\gamma \cdot \dot\gamma)\frac{1}{2}\upsilon_2 (\dot\gamma : \dot\gamma)\,\delta\right). \tag{6.11}$$

The quantities $\mu_0, \mu_1, \mu_2, \nu_1$, and ν_2 are time constants. The rate of strain tensor $\dot\gamma$ and vorticity tensor ω are defined by

$$\dot\gamma = \nabla U + (\nabla U)^T \tag{6.12}$$

$$\omega = \nabla U - (\nabla U)^T. \tag{6.13}$$

The corotational derivative in eq 6.11 is defined by

$$\frac{D\tau}{Dt}=\frac{\partial\tau}{\partial t}+(U\cdot\nabla)\,\tau+\frac{1}{2}(\omega\cdot\tau-\tau\cdot\omega)\tag{6.14}$$

$$\frac{D\dot\gamma}{Dt}=\frac{\partial\dot\gamma}{\partial t}+(U\cdot\nabla)\,\dot\gamma+\frac{1}{2}(\omega\cdot\dot\gamma-\dot\gamma\cdot\omega)\tag{6.15}$$

To carry out a linear stability analysis, the basic flow is disturbed by a small disturbance. The dependent variables for the velocities, stress tensor can be presented as the sum of the value found in the undisturbed steady state plus the unsteady perturbation,

$$\tilde U=U+U'=(U_z,0)+(U_z,U_r)\tag{6.16}$$

$$\tilde\tau=\tau+\tau'=0+\tau'.\tag{6.17}$$

In the undisturbed state $U=(U_z,\,U_r)$, the undisturbed strain tensor $\dot\gamma$ and vorticity tensor ω are both 0. Similarly, the strain tensor and the vorticity tensor are expressed below. In addition, the strain tensor and the vorticity tensor of the perturbation are in terms of the velocity vector of the perturbation according to eqs 6.12 and 6.13:

$$\tilde{\dot\gamma}=\dot\gamma+\dot\gamma'=0+\dot\gamma'$$
$$\tilde\omega=\omega+\omega'=0+\omega'\tag{6.18}$$

$$\dot\gamma'=\nabla U'+(\nabla U')^T$$
$$\omega'=\nabla U'-(\nabla U')^T.\tag{6.19}$$

Substituting eqs 6.18 and 6.19 for eq 6.11, and considering that the coordinate system is moving with the liquid jet, the constitutive eq 6.11, can be linearized as below,

$$\left(1+\lambda_1\frac{\partial}{\partial t}\right)\tau'_{ij}=-\mu_0\left(1+\lambda_2\frac{\partial}{\partial t}\right)\dot\gamma'_{ij}\tag{6.20}$$

where τ'_{ij} and $\dot\gamma'_{ij}$ denote the shear stress and the (i,j) component of the rate of deformation tensor, in cylindrical coordinates $\dot\gamma'_{rz}=\frac{\partial u_z}{\partial r}+\frac{\partial u_r}{\partial z}$. Equation 6.20 is a somewhat more versatile model, with three constants $(\mu_0,\lambda_1,\lambda_2)$ put forward by Jeffreys in 1929. The ratio of λ_2/λ_1 is defined as the time constant ratio. For gel $\lambda_2<\lambda_1$ as gel is the shear-thinning fluid.

According to the literature, assuming that variables $u_z; u_r$ can be written in the general normal-mode as at z form $(u_z, u_r, \tau') = [U'_z(r), U'_r(r), T(r)] e^{ikz+\alpha t}$, which is referred to as the temporally growing and spatially harmonic wave form, orinfinite extent-initial value problem and has also been used by Gordon, and Sadik and Zimmels,[17] then eq 6.20 can be written into

$$(1+\alpha\lambda_1)T_{rz} = -\mu_0(1+\alpha\lambda_2)\left(ikU'_r + \frac{dU'_z}{dr}\right). \tag{6.21}$$

In this relation, the wave number is $k = \dfrac{2\pi}{\lambda}$. The most unstable disturbance has the largest value of α, denoted by α_{max} in the present work, and is assumed to be responsible for the breakup. The resulting ligament size is related to the most unstable wave length $\lambda_s = \dfrac{2\pi}{k_d}$, where k_d is the wavenumber corresponding to the maximum growth rate α_{max}, that is, the dominant wavenumber. Equation 6.21 reduces to

$$T_{rz} = -\mu_1\left[ikU'_r + \frac{dU'_z}{dr}\right] \tag{6.22}$$

where

$$\mu_1 = \mu_0 \frac{1+\alpha\lambda_2}{1+\alpha\lambda_1}. \tag{6.23}$$

The shear stress of Newtonian fluid in the linearized form can be expressed as

$$\tau'_{rz} = -\mu\left[\frac{\partial u_r}{\partial z} + \frac{\partial u_z}{\partial r}\right]. \tag{6.24}$$

Considering the general form $(u_z, u_r, \tau') = [U'_z(r), U'_r(r), T(r)] e^{ikz+\alpha t}$, eq 6.25 should be rewritten as

$$T_{rz} = -\mu\left[ikU'_r(r) + \frac{dU'_z(r)}{dr}\right]. \tag{6.25}$$

6.4 MODELS BASED ON INTEGRAL FORMULATIONS

The equations that govern the fluid dynamics of incompressible, inviscid, irrotational, axisymmetric, annular liquid jets are the continuity and Euler equations and their rotationality condition, that is,

$$\frac{\partial u^*}{\partial z^*} + \frac{1}{r^*}\frac{\partial\left(v^* r^*\right)}{\partial r^*} = 0 \tag{6.26}$$

$$\rho^*\left(u^*\frac{\partial u^*}{\partial z^*} + v^*\frac{\partial u^*}{\partial r^*}\right) = -\frac{\partial p^*}{\partial z^*} + \rho^* g^* \tag{6.27}$$

$$\rho^*\left(u^*\frac{\partial v^*}{\partial z^*} + v^*\frac{\partial v^*}{\partial r^*}\right) = -\frac{\partial p^*}{\partial r^*} \tag{6.28}$$

$$\frac{\partial u^*}{\partial r^*} = \frac{\partial v^*}{\partial z^*}. \tag{6.29}$$

Equations 6.26–6.29 are subject to the following boundary conditions:

$$v^*\left(R_i^*, z^*\right) = u^*\left(R_i^*, z^*\right)\frac{\mathrm{d}R_i^*}{\mathrm{d}z^*} \quad \in (i = 1, 2) \tag{6.30}\backslash$$

$$p_1^* - p^*\left[R_1^*\left(z^*\right), z^*\right] = \sigma^* J_1^*$$
$$p^*\left[R_2^*\left(z^*\right), z^*\right] - p_2^* = \sigma^* J_2^*. \tag{6.31}$$

If the velocity components, coordinates, and pressure are nondimensionalized with respect to U^*, R_0^*, and $\rho^* U^{*2}$, respectively, where U^* is a constant reference velocity at the nozzle exit or the capillary velocity for inertia-or surface tension-dominated jets, respectively, then eqs 6.26–6.31 become

$$\frac{\partial u}{\partial z} + \frac{1}{r}\frac{\partial\left(vr\right)}{\partial r} = 0 \tag{6.32}$$

$$u\frac{\partial u}{\partial z} + v\frac{\partial u}{\partial r} = -\frac{\partial p}{\partial z} + \frac{1}{F} \tag{6.33}$$

$$u\frac{\partial v}{\partial z} + v\frac{\partial v}{\partial r} = -\frac{\partial p}{\partial r} \tag{6.34}$$

$$\frac{\partial u}{\partial r} = \frac{\partial v}{\partial z} \tag{6.35}$$

$$v(R_i, z) = u(R_i, z)\frac{\mathrm{d}R_i}{\mathrm{d}z} \quad \in (i = 1, 2) \tag{6.36}$$

REFERENCES

1. Özgen, S.; Uzol, O. Investigation of The Linear Stability Problem of Electrified Jets, Inviscid Analysis. *J. Fluids Eng.* **2012,** *134* (9), 091201-1–091201-9.
2. Reitz, R. D.; Bracco, F. V. Mechanism of Atomization of a Liquid Jet. *Phys. Fluids (1958–1988)* **1982,** *25* (10), 1730–1742.
3. Basset, A. B. Waves and Jets in a Viscous Liquid. *Am. J. Math.* **1894,** *16* (1), 93–110.
4. Brenn, G.; Liu, Z.; Durst, F. Linear Analysis of the Temporal Instability of Axisymmetrical Non-Newtonian Liquid Jets. *Int. J. Multiphase Flow* **2000,** *26* (10), 1621–1644.
5. Ibrahim, E. A.; Marshall, S. O. Instability of a Liquid Jet of Parabolic Velocity Profile. *Chem. Eng. J.* **2000,** *76* (1), 17–21.
6. Lasheras, J. C.; Hopfinger, E. J. Liquid Jet Instability and Atomization in a Coaxial Gas Stream. *Annu. Rev. Fluid Mech.* **2000,** *32* (1), 275–308.
7. Schneider, J. M.; Lindblad, N. R.; Hendricks, Jr., C. D.; Crowley, J. M. Stability of an Electrified Liquid Jet. *J. Appl. Phys.* **1967,** *38* (6), 2599–2605.
8. Eliseev, Y. G. Flow Stability of a Liquid Dielectric Cylinder in a Transverse Electrostatic Field. *Sov. Phys. J.* **1969,** *12* (12), 1536–1540.
9. Saville, D. A. Electrohydrodynamic Stability: Effects of Charge Relaxation at the Interface of a Liquid Jet. *J. Fluid Mech.* **1971,** *48* (04), 815–827.
10. Turnbull, R. J. On the Instability of an Electrostatically Sprayed Liquid Jet. *Ind. Appl., IEEE Trans.* **1992,** *28* (6), 1432–1438.
11. Artana, G.; Romat, H.; Touchard, G. Study of a High-Velocity Liquid Jet Stressed by an Electric Field. *Phys. Fluids (1994–present)* **1998,** *10* (11), 2922–2931.
12. Artana, G.; Romat, H.; Touchard, G. Theoretical Analysis of Linear Stability of Electrified Jets Flowing at High Velocity Inside a Coaxial Electrode. *J. Electrostat.* **1998,** *43* (2), 83–100.
13. Priol, L.; Baudel, P.; Louste, C.; Romat, H. Theoretical and Experimental Study (Linear Stability and Malvern Granulometry) on Electrified Jets of Diesel Oil in Atomization Regime. *J. Electrostat.* **2006,** *64* (7), 591–596.
14. Renardy, M. Similarity Solutions for Jet Breakup for Various Models of Viscoelastic Fluids. *J. Non-Newtonian Fluid Mech.* **2002,** *104* (1), 65–74.
15. Yarin, A. L. *Free Liquid Jets and Films: Hydrodynamics and Rheology.* Longman Publishing Group: Harlow, 1993.
16. Ramos, J. I. One-Dimensional Models of Steady, Inviscid, Annular Liquid Jets. *Appl. Math. Model.* **1996,** *20* (8), 593–607.
17. Yang, L. J.; Liu, Y. N.; Fu, Q. F. Linear Stability Analysis of an Electrified Viscoelastic Liquid Jet. *J. Fluids Eng.* **2012,** *134* (7), 071303-1–071303-13.

CHAPTER 7

THE ELECTRICALLY FORCED JET AND INSTABILITIES

SHIMA MAGHSOODLOU* and S. PORESKANDAR

Textile Engineering, University of Guilan, Rasht, Iran

Corresponding author. E-mail: sh.maghsoodlou@gmail.com

CONTENTS

ABSTRACT

In the electrospinning process, polymer nanofibers are formed by subjecting a fluid jet to a high electric field. An investigation of electrically forced jet and its instability will be studied in this chapter. Frame work is based on a developed theory for electrified fluid jets. The process can be described by a small set of operating parameters and summarized through their relationship equations. The nonaxisymmetric dispersion relation has also been compared to special cases, for example, inviscid jets without surface charge or perfect conductors.

7.1 INTRODUCTION

As mentioned in previous chapters, after a small distance of stable traveling, the jet will start unstable behavior because of axisymmetric and nonaxisymmetric instabilities and sprays or separates into many fibers.[1-3]

The polymer jet is influenced by these instabilities. Recently instability in electrospinning has received much attention.[1] Each instability grows at different rates.[4,5]

These instabilities arise owing to the charge–charge repulsion between the excess charges present in the jet, which encourages the thinning and elongation of the jet.[6] Bending or stretching of the jet is the key role in reduction of the jet diameter from micrometer to nanometer. When the polymer jet becomes very long and thin, the time required for the excess charge to redistribute itself along the full length of the jet becomes longer. The location of the excess charge then tends to change with the elongation. The repulsive Coulomb forces between the charges carried with the jet elongate the jet in the direction of its axis until the jet solidifies. This leads to an incredibly high velocity at the thin leading end of the straight jet. As a result, the jet bends and develops a series of lateral excursions that grow into spiraling loops. Each of these loops grows larger in diameter as the jet grows longer and thinner.[7]

At high field, axisymmetric and nonaxisymmetric instabilities are due to fluctuations in the dipolar component of the charge distribution. These conducting modes are electrically driven and essentially independent of surface tension. The axisymmetric conducting mode arises as a consequence of the finite, nonzero conductivity of the fluid, which produces an extra root to the growth rate equations. The whipping instability can occur through either: (a) small lateral fluctuations in the centerline of the jet result in the induction of a dipolar charge distribution, as the free charge adjusts to screen out the field inside the jet. The dipoles interact with the external electric field, producing a torque that further bends the jet; and (b) mutual repulsion of surface charges carried by jet causes the centerline to bend.[8]

At high electric forces, the jet is dominated by bending (axisymmetric) and whipping instability (nonaxisymmetric), causing the jet to move around and produces wave in the jet. At higher electric fields and at enough charge density in the jet, the axisymmetric (i.e., Rayleigh and bending) instabilities are suppressed and the nonaxisymmetric instability is increased.[6] Whipping of the jet dominates for a given set of conditions depends on the jet radius and surface charge experienced by a fluid element as it travels downstream. In general, the whipping mode dominates at high-charge density, whereas the axisymmetric mode dominates at low-charge density. As a jet thins away from the nozzle, both the charge density and the radius change.[8]

A significant part of our information of the electrospinning process comes from empirical observations but the complexity of the process makes an empirical determination of parameter effects very difficult if not impractical.[9] Of crucial importance to understanding the electrospinning process is the competition between modes of instability. For prediction the final properties of this process, a suitable model can be used.[10] The aim of this chapter is to use a suitable model for investigating instabilities behavior of this jet during pathways toward the collector. In the next part, more detailed will be investigated about this model.

7.2   MATHEMATICAL MODEL

7.2.1   THE ELECTROHYDRODYNAMIC EQUATIONS

The theoretical approach presented here is based on our empirical observation that the instabilities occur on a length scale much longer than the jet radius. The jet can express as a long, slender object. As a first step, it has assumed that the fluid is Newtonian and incompressible. Treatment of more complexes, for example, viscoelastic fluid behavior should also be possible without major revision of the basic theoretical approach. A more detailed exposition of the mathematical development is presented elsewhere,[11] and only the key aspects are summarized here.

The relevant three-dimensional variables—radial velocity (v_r), axial velocity (v_z), radial electric field (E_r), and axial electric field (E_z) are expanded as Taylor series in the jet radius. These expansions are substituted into the full three-dimensional equations for conservation of mass, conservation of charge, and differential momentum balance, and only the leading order terms are retained. Due to viscous dissipation and external forcing by both gravity and the electric field, the isolated system of the jet does not conserve energy.

The resulting hydrodynamic equations are made nondimensional by choosing a length scale r_0, where r_0 is the diameter of capillary; a time scale $t_0 : \sqrt{\rho r_0^3 / \gamma}$, where γ is the surface tension and ρ is the density of the fluid; an electric field strength $E_0 : \sqrt{\gamma(\varepsilon - \bar{\varepsilon}) r_0}$, where $\varepsilon, \bar{\varepsilon}$ is the permittivity of fluid and air; and a surface charge density $\sqrt{\gamma \bar{\varepsilon} / r_0}$. The dimensionless asymptotic field is $\Omega_0 = E_\infty / \sigma_0$. The material properties of the fluid are characterized by four dimensionless parameters: $\beta = \varepsilon / \bar{\varepsilon} - 1$; the dimensionless viscosity $\dot{v} = \sqrt{\dot{v}^2 / (\rho \gamma r_0)}$; the dimensionless gravity $g^* = g \rho r_0^2 / \gamma$ and the dimensionless conductivity $k^* = k \sqrt{\rho r_0^3 / (\beta \gamma)}$. The nondimensionalized equations for conservation of mass and charge, and the Navier–Stokes equation are shown in eqs 7.1–7.3, respectively:

$$\partial_t \left(h^2 \right) + \left(h^2 v \right)' = 0 \tag{7.1}$$

$$\partial_t \left(\sigma h \right) + \left(\sigma h v + \frac{k^*}{2} h^2 E \right)' = 0 \tag{7.2}$$

$$\partial_t v + vv' = -\left(\frac{1}{h} - h'' - \frac{E^2}{8\pi} - 2\pi\sigma^2 \right)' + \frac{2\sigma E}{\sqrt{\beta h}} + g^* + \frac{3 v^{*2}}{h^2} \left(h^2 v \right)'. \tag{7.3}$$

Here, $h(z)$ is the radius of the jet at axial coordinate z; $v(z)$ is the axial velocity of the jet and is constant across the jet cross-section to leading order; $\sigma(z)$ is the surface charge density; and $E(z)$ is the electric field in the axial direction. The prime (') denotes differentiation with respect to z.

Then nondimensionalized tangential field inside the jet is derived from Coulomb's law

$$E = E_\infty + \int ds \frac{\lambda(s)}{|x - r(s)|} \approx E_\infty + \ln \frac{r}{L} \left(\frac{\beta}{2} \left(h^2 E \right)' - \frac{4\pi}{\bar{\varepsilon}} h \sigma \right) \tag{7.4}$$

7.2.2   LINEAR STABILITY ANALYSIS

For the linear stability analysis, axisymmetric perturbation at the form $h^*/h = 1 + h_e e^{\alpha t + ikz}$ were applied to the radius, where h_z is assumed to be small. Similar perturbations in velocity, surface charge density, and the local electric field strength were also introduced. Substituting these perturbations into the governing hydrodynamic equations yields a dispersion relation for axisymmetric instabilities, which is shown in the following equation:

$$
\omega^3 + \omega^2 \left[\frac{4\pi k^* \Lambda}{\delta\sqrt{\beta}} + 3 v \times k^2 \right] + \omega \left[3 v \times k^2 \frac{4\pi k^* \Lambda}{\delta\sqrt{\beta}} + \frac{k^2}{2}(k^2 - 1) + 2\pi\sigma^2 - k^2\left(\frac{8l}{\delta} - 1\right) + \frac{\Lambda}{\delta}\frac{\Omega_0^2}{4\pi}k^2 \right]
$$

$$
+ \frac{4\pi k^* \Lambda}{\delta\sqrt{\beta}} \left[\frac{k^2}{2}(k^2 - 1) + 2\pi\sigma_0^2 k^2 + \frac{\delta}{\Lambda}\frac{\Omega_0^2}{4\pi}k^2 + \frac{E_0\sigma_0}{\sqrt{\beta}} ik\left(\frac{1}{l} - 4\right) \right] = 0
$$

(7.5)

$1/\chi$ is the local aspect ratio of the jet, indicative of the ratio of the jet diameter to length scale over which the charge various; its value is assumed to be small. In accord with the assumption of long wavelength instabilities, $l = \ln(1/\chi)$, $\Lambda = \beta \ln \chi(k^2)$, and $\delta = 2 \cdot \Lambda$. The axisymmetric dispersion relation works for arbitrary values of conductivity, dielectric constant viscosity, and field strength, as long as the condition that the tangential electric stress be smaller than the radial viscous stress is satisfied. It therefore permits comparison to experimental results. By taking special limits of eq 7.5, for example, zero viscosity and infinite conductivity, excellent agreement with Saville's results,[12] has been obtained.[11]

Nonaxisymmetric disturbances can be modeled by considering long wavelength modulations of the jet centerline. The equations of motion are more complicated than eqs 7.1–7.3, due to allowance for bending. However, they are structurally similar to the equations for a slender elastic rod under slight bending. When the jet bends, the charge density along the jet is no longer uniform around the circumference of the jet but now also contains a dipolar component, $P(z)$, oriented perpendicular to the jet axis, as the internal charges adjust to screen the external field. These dipoles set up a localized torque that bends the jet, and oscillations of the bending instability account for the whipping motion of the jet. This is illustrated in Figure 7.1(b).

To derive the dispersion relation for nonaxisymmetric disturbances, the centerline of the jet is described in nondimensional coordinates by $r(z,t) = z\hat{z} + \varepsilon e^{\alpha t + ikz}\hat{x}$.

The final form of the dispersion relation for nonaxisymmetric disturbances is shown in the following equation:

$$
\omega^2 + \frac{3}{4}\dot{v}\,\alpha k^4 + 4\pi\sigma_0^2 k^2 \ln(k) + ik\frac{2\sigma_0\Omega_0}{\sqrt{\beta}} + \frac{ik\Omega_0}{\sqrt{\beta}}(\sigma_0 k^2 + c) + k^2 - \left[\Omega_0\left(\frac{1}{4\pi} + \frac{(\beta+1)k^2}{16\pi\beta}\right) + \frac{ik\sigma_0}{\beta} \right]
$$

$$
\times \frac{\sqrt{\beta}}{\beta + 2}\left[k^2 \sqrt{\beta}\Omega_0 + ik(4\pi c + 2\pi\beta\sigma_0 k^2) \right] = 0
$$

(7.6)

REFERENCES

1. Ghochaghi, N. Experimental Development of Advanced Air Filtration Media based on Electrospun Polymer Fibers. In *Mechnical and Nuclear Engineering*; Virginia Commonwealth: Richmond, VA, 2014; pp 1–165.
2. Brooks, H.; Tucker, N. Electrospinning Predictions Using Artificial Neural Networks. *Polymer* **2015,** *58,* 22–29.
3. Reneker, D. H.; Chun, I. Nanometre Diameter Fibres of Polymer, Produced by Electrospinning. *Nanotechnology* **1996,** *7* (3), 216–223.
4. Zhang, S. *Mechanical and Physical Properties of Electrospun Nanofibers*, 2009; pp 1–83.
5. Wu, Y.; Yu, J. Y.; He, J. H.; Wan, Y. Q. Controlling Stability of the Electrospun Fiber by Magnetic Field. *Chaos, Solit. Fract.* **2007,** *32* (1), 5–7.
6. Baji, A.; Mai, Y. W.; Wong, S. C.; Abtahi, M.; Chen, P. Electrospinning of Polymer Nanofibers: Effects on Oriented Morphology, Structures and Tensile Properties. *Compos. Sci. Technol.* **2010,** *70* (5), 703–718.
7. Reneker, D. H.; Yarin, A. L.; Fong, H.; Koombhongse, S. Bending Instability of Electrically Charged Liquid Jets of Polymer Solutions in Electrospinning. *J. Appl. Phys.* **2000,** *87,* 4531–4547.
8. Shin, Y. M.; Hohman, M. M.; Brenner, M. P.; Rutledge, G. C. Electrospinning: A Whipping Fluid Jet Generates Submicron Polymer Fibers. *Appl. Phys. Lett.* **2001,** *78* (8), 1149–1151.
9. Thompson, C. J.; Chase, G. G.; Yarin, A. L.; Reneker, D. H. Effects of Parameters on Nanofiber Diameter Determined from Electrospinning Model. *Polymer* **2007,** *48* (23), 6913–6922.
10. Thompson, C. J. *An Analysis of Variable Effects on A Theoretical Model of the Electrospin Process for Making Nanofibers*, University of Akron: Lakewood, 2006.
11. Hohman, M. M.; Shin, M.; Rutledge, G.; Brenner, M. P. Electrospinning and Electrically Forced Jets. I. Stability Theory. *Phys. Fluids (1994–present)* **2001,** *13* (8), 2201–2220.
12. Saville, D. A. Electrohydrodynamic Stability: Effects of Charge Relaxation at the Interface of a Liquid Jet. *J. Fluid Mech.* **1971,** *48* (04), 815–827.

CHAPTER 8

UPDATE ON DEFORMATION OF NEWTONIAN AND NON-NEWTONIAN CONDUCTING DROPS IN THE ELECTRIC FIELD

SHIMA MAGHSOODLOU* and S. PORESKANDAR

Textile Engineering, University of Guilan, Rasht, Iran

Corresponding author. E-mail: sh.maghsoodlou@gmail.com

CONTENTS

ABSTRACT

Jets occur from the micro scale up to the large-scale structure of the universe. Liquid jets thus serve as a paradigm for free-surface motion, hydrodynamic instability, and singularity formation leading to drop breakup. In this chapter, the deformation and breakup of conducting drops will be investigated. For the Newtonian will be described by the electrohydrostatic theory, especially with regard to the prediction of the critical point. The non-Newtonian effect on the drop deformation and breakup will be studied for highly conducting drops which become stable in the weak or moderate field strength.

8.1 INTRODUCTION

When a fluid drop is suspended in another immiscible fluid, it can be influenced by an externally applied uniform electric field in many ways. Apart from its basic scientific interest, understanding of the behavior of drops in an electric field is playing an increasingly important role in practical applications.[1] Part of the interest stemmed also from applications including the deformation and breakup of raindrops in thunderstorms and the breakdown of dielectric liquids due to contaminants such as the water droplets with a substantial conductivity.[2,3] The electric-field-driven flow is of practical significance in the processes in which enhancement of the rate of mass or heat transfer between the drops and their surrounding fluid is desired.[4,5]

The initial stages of breakup are governed by linear theory. Since growth is exponential, this gives a dependable estimate of the total time to breakup, and thus of the breakup length of a jet. Yet the growth of sinusoidal modes cannot even explain basic features such as satellite drops, so the inclusion of nonlinear effects is essential.

There are two simple limits when considering the effect of an external electric field on the drop dynamics. If the two contiguous fluids are considered as perfectly insulating dielectrics and no free charges are present, or if, on the other hand, the drop contains a fluid that is highly conducting while the surrounding fluid is perfectly insulating, an electric field induces a net force at the fluid interface as a result of discontinuity in the electric stresses.[6] Because this electric surface force acts only in the direction normal to the interface, it can be balanced by the interfacial tension. Therefore, the drop is always elongated into a prolate shape with its axis of symmetry parallel to the direction of the applied field, while the fluids remain motionless at a steady state. These ideal situations of electrified drops have been extensively studied within the well-established theoretical framework of electrohydrostatics.[7]

When finite conductivity of fluid is considered, the problem becomes more interesting yet complicated. One of the most noteworthy phenomena is that, in addition to the prolate shape predicted by the electrohydrostatic theory, a drop in an electric field may remain in a spherical shape or even be deformed into an oblate spheroid as observed in experiments.[5]

The initial stages of breakup are governed by linear theory. Since growth is exponential, this gives a good estimate of the total time to breakup, and thus of the breakup length of a jet. Yet the growth of sinusoidal modes cannot even explain basic features such as satellite drops, so the inclusion of nonlinear effects is essential.

8.2   BREAKUP CONCEPTS

In this part, what actually takes place as the radius of the fluid neck goes to zero were investigated to produce a theory of drop formation from the continuum perspective. Near breakup, the typical size of the solution goes to zero, and therefore does not possess a characteristic scale. It is therefore natural to look for solutions that are invariant under an appropriate scale transformation, that is, to look for similarity solutions. We choose the intrinsic scales as units of length and time, and the point z_0, t_0 where the singularity occurs as the origin of our coordinate system. The dimensionless coordinates are[8]

$$z' = \frac{(z - z_0)}{l_v} \quad \text{and} \quad t' = \frac{(t - t_0)}{t_v}. \tag{8.1}$$

In the simplest case that a one-dimensional description is appropriate, we expect a similarity description

$$h(z,t) = l_v |t'|^\alpha \, \Phi(\xi) \tag{8.2}$$

$$v(z,t) = \left(\frac{l_v}{t_v}\right)|t'|^{\alpha_2} \, \Psi(\xi) \tag{8.3}$$

$$\xi = \frac{z'}{|t'|^\beta}. \tag{8.4}$$

The plus sign refers to the time before breakup. To calculate the values of the exponents, one must know the balance of terms in the original equation, or its one-dimensional counterpart. We will see below that in the absence of an outer fluid, asymptotically breakup always proceeds according to balances surface tension, viscous, and inertial forces. However, for extreme values of the Ohnesorge number, other transient regimes are possible. For example, if the viscosity is small ($Oh \ll 1$), it must initially drop out of the description.[9]

It is easy to confirm that with the choice $\alpha_1 = 1$, $\alpha_2 = 1/2$, $\beta_2 = 1/2$ the self-similar profile solves the slender-jet equations, in which all terms are balanced as $t' \to 0$. The scaling implies that $\mathcal{E}$ is proportional to $|t'|^{1/2}$; thus, higher order terms in the expansion in $\mathcal{E}$ go to zero as the singularity is approached. It is now transformed into a similarity equation for the scaling functions ϕ, Ψ[8]:

$$\pm\left(-\varphi + \frac{\xi\varphi'}{2}\right) + \Psi\varphi' = -\frac{\Psi\varphi'}{2} \tag{8.5}$$

$$\pm\left(\frac{\Psi}{2} + \frac{\xi\Psi'}{2}\right) + \Psi\Psi' = \frac{\varphi'}{\varphi^2} + \frac{3\left(\Psi'\varphi^2\right)'}{\varphi^2}. \tag{8.6}$$

Taylor recognized that finite conductivity enables electrical charge to accumulate at the drop interface, permitting a tangential electric stress to be generated. The tangential electric stress drags fluid into motion and thereby generates hydrodynamic stress at the drop interface. The complicated interplay between the electric and hydrodynamic stresses causes either oblate or prolate drop deformation and, in some special cases, keeps the drop from deforming. This electrohydrodynamic theory proposed by Taylor is referred to as the leaky dielectric model and is capable of predicting the drop deformation in qualitative agreement with the previous experimental observations. Although Taylor's leaky dielectric theory provides a good qualitative description for the deformation of a Newtonian drop in an electric field, the validity of its analytical results is strictly limited to the drop experiencing small deformation in an infinitely extended domain (Fig. 8.1).

While the asymptotic leaky dielectric theory gives a satisfactory description for the behavior of a conducting Newtonian drop in an insulating medium within the small deformation limit, the electrohydrostatic theory predicts more readily the critical electric field strength above which the drop loses its stable configuration. It has been shown that Taylor's spheroidal approximation is in good agreement with more rigorous numerical methods if the drop is perfectly conducting.[10]

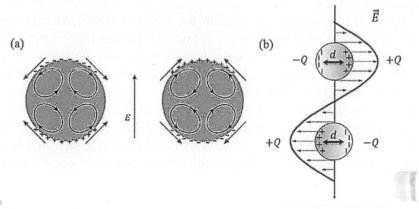

FIGURE 8.1   Electrohydrodynamic model of drop deformation in (a) AC electric field and (b) DC electric field.

Unlike the Newtonian fluids, there are few available theoretical investigations which pertain to the deformation and stability of the interface between the non-Newtonian fluids in an electric field. This is mostly because of the anticipated uncertainties in selecting an appropriate constitutive model for non-Newtonian fluids, as well as the obvious difficulty in solving the equations of motion after the choice has been made. Instead, in the present study, we consider experimentally the effect of non-Newtonian properties on the drop deformation and breakup and identify the role of the non-Newtonian response by comparing the deviations from the well-known theories or experimental observations for a Newtonian-fluid pair. The aim of this chapter is to explain what actually takes place as the radius of the fluid neck goes to zero, to produce a theory of drop formation from the continuum perspective.

8.3   THE EQUATIONS OF DROP DEFORMATION

The degree of drop deformation can be expressed conveniently by the parameter defined by

$$D = \frac{(L - B)}{(L + B)}. \tag{8.7}$$

This parameter is especially sensitive to small deviations from sphericity. In the present experimental study, none of the steady-state drop shapes were highly elongated and D was used exclusively to define the deformed drop shapes.

One of the difficulties involved in the study of non-Newtonian fluids arises from the fact that the viscosity is not a unique quantity.[11] Specifically, the shear viscosity differs from the extensional viscosity, both of which are usually influenced by the rate of deformation. Therefore, the shear viscosity of a fully viscoelastic fluid exhibits shear-thinning behavior. Herein, we defined the viscosity ratio λ in terms of the steady shear viscosity instead of the extensional viscosity. For the shear-thinning fluids, we defined the ratio of the viscosity, λ, of the drop phase to that of the ambient fluid in terms of the zero-shear-rate viscosities exclusively. As we shall see later, the neglect of shear dependence of the viscosity does not oversimplify the problem, especially for the conducting drops, since the electric-field-induced flow is too weak to induce an appreciable rheological effect. Under these circumstances, the role of non-Newtonian effects arises from the fluid elasticity. This is especially true when the field strength is below its critical value or prior to the instability. The determination of the interfacial tension, γ, is not easy because the fluids used in this study are very viscous and have almost the same densities. Therefore, conventional techniques for interfacial tension measurements such as the pendant drop, ring or plate methods are not appropriate. One simple way to estimate the interfacial tension is based on the small deformation theory describing the relationship between the degrees of drop deformation and electric field strength, that is,

$$D = \frac{9}{16(2R+1)^2}\left[\left(1+R^2-2SR^2\right)+3R(1-SR)\frac{2+3\lambda}{5+5\lambda}\right]\frac{a \in E^2}{\gamma} \qquad (8.8)$$

Thus, Taylor's theory for the drop deformation in a weak electric field may offer a convenient measure of the interfacial tension. In the present study, Taylor's small deformation theory was used to obtain the interfacial tension. Therefore, as we shall see shortly, the experimental deformation data and theory coincide at small deformations. The dimensionless group, $a \in E^2/\gamma$ is sometimes referred to as the dimensionless electric field strength or the electric capillary number y, due to the analogy with the capillary number, $\mu u_c/\gamma$, which is frequently used in studies of drop dynamics under external flow fields in the low Reynolds number limit. In fact, the characteristic velocity u_c is given by $a \in E^2/\mu$ for electrohydrodynamic flow. Meanwhile, although Taylor's theory is valid strictly for the case of Newtonian fluids, it is not applicable in general when at least one of the fluids is viscoelastic. However, the first correction to Taylor's theory taking into account the viscoelastic effects occurs at the order of Deborah number. In the small deformation limit, therefore, the viscoelastic correction is expected to be small because the electric-field driven flow is very weak at modest electric field strength.

8.4 CONCLUDING REMARK

In the present study, the deformation and breakup of Newtonian or non-Newtonian fluid drops in an external electric field have been studied. For Newtonian-fluid pairs, the electrohydrostatic assumption was valid when $R < O(10^{-5})$. The breakup mode of Newtonian drops was slightly influenced by the resistivity and viscosity ratios. In this chapter, the deformation and breakup of Newtonian and non-Newtonian conducting drops in surrounding fluid subjected to a uniform electric field were investigated. The results were on the Newtonian fluids was described well by the electrohydrostatic theory, especially with regard to the prediction of the critical point. Then, the non-Newtonian effect on the drop deformation and breakup was studied for highly conducting drops which satisfied the condition $R < O(10^{-5})$. The highly conducting drop became stable in the weak or moderate field strength when either the drop or the continuous phase was non-Newtonian.

KEYWORDS

- Newtonian
- non-Newtonian
- fluid
- breakup modes
- drop formation

REFERENCES

1. O'Konski, C. T., Harris, F. E. Electric Free Energy and the Deformation of Droplets in Electrically Conducting Systems. *J. Phys. Chem.* **1957,** *61* (9), 1172–1174.
2. Sartor, J. D. Electricity and Rain. *Phys. Today* **1969,** *22,* 45.

3. Beard, K. V., Ochs, H. T., Kubesh, R. J. Natural Oscillations of Small Raindrops. *Nature* **1989,** *342*, 408–410.
4. Ptasinski, K. J., Kerkhof, P. J. A. M. Electric Field Driven Separations: Phenomena and Applications. *Separ. Sci. Technol.* **1992,** *27* (8–9), 995–1021.
5. Ha, J. W., Yang, S. M. Effects of Surfactant on the Deformation and Stability of a Drop in a Viscous Fluid in an Electric Field. *J. Colloid Interface Sci.* **1995,** *175* (2), 369–385.
6. Melcher, J. R., Taylor, G. I. Electrohydrodynamics: A Review of the Role of Interfacial Shear Stresses. *Annu. Rev. Fluid Mech.* **1969,** *1* (1), 111–146.
7. O'Konski, C. T., Thacher Jr., H. C. The Distortion of Aerosol Droplets by an Electric Field. *J. Phys. Chem.* **1953,** *57* (9), 955–958.
8. Ting, L., Keller, J. B. Slender Jets and Thin Sheets with Surface Tension. *SIAM J. Appl. Math.* **1990,** *50* (6), 1533–1546.
9. Hirt, C. W., Nichols, B. D. Volume of Fluid (VOF) Method for the Dynamics of Free Boundaries. *J. Comput. Phys.* **1981,** *39* (1), 201–225.
10. Eggers, J., Villermaux, E. Physics of Liquid Jets. *Rep. Progr. Phys.* **2008,** *71* (3), 036601.
11. Beard, K. V., Ochs, H. T., Kubesh, R. J. *Natural Oscillations of Small Raindrops,* 1989.

CHAPTER 9

UPDATE ON BEHAVIOR OF AN ELECTROSPUN FLUID JET SIMULATION USING A MICROSCOPIC MODEL

SHIMA MAGHSOODLOU* and S. PORESKANDAR

Textile Engineering, University of Guilan, Rasht, Iran

Corresponding author. E-mail: sh.maghsoodlou@gmail.com

CONTENTS

ABSTRACT

In the recent years, electrospinning is used as an efficient and simple method for producing nanofibers. The efforts of researchers have mainly been in producing nanofibers from varied polymers and their applications. But, little attention has been given in the process of electrospinning. The most important part of electrospinning is how to control the process. Studying the dynamics of the jet becomes easier and faster if it can be simulated, rather than doing experiments. A microscopic model can be expressed by the dynamic behavior of electrospun fluid jet (the whipping instability) by momentum equation. This chapter focuses on simulating the electrospinning process by studying the dynamics of the jet. An existing mathematical model, which describes the jet as a mechanical system of masses, is interconnected with viscoelastic elements, will be used to build a Matlab script.

9.1 INTRODUCTION

The dynamics of charged polymer jets under the effect of an external electrostatic field stands out as a major challenge in nonequilibrium thermodynamics, with numerous applications in micro- and nano-engineering and life sciences as well.[1]

As mentioned in previous chapters, electrospinning is an electrohydrodynamic phenomenon in which a thin fiber is ejected from the surface of a polymer solution destabilized by the influence of a strong external electric field of intensity exceeding a certain critical threshold value.[2]

Indeed, charged liquid jets may develop several types of instabilities depending on the relative strength of the various forces acting upon them, primarily electrostatic Coulomb self-repulsion, viscoelastic drag, and surface tension effects.[3-5]

As the jet spirals toward the collector, higher order instabilities have resulted in spinning distance.[6]

The polymer jet is influenced by these instabilities. These instabilities arise owing to the charge–charge repulsion between the excess charges present in the jet, which encourages the thinning and elongation of the jet. At high electric forces, the jet is dominated by bending (axisymmetric) and whipping instability (nonaxisymmetric), causing the jet to move around and produce waves. At higher electric fields and at enough charge density in the jet, the axisymmetric (i.e., Rayleigh and bending) instabilities are suppressed and the nonaxisymmetric instability is increased.[7]

Recently, instability in electrospinning has received much attention.[3] Each instability grows at different rates.[8,9]

These instabilities vary and increase with distance, electrical field, and fiber diameter at different rates depending on the fluid parameters and performing conditions.[7]

Based on the above, it is clear that an accurate control of the effects of the bending and whipping instabilities on the morphological features of the resulting spirals is key to an efficient design of the electrospinning process. The fundamental physics of the electrospinning process is governed by the competition between Coulomb repulsion and the stabilizing effects of viscoelastic drag and surface tension.[1]

Suitable methods selection for analyzing these behaviors is really important. Modeling and simulating can be used for prediction electrospun nanofiber properties and morphology.[10] In addition, they can be useful for investigating the effects of factors perception that cannot be measured experimentally.[11]

Broadly speaking, these models fall within two general classes: continuum and discrete. The former treats the polymer jet as a charged fluid, obeying the equations of continuum mechanics, while the latter represents the jet as a discrete collection of charged particles (beads), subject to four type of interactions: Coulomb repulsion, viscoelastic drag, curvature-driven surface tension, and, finally, the external electric field.[1]

Results from modeling also explain how processing parameters and fluid behavior lead to nanofibers with appropriate properties.[12,13] A suitable theoretical model of the electrospinning process is one that can show a strong–moderate–minor rating effects of these parameters on the nanofiber diameter.[14]

The main focus of this chapter is to investigate microscopic model to give a better understanding of unstable behavior of the electrospinning process. In the next part, more details are investigated about this model.

9.2 MICROSCOPIC MODELS

One of the aims of computer simulation is to reproduce experiments to elucidate the invisible microscopic details and further explain the experiments. Physical phenomena occurring in complex materials cannot be encapsulated within a single numerical paradigm. In fact, they should be described within hierarchical, multilevel numerical models in which each submodel is responsible for different spatial–temporal behavior and passes out the averaged parameters of the model, which is next in the hierarchy. The understanding of the nonequilibrium properties of complex fluids such as the viscoelastic behavior of polymeric liquids, the rheological properties of fluids and liquid crystals subjected to magnetic fields, based on the architecture of their molecular constituents, is useful to get a comprehensive view of the process. The analysis of simple physical particle models for complex fluids has developed from the molecular computation of basic systems (atoms, rigid molecules) to the simulation of macromolecular "complex" system with a large number of internal degrees of freedom exposed to external forces.[15,16]

With the advance of computer technology and the development of efficient simulation techniques, molecular simulation methodology has made it possible to study increasingly complex problems and this approach has become commonplace in many such fields, including mechanical engineering and material design, biology, and even medicine, often shedding light on problems that are inaccessible by other methods.[2]

The most widely used simulation methods for molecular systems are Monte Carlo, Brownian dynamics, and molecular dynamics (MD). The microscopic approach represents the microstructural features of material by means of a large number of micromechanical elements (beads, platelet, rods) obeying stochastic differential equations. The evolution equations of the microelements arise from a balance of momentum at the elementary level. The Monte Carlo method is a stochastic strategy that relies on probabilities. The Monte Carlo sampling technique generates large numbers of configurations or microstates of equilibrated systems by stepping from one microstate to the next in a particular statistical ensemble. Random changes are made to the positions of the species present, together with their orientations and conformations where appropriate. Brownian dynamics are an efficient approach for simulations of large polymer molecules or colloidal particles in a small molecule solvent. MD is the most detailed molecular simulation method which computes the motions of individual molecules. MD efficiently evaluates different configurational properties and dynamic quantities which cannot generally be obtained by Monte Carlo.[17,18]

A working definition of MD simulation is technique by which one generates the atomic trajectories of a system of N particles by numerical integration of Newton's equation of motion, for a specific interatomic potential, with certain initial condition (IC) and boundary condition (BC). One can often treat a MD simulation like an experiment.[19] Figure 9.1 shows a common flowchart of an ordinary MD run.

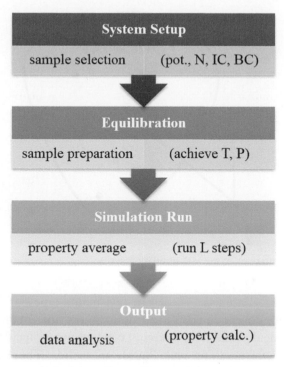

FIGURE 9.1   A common flowchart of an ordinary MD run.

In the figure, we fine-tune the system until it reaches the desired condition (e.g., temperature T and pressure P), and then perform property averages, for instance, calculating the function. Central idea of MD simulations can be expressed as

- Biological activity is the result of time-dependent interactions between molecules and these interactions occur at the interfaces such as protein–protein, protein–NA, protein–ligand.
- Macroscopic observables (laboratory) are related to microscopic behavior (atomic level).
- Time dependent (and independent) microscopic behavior of a molecule can be calculated by MD simulations.

There are five key ingredients to a MD simulation, which are BC, IC, force calculation, integrator/ensemble, and property calculation. A brief overview of them is given below, followed by more specific discussions (Fig. 9.2).[19,20]

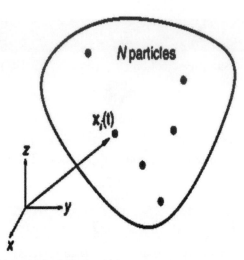

FIGURE 9.2   Illustrations of the MD simulation system.

a) *Boundary condition.* There are two major types of BCs: isolated boundary condition (IBC) and periodic boundary condition (PBC). IBC is ideally suited for studying clusters and molecules, while PBC is suited for studying bulk liquids and solids. There could also be mixed BCs such as slab or wire configurations for which the system is assumed to be periodic in some directions but not in the others. In IBC, the N-particle system is surrounded by vacuum; these particles interact among themselves but are presumed to be so far away from everything else in the universe that no interactions with the outside occur except perhaps responding to some well-defined "external forcing." In PBC, one explicitly keeps track of the motion of N particles in the so-called supercell, but the supercell is surrounded by infinitely replicated, periodic images of itself. Therefore, a particle may interact not only with particles in the same supercell but also with particles in adjacent image supercells (Fig. 9.3).[21,22]

b) *Initial condition.* Since Newton's equations of motion are second-order ordinary differential equations (ODEs), IC basically means x^{3N} ($t = 0$) and y^{3N} ($t = 0$), the initial particle positions, and velocities. Generating the IC for crystalline solids is usually quite easy, but IC for liquids needs some work, and even more so for amorphous solids. A common strategy to create a proper liquid configuration is to melt a crystalline solid. And if one wants to obtain an amorphous configuration, a strategy is to quench the liquid during a MD run.[19]

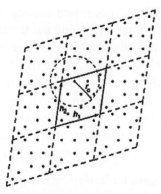

FIGURE 9.3 Illustration of periodic boundary condition (PBC). We explicitly keep track of trajectories of only the atoms in the center cell called the supercell (defined by edge vectors h_1, h_2, and h_3), which is infinitely replicated in all three directions (image supercells). An atom in the supercell may interact with other atoms as well as atoms in the surrounding image supercells. r_c is a cut-off distance of the interatomic potential, beyond which interaction may be safely ignored.

c) *Force calculation.* Consider the classical equation of motion to describe the atoms: $m_i (d^2x_i (t)/dt^2) = f_i$. The evaluation of the equation is the key step that usually consumes the computational time in a MD simulation, so its efficiency is crucial. For long-range Coulomb interactions, special algorithms exist to break them up into two contributions: a short-ranged interaction, plus a smooth, field-like interaction, both of which can be computed efficiently in separate ways (Fig. 9.4).[23]

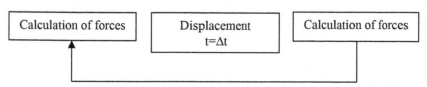

FIGURE 9.4 Force calculation steps in MD.

d) *Integrator/ensemble.* Equation is a set of second-order ODEs, which can be strongly nonlinear. By converting them to first-order ODEs in the 6N dimensional space of (xN, yN), general numerical algorithms for solving ODEs such as the Runge–Kutta method can be applied. However, these general methods are rarely used in MD, because the existence of a Hamiltonian allows for more accurate integration algorithms, prominent among which are the family of predictor–corrector integrators and the family of symplectic integrators. A section in this chapter gives a brief overview of integrators.[21]

e) *Property calculation.* A great value of MD simulation is that it is "omnipotent" at the level of classical atoms. All properties that are well posed in classical mechanics and statistical mechanics can in principle be computed. The issues remaining are accuracy (the error comes from the interatomic potential) and computational efficiency. The properties can be roughly grouped into four categories[23]:

1. *Structural characterization.* Examples include radial distribution function, dynamic structure factor, etc.
2. *Equation of state.* Examples include free-energy functions, phase diagrams, static response functions like thermal expansion coefficient, etc.
3. *Transport.* Examples include viscosity, thermal conductivity (electronic contribution excluded), correlation functions, diffusivity, etc.
4. *Nonequilibrium response.* Examples include plastic deformation, pattern formation, etc.

Now, the applied models for the polymers MD simulation are investigated. The first computer simulation of liquids was carried out in 1953. The model was an idealized two-dimensional representation of molecules as rigid disks. For macromolecular systems, the coarse-grained approach is widely used as the modeling process is simplified, hence becomes more efficient, and the characteristic topological features of the molecule can still be maintained. The level of detail for a coarse-grained model varies in different cases. The whole molecule can be represented by a single particle in a simulation and interactions between particles incorporate average properties of the whole molecule. With this approach, the number of degrees of freedom is greatly reduced.[24]

On the other hand, a segment of a polymer molecule can also be represented by a particle (bead). The first coarse-grained model, called the "dumbbell" model, was introduced in the 1930s. Molecules are treated as a pair of beads interacting via a harmonic potential. However by using this model, it is possible to perform kinetic theory derivations and calculations for nonlinear rheological properties and solve some flow problems. The analytical results for the dumbbell models can also be used to check computer simulation procedures in MD and Brownian dynamics (Fig. 9.5).[25,26]

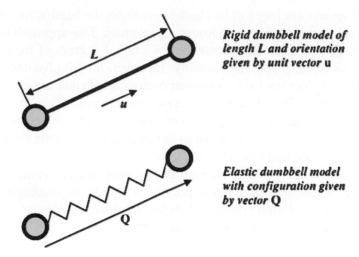

FIGURE 9.5 The first coarse-grained models—the rigid and elastic dumbbell models.

The bead-rod and bead-spring models were introduced to model chain-like macromolecules. Beads in the bead-rod model do not represent the atoms of the polymer chain backbone, but some portion of the chain, normally 10–20 monomer units. These beads are connected by rigid and mass-less rods. While in the bead-spring model, a portion of the chain containing several hundreds of backbone atoms are replaced by a "spring" and the masses of the atoms are concentrated on the mass of beads (Fig. 9.6).[27]

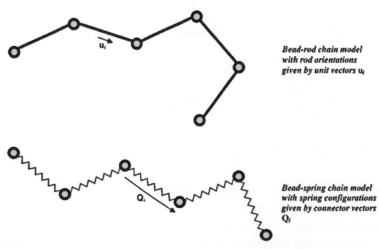

FIGURE 9.6 The freely jointed bead-rod and bead-spring chain models.

If the springs are taken to be Hookean springs, the bead-spring chain is referred to as a Rouse chain or a Rouse-Zimm chain. This approach has been applied widely as it has a large number of internal degrees of freedom and exhibits orientability and stretchability. However, the disadvantage of this model is that it does not have a constant contour length and can be stretched out to any length. Therefore in many cases finitely extensible springs with two more parameters, the spring constant and the maximum extensibility of an individual spring can be included so the contour length of the chain model cannot exceed a certain limit (Fig. 9.7).[28,29]

The understanding of the nonequilibrium properties of complex fluids such as; the viscoelastic behavior of polymeric liquids, the rheological properties of fluids and liquid crystals subjected to magnetic fields, are based on the architecture of their molecular constituents.[15]

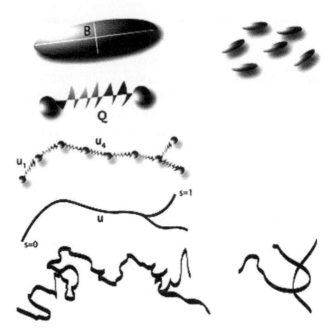

FIGURE 9.7   Simple microscopic models for complex fluids by using dumbbell model

Dumbbell models are very crude representations of polymer molecules, too crude to be of much interest to a polymer chemist, since it in no way accounts for the details of the molecular architecture. It certainly does not have enough internal degrees of freedom to describe the very rapid motions that contribute, for example, to the complex viscosity at high frequencies. On the other hand, the elastic dumbbell model is orientable and stretchable, and these two properties are essential for the qualitative description of steady-state rheological properties and those involving slow changes with time. For dumbbell models, one can go through the entire program of endeavor—from molecular model for fluid dynamics—for illustrative purposes, to point the way toward the task that ultimately needs to be performed for more realistic models. According to the research, dumbbell models must, to some extent then, be regarded as mechanical playthings, somewhat disconnected from the real world of polymers. When used intelligently, however, they can be useful pedagogically and very helpful in developing a qualitative understanding of rheological phenomena.[15,30]

The simplest model of flexible macromolecules in a dilute solution is the elastic dumbbell (or bead-spring) model. This has been widely used for purely mechanical theories of the stress in electrospinning modeling which are investigated as following.[31]

9.3 MICROSCOPIC MODEL FOR INVESTIGATING DYNAMICAL BEHAVIOR OF ELECTROSPINNING PROCESS

The aim of mathematical analysis of engineering process is to obtain a mathematical equation to determine the relations between variables, input and output parameters. Also, several samples can deeply help for investigating behaviors of physical phenomena.[15,16] The simplest model of flexible macromolecules in a dilute solution is the bead-spring model. This has been widely utilized in electrospinning modeling which are investigated as following.[31]

9.3.1 THE GOVERNING EQUATIONS IN MICROSCOPIC MODEL

To understand the mathematical model, first some details are given about the modeling of the system. The model of the fluid jet is described by a sequence of discrete charged particles (beads), obeying Maxwell fluid mechanics. The polymer jet is represented by means of a Maxwellian liquid, coarse-grained into a viscoelastic bead-spring model with a charge associated to each bead for investigating the behavior of whipping instability in electrospinning process.[32–34]

Each bead has a mass m and possesses a charge e. According to Newton's second law, each bead acts on by Coulomb forces, electric field and viscoelastic forces. So, the equation of motion of the beads is[32–34]

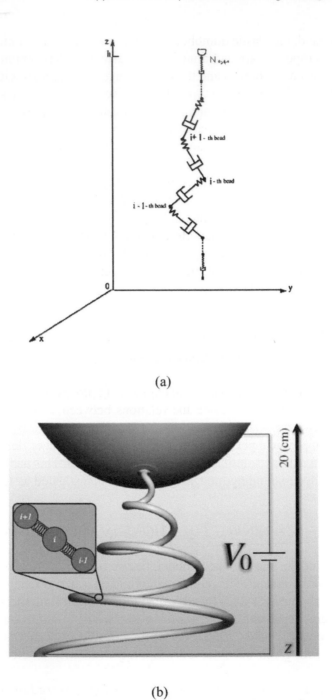

(a)

(b)

FIGURE 9.8 (a) Jet path in three dimensions with beads and springs and (b) schematic representation of the model for the polymer jet.

$$m\frac{dV}{dt} = \frac{-e^2}{L^2} - \frac{-eV0}{h} + \pi a^2 \sigma \qquad (9.1)$$

The total number of beads, N, increases over time as new electrically charged beads are inserted at the top of Figure 9.8 to represent the flow of solution into the jet. The net Coulomb force acting on bead i from all the other beads is[32–34]

$$F_C = \sum_{\substack{j=1 \\ j \neq i}}^{N} \frac{e^2}{R_{ij}^2} \left[i\frac{X_i - X_j}{R_{ij}} + \frac{Y_i - Y_j}{R_{ij}} + \frac{Z_i - Z_j}{R_{ij}} \right] \qquad (9.2)$$

R_{ij} is the distance between bead i and bead j and is expressed as[32–34]

$$R_{ij} = \left[\left(X_i - X_j \right)^2 + \left(Y_i - Y_j \right)^2 + \left(Z_i - Z_j \right)^2 \right]^{1/2} \qquad (9.3)$$

The electric force imposed on the ith bead by the electric field is[32–34]

$$FE_i = -e\frac{V_0}{h} K \qquad (9.4)$$

The net viscoelastic force acting on the ith bead is[32–34]

$$F_{Ve} = \pi a_{ui}^2 \sigma_{ui} \left[i\frac{X_{i+1} - X_i}{L_{ui}} + j\frac{Y_{i+1} - Y_i}{L_{ui}} + k\frac{Z_{i+1} - Z_i}{L_{ui}} \right] - \pi a_{di}^2 \sigma_{di} \cdot$$
$$\left[i\frac{X_i - X_{i-1}}{L_{di}} + j\frac{Y_i - Y_{i-1}}{L_{di}} + k\frac{Z_i - Z_{i-1}}{L_{di}} \right] \qquad (9.5)$$

These stresses can be calculated by integrating the following equations[32–34]:

$$\frac{d\sigma_{ui}}{dt} = G\frac{1}{l_{ui}}\frac{dL_{ui}}{dt} - \frac{G}{\mu}\sigma_{ui}$$

$$\frac{d\sigma_{di}}{dt} = G\frac{1}{l_{di}}\frac{dL_{di}}{dt} - \frac{G}{\mu}\sigma_{di} \qquad (9.6)$$

where L_{ui} and L_{di} can be calculated as

$$L_{ui} = \left[\left(X_{i+1} - X_i \right)^2 + \left(Y_{i+1} - Y_i \right)^2 + \left(Z_{i+1} - Z_i \right)^2 \right]^{1/2}$$
$$L_{di} = \left[\left(X_i - X_{i-1} \right)^2 + \left(Y_i - Y_{i-1} \right)^2 + \left(Z_i - Z_{i-1} \right)^2 \right]^{1/2} \qquad (9.7)$$

When bead i is deviated from its equilibrium position, the jet segments (i − 1, i), and (i, i + 1) are bent. The surface tension force exerting on the ith bead tends to restore the rectilinear shape of the bending part of the jet and is given as[32–34]

$$F_{cap} = -\frac{\alpha\pi(a)_{av}^2 K_i}{\left(X_i^2 + Y_i^2\right)^{1/2}}\left[i|X_i|\text{sign}(X_i) + j|Y_i|\text{sign}(Y_i)\right] \tag{9.8}$$

The average jet radius is defined as follows[32–34]:

$$(a)_{av}^2 = \frac{\left(a_{ui} + a_{di}\right)^2}{4} \tag{9.9}$$

The meaning of "sign" in eq 9.8 is as follows[32–34]:

$$\text{sign}(\text{function}) = \begin{cases} 1 & x > 0 \\ 0 & x = 0 \\ -1 & x < 0 \end{cases} \tag{9.10}$$

Additionally, in the calculation, the air drag force and gravity force are neglected as both space and time are dependent perturbations. Therefore, the development of whipping instability occurs. To model this, a single perturbation is added by inserting an initial bead of i by[32–34]

$$\begin{bmatrix} X_i \\ Y_j \end{bmatrix} = 10^{-3} L \begin{bmatrix} \sin(\alpha t) \\ \cos(\alpha t) \end{bmatrix} \tag{9.11}$$

In this formula, L can be defined as

$$L = \left(\frac{4e^2}{\pi d_0^2 G}\right)^{1/2} \tag{9.12}$$

Therefore, with the second law of Newton, the momentum equation for the motion of the ith bead is[32–34]

$$\sum F_i = \sum F_{total} \tag{9.13}$$

$$ma = m\frac{d^2 R_i}{dt^2} \tag{9.14}$$

$$R_i = iX_i + jY_i + kZ_i \tag{9.15}$$

$$m\frac{d^2R_i}{dt^2} = \sum_{\substack{j=1 \\ j \neq i}}^{N} \frac{e^2}{R_{ij}^3}(R_i - R_j) - e\frac{V_0}{h}\hat{k} + \frac{\pi a_{ui}^2\left(\overline{\sigma}_{ui} + G\ln(l_{ui})\right)}{L_{ui}}(R_{i+1} - R_i)$$

$$-\frac{\pi a_{di}^2\left(\overline{\sigma}_{di} + G\ln(l_{di})\right)}{L_{di}}(R_i - R_{i-1}) - \frac{\alpha\pi a_{av}^2 k_i}{\sqrt{\left(x_i^2 + y_i^2\right)}}\left[i|x_i|\text{sign}(x_i) + j|y_i|\text{sign}(y_i)\right]$$

$$(9.16)$$

9.4 THE CONSTITUTIVE RELATIONS FOR INITIAL PARAMETER OF SIMULATION ELECTROSPUN NANOFIBER JET

Initial parameters were needed for running simulation program. In addition, some of these parameters were related to each other (i.e., viscosity and elastic modulus, viscosity and concentration, the charge of solution and elastic modulus). For this reason, a correct equation is needed for each parameter. In this work, surface tension, viscosity, the mass of consumable polymer, elastic modulus, specific density, length and radius of the needle, and charge of the solution should be calculated for running the simulation program.

9.4.1 VISCOSITY OF THE SOLUTION

One of the important parameters for giving a simulation program is viscosity. This parameter offers a wealth of information relating to the size of the polymer molecule in solution, including the effects upon chain dimensions of polymer, structure, molecular shape, degree of polymerization, and polymer–solvent interactions. Most commonly, however, it is applied to estimate the molecular weight of a polymer. This involves the utilization of semi-empirical equations which have to be established for each polymer/solvent/temperature system analysis of samples whose molecular weights are known. Absolute measurements of viscosity are not essential in dilute solution viscosity, since it is exclusively necessary to determine the viscosity of a polymer relative to that of the pure solvent. The limiting or intrinsic viscosity, quantity is related to the molecular weight of polymer by the semi-empirical Mark–Houwink equation[35]:

$$[\eta] = K \cdot \overline{M}_v^a$$

$$(9.17)$$

where k and α are constants for a given polymer, solvent, and temperature. Generally, for polyvinyl alcohol by ranged molecular weights between

69,000 and 690,000 are 6.51 and 0.628. In addition, the intrinsic viscosity of this polymer with 72,000 g/L molecular weights was 0.73. Also, the relation between relative viscosity and specific viscosity were calculated from this equation.[35]

$$\eta_{sp} = \eta_r - 1 = \frac{\eta_s - \eta_0}{\eta_0}.$$ (9.18)

In this work, Ram Mohan Rao and Yaseen equations were utilized for calculation of specific viscosity[35]:

$$[\eta] = \left[\ln \eta_r + \eta_{sp}\right](2c)^{-1}$$ (9.19)

Equation 9.19 was solved by Matlab program and the values of specific viscosity were obtained as initial parameters for viscosity in Matlab script.

9.4.2   SURFACE TENSION OF THE SOLUTION

The other important parameter in the simulation of the electrospun jet is surface tension. It is a property of fluids that causes the outer layer to act as an elastic sheet. In other words, the surface tension is equal to the force of section per unit length. In addition, a govern equation was obtained between surface tension and voltage. As mentioned before, electrospinning solution is usually an ionic solution that contains charged ions. The amounts of positive and negative charged particles are equal; therefore, the solution is electrically neutral. When an electrical potential difference is given between needle and collector, a hemispherical surface of the polymeric droplet at the orifice of the needle is gradually expanded. When potential came into a critical value, a flow of jet starts formation to drop.[3] As mentioned in eq 9.20, the relation between surface tension and critical voltage was observed. This behavior may arise as a result of surface tension differences between several solutions. It has been demonstrated that the surface tension of polyvinyl alcohol solutions is proportional to the degree of polymerization and to the extent of hydrolysis. The surface tension decreases with increasing concentration of the polymer in the solution. A low surface tension is desirable in electrospinning as it thins out the critical voltage V_c needed for the ejection of the jet from the Taylor's cone as shown below[36,37]:

$$V_c^2 = 4\frac{H^2}{L^2}\left(\ln \frac{2L}{R} - \frac{3}{2}\right)(0.117\,\pi\gamma R).$$ (9.20)

9.4.3 CHARGE OF THE SOLUTION

The amount of force between two charges that has a certain distance from each other was named as the electrical force. For determining the amount of the charge of solving in simulation program, the dimensionless amount of the charge was used for this parameter:

$$e = \sqrt{\frac{mL_{el}^3 G^2}{\eta_{sp}^2}} \qquad (9.21)$$

The amount of mass of polymer powder for preparing $v = 20$ mL of polyvinyl alcohol solution was determined by this equation:

$$\%c = \frac{m}{v} \qquad (9.22)$$

9.4.4 ELASTIC MODULUS

Elastic modulus is defined as the ratio of stress (force per unit area) along an axis to strain (ratio of deformation over initial length) along that axis. During electrospinning, the stable jet ejected from Taylor's cone is subjected to tensile stresses and may undergo significant elongational flow. The characteristics of this elongation flow can be determined by examining the elasticity of the solution. The longest relaxation time (λ) of the molecules in solution can be estimated from the Rouse model[37]:

$$\lambda = \frac{6\eta_s [\eta] M_w}{\pi^2 RT} \qquad (9.23)$$

In summation, a relation exists between the viscosity η and the longest relaxation time λ were defined[38]:

$$\eta = G \cdot \lambda \qquad (9.24)$$

By combination of eqs (9.23) and (9.24), new longest relation time was utilized for determining the elastic modulus:

$$\eta_{sp} = G \frac{6\eta_s [\eta] M_w}{R \cdot T \pi^2} \qquad (9.25)$$

At last, initial parameters were calculated from constitute equations for running the simulation program.

9.5 SIMULATION ANALYSIS

According to the mathematical model described above, the time evolution of the jet whipping instability is determined by the following procedure: At $t = 0$, the initial whipping jet includes two beads, bead 1 and bead 2. The distance $l_{1,2}$ is set to be a small distance, say, $H/10,000$. Other ICs, including the stresses $\sigma_{i-1,i}$ and $\sigma_{i,i+1}$ and the initial velocity of bead i, dR_i/dt, are set to be zero. For a given time, t, eq 9.16 is solved numerically, using 4-order Runge–Kutta algorithm. For the numerical solution, the equations are made dimensionless. All the variables related to bead i, including the stresses $\sigma_{i-1,i}$ and $\sigma_{i,i+1}$, the position $\mathbf{r}_i$, the length of the jet segment $l_{i-1,i}$ are obtained simultaneously. The new values of all the variables at time $t + \Delta t$ are calculated numerically. We denote the last bead pulled out of the spinneret by $i = N$. When the distance between this bead and the spinneret becomes long enough, say $H/5000$, a new bead $i = N + 1$ is inserted at a small distance, $H/10,000$, from the previous bead. With this work, we can follow the positions of all beads and obtain the path of the jet during the time. As the jet arrives at the collector, the calculation stops. The dates for simulating electrospinning process are summarized in Table 9.1.

TABLE 9.1 Calculation Parameters Using in This Model.

Calculation parameters	Value	Calculation parameters	Value
a_0	150×10^{-4}	Alpha	700
e	8.48	mu	10^5
G	10^6	r_0	1.21×10^{-3}
h	20	N	10
m	0.283×10^{-5}	t period	$0–10^{-7}$
V_0	10,000	step time period	10^{-12}

The jet segment length increases as time develops. It demonstrates that the jet is stretched as it moves downward from the initial position to the collector. The results of simulating electrospinning process are shown in Figures 9.9–9.19.

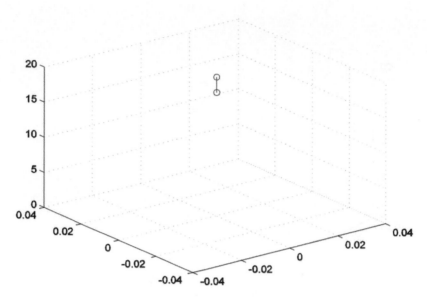

FIGURE 9.9   Jet path calculated for $N = 10$ at time $= 0$–10^{-7} s, with step time period 10^{-12}.

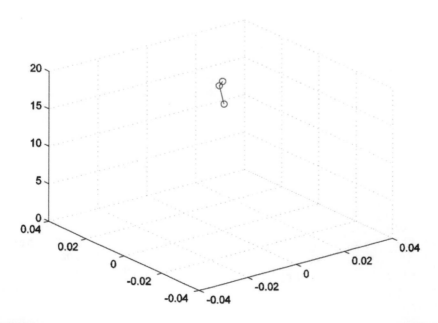

FIGURE 9.10   Jet path calculated for $N = 10$ at time $= 0$–10^{-7} s, with step time period 10^{-12}.

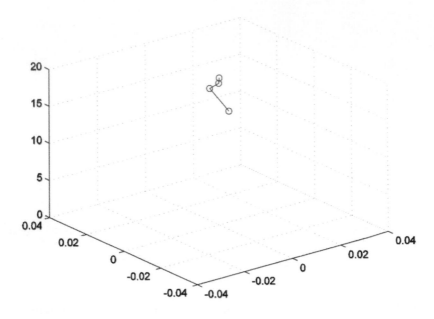

FIGURE 9.11   Jet path calculated for $N = 10$ at time $= 0\text{–}10^{-7}$ s, with step time period 10^{-12}.

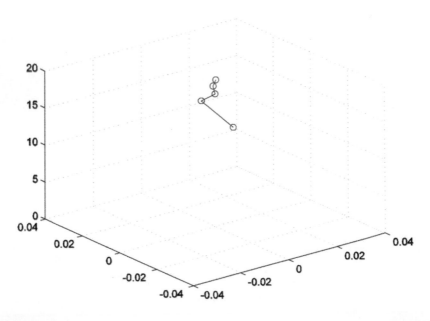

FIGURE 9.12   Jet path calculated for $N = 10$ at time $= 0\text{–}10^{-7}$ s, with step time period 10^{-12}.

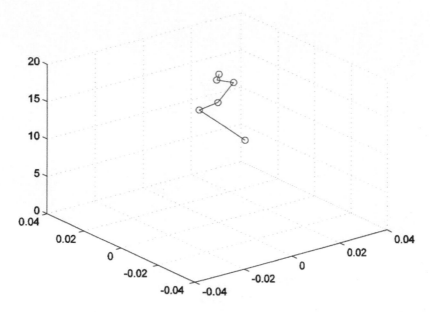

FIGURE 9.13 Jet path calculated for $N = 10$ at time $= 0$–10^{-7} s, with step time period 10^{-12}.

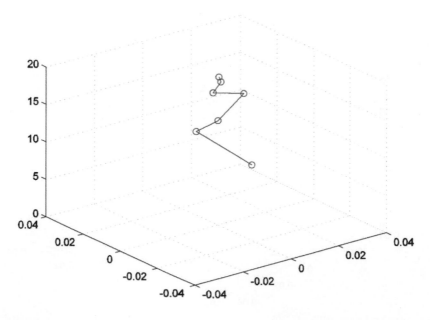

FIGURE 9.14 Jet path calculated for $N = 10$ at time $= 0$–10^{-7} s, with step time period 10^{-12}.

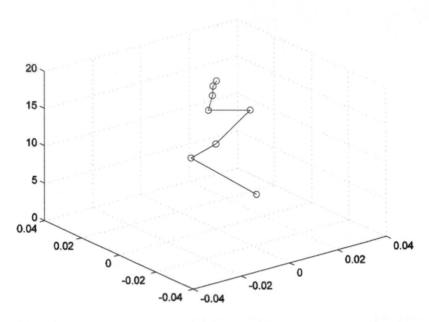

FIGURE 9.15   Jet path calculated for $N = 10$ at time $= 0-10^{-7}$ s, with step time period 10^{-12}.

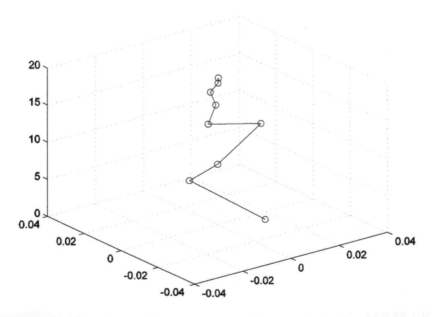

FIGURE 9.16   Jet path calculated for $N = 10$ at time $= 0-10^{-7}$ s, with step time period 10^{-12}.

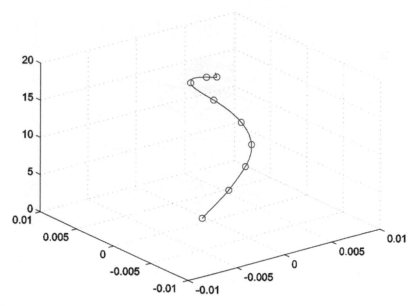

FIGURE 9.17   Path of a single bead calculated for $N = 10$ at times ranging from 0 to 10^{-7} s, with step time period 10^{-12}.

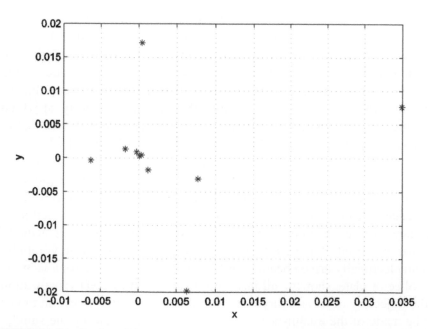

FIGURE 9.18   Beads path in x and y directions during electrospinning because perturbations for better understanding about electrospinning.

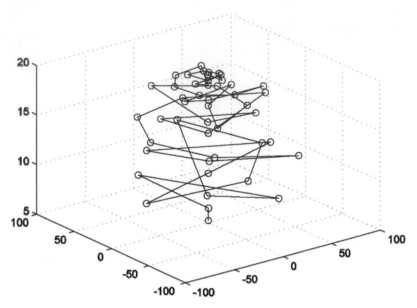

FIGURE 9.19   Path of a single bead calculated for $N = 50$ at times ranging from $= 0–10^{-7}$ s, with step time period 10^{-12}.

Figures 9.9–9.19 illustrate the results of the model data output at various times throughout the calculation of the jet path. As described earlier, new beads are inserted into the calculation when the distance between the last bead and nozzle (located at 20 cm in this case) exceeds the set value. As the beads progress downward, the perturbation added to the x and y coordinates begins to grow as it fully develops into the bending instability, at which point the loops continue to grow outward as the jet moves down. Figure 9.17 shows the path of a single bead at period times (in Table 9.1). The bead does not follow a spiral in the motion. This type of behavior corresponds to observed path of a jet during electrospinning. It shows that the longitudinal stress caused by the external electric field acting on the charge carried by the jet stabilized the straight jet for some distance. Then, a lateral perturbation grew in response to the repulsive forces between adjacent elements of the charge carried by the jet. The motion segment of the jet grew rapidly into an electrically driven bending instability. Figure 9.18 shows a view on the collector plate, showing all incoming beads, next to a 3D visualization of the trajectory of an arbitrary bead in the system. Furthermore, an image can be made of the established jet at a certain moment during the simulation. In Figure 9.19, simulation running can be seen for 50 bead-viscoelastic elements. By considering more beads in lower time, the accuracy of simulation can be increased.

9.6 CONCLUDING REMARKS

Electrospinning is a very simple technique for producing nanofibers by applying electrostatic forces that overcomes the fluid surface tension. It is necessary to control the morphology of these nanofibers and also the electrospinning process. For achieving this goal, using simple models would be useful. In this chapter, microscopic model was reviewed for developing the electrospinning process. This model can be used to describe the dynamic behavior of the electrospun jet in instability part. The jet is described as a bead-viscoelastic element chain. The forces acting on the jet in this model are Coulomb, electric field, viscoelastic, and surface tension forces. The results of the bending instability phenomenon with simulated model were presented. Also, the importance of perturbation could obviously be seen in the results of simulations.

KEYWORDS

- electrospun nanofiber
- microscopic model
- simulation dynamical behavior
- viscoelastic elements
- dynamics of the jet

REFERENCES

1. Coluzza, I.; Pisignano, D.; Gentili, D.; Pontrelli, G.; Succi, S. Ultrathin Fibers from Electrospinning Experiments under Driven Fast-Oscillating Perturbations. *Phys. Rev. Appl.* **2014,** *2* (5), 054011-1–054011-10.
2. Jirsák, J.; Moučka, F.; Nezbeda, I. Insight into Electrospinning via Molecular Simulations. *Ind. Eng. Chem. Res.* **2014,** *53* (19), 8257–8264.
3. Ghochaghi, N. Experimental Development of Advanced Air Filtration Media Based on Electrospun Polymer Fibers. In *Mechnical and Nuclear Engineering*, Virginia Commonwealth, 2014; pp 1–165.

CHAPTER 10

UPDATE ON CONTROLLING INSTABILITY OF ELECTROSPUN NANOFIBERS (PART I)

SHIMA MAGHSOODLOU* and S. PORESKANDAR

Textile Engineering, University of Guilan, Rasht, Iran

Corresponding author. E-mail: sh.maghsoodlou@gmail.com

CONTENTS

ABSTRACT

Recently, the instability behavior of the electrospinning process has received much attention. In this chapter, a model for magnetic electrospinning will be reviewed at first. Then a discrete mathematical model of the magnetic electrospinning process will be investigated and the moving behavior of the jet will be analyzed. The jet can be simulated as many discrete electrified particles, which are joined by viscous flexible materials.

10.1 INTRODUCTION

As mentioned before, electrospinning has been recognized as an efficient technique for the fabrication of polymer nanofibers and has received much attention in recent years. The arrangement of fibers obtained from electrospinning are generally random, but some special nanofibers with well-aligned and highly ordered architectures are greatly necessary when they are used as field effect transistors, gas and optical sensors, fiber-reinforced composite materials, tissue engineering, etc.[1-3]

The charged jet is unstable during the electrospinning process, which leads to uneven construction of nanofibers and waste of energy, as well as other unwanted consequences. Therefore, many researchers have studied instability and tried to control the instability of the jet. For controlling the instability, magnetic electrospinning is proposed, using a magnetic field to control the jet for making ordered nanofibers,[4] as illustrated in Figure 10.1.

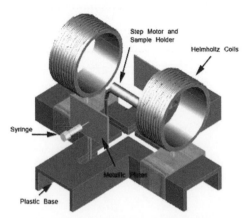

FIGURE 10.1 Magnetic electrospinning setup: pump, nozzle, high voltage supply, excitation coil, collecting plate, and resistance.

The branching of electrospun fibers is due to the electrostatic interactions between different segments of the charged jet. When the repulsive forces on the charged fiber jet generated by magnetic field surpassed the electrostatic interactions, straight fibers without branching are obtained.[5]

This set up can have many advantages such as[5,6]

- The fibers fabricated using the magnetic field-assisted electrospinning method are substantially more uniform and with much less, if any, splitting than those without the field.
- The distribution of diameters also became narrower for fibers made with the magnetic field than those without (Fig. 10.2). The large improvement in fiber uniformity could be attributed to the reducing of instability of spinning jet by applying magnetic field during the electrospinning.
- These ordered arrays of nanofibers can be made over large areas and as thick matrix films. As they were suspended over the gap between the two magnets, the fibers could easily be removed from the substrates and used as scaffolds or transferred onto other surfaces for subsequent treatment or characterization.

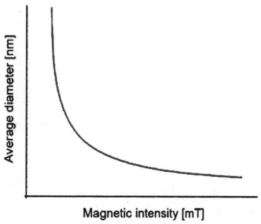

FIGURE 10.2   The relationship between the magnetic intensity and diameters of jets.

The resultant force of electric force and the viscous force of the jet flow are illustrated in Figure 10.3(c).

They cause the whipping circle increasing larger and larger. If a magnetic field were applied in the electrospinning, the problem can be completely overcome. The current in the jet, under the magnetic field, produces a centripetal force, that is, the direction of the ampere force is always toward the initial equilibrium point (Fig. 10.3(b) and (c)), leading to the shrinking of the radius of whipping circle.

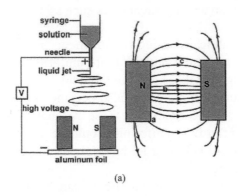

(a)

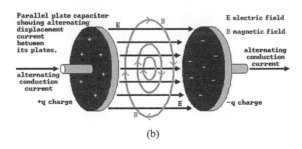

(b)

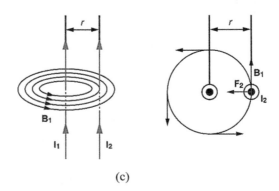

(c)

FIGURE 10.3 Effect of magnetic field on electrospinning: (a) Application of a magnetic field to the electrospinning, (b) mechanics analysis of electric field with magnetic field, and (c) ampere force in the electronic jet induced by the magnetic field.

In this chapter, a discrete mathematical model of the magnetic electrospinning process is established and the moving behavior of the jet is analyzed. The jet can be simulated as many discrete electrified particles, which are joined by viscous flexible materials. The result indicates that the magnetic field creates a radial Lorenz force on the jet and so the jet shrinks in the falling process.

The relationships of the swing of the jet and its distance with different excitation currents applied are obtained. From the relationships, we know that the swing scope with a magnetic field is smaller than that without a magnetic field. The results show that the magnetic field can influence the jet at an appropriate distance.

10.2 MAGNETIC ELECTROSPINNING MODEL

If a magnetic field were applied in the electrospinning process, as illustrated in Figure 10.3, an Ampere force, $d\vec{F}$, acting on the elementary length dL is generated due to the current in the polymer jet:

$$d\vec{F} = dL\vec{J} \times \vec{B}.$$ (10.1)

This set up produces a homogeneous magnetic field B in the mid-plane between the two circular coils, given by[7]

$$B = \frac{N\mu_0 J}{2R}$$ (10.2)

A high-voltage power supply (ES30, 0–30 kV) was used at 15 kV to create an electric field.

In the electrospinning process, the charged jet lands the gap of the permanent magnets, and a couple of Ampere forces are generated due to the current in the jet,[8,9] as illustrated in Figure 10.4.

$$F_m = e\frac{1}{c}u \times B + \frac{1}{c}J \times B + (\nabla B)M + \frac{1}{c}(P \times Bu) + \frac{1}{c}\frac{\partial}{\partial t}(P \times B)$$ (10.3)

$$J = k\left(E + \frac{1}{c}uB\right) + eu + \sigma_T\nabla T$$ (10.4)

$$P = \varepsilon_P\left(E + \frac{1}{c}uB\right)$$ (10.5)

$$e = \nabla D, \quad D = E + P \tag{10.6}$$

The current in the jet, under the magnetic field, produces a couple of Ampere forces. The direction of the Ampere force is always parallel to the velocity of the charged jet (Fig. 10.4), leading to straightening the whipping circle. The straightened circle means less energy waste in the electrospinning process, and the saved energy is used to increase the kinetic energy of the moving jet. As a result the stability condition is enormously improved. More kinetic energy of the moving jet means a higher velocity of the jet.

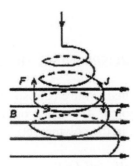

FIGURE 10.4   A couple of Ampere forces in the charged jet induced by the magnetic field.

According to conservation of mass $r^2 \sim 1/u$, the radius becomes much smaller than that without magnetic field.

In addition, according to the mathematical model of the charged jet under coupled multifield, an equation determining the quantity of heat (q) has been written[9]:

$$q = k\nabla T + k_E \left(E + \frac{1}{c} u \times B \right) \tag{10.7}$$

Above equation shows the effect of magnetic field on the quantity of heat in the charged jet. It can be seen that with increase in the magnetic intensity, the jet will obtain a larger quantity of heat.

10.3   A MULTIPHASE FLOW MODEL FOR ELECTROSPINNING PROCESS

Electrospinning is a multiphase and multiphysics process involving electrohydrodynamics, mass and heat diffusion and transfer, evaporation, etc.

Almost the models are single-phase models and can't offer in-depth insight into physical understanding of many complex phenomena which cannot be fully explained experimentally.

The multiphase flow model takes into account solvent evaporation and dispersion of additive particles, which play pivotal roles in determining the internal fiber morphology of the electrified jet.[10]

The modified Navier–Stokes equations governing heat and a jet under the influence of electric field and magnetic field are[8]

Maxwell's equations:

$$\frac{\partial e}{\partial t} + \nabla J = 0 \tag{10.8}$$

$$\nabla \times E + \frac{1}{c}\frac{\partial B}{\partial t} = 0 \tag{10.9}$$

$$\nabla \times H - \frac{1}{c}\frac{\partial D}{\partial t} = \frac{1}{c}J \tag{10.10}$$

Continue equation:

$$\frac{\partial \rho}{\partial t} + \nabla(\rho u) = 0 \tag{10.11}$$

Momentum equation:

$$\rho\frac{Du}{Dt} = \nabla t + \rho f + eE + (\nabla E)P + F_m \tag{10.12}$$

Energy equation:

$$\rho c_p \frac{DT}{Dt} = Q_h + \nabla q + (J - eu)\left(E + \frac{1}{c}u \times B\right) - \left(E + \frac{1}{c}u \times B\right)$$
$$\frac{DP}{Dt} - \left(M + \frac{1}{c}u \times P\right)\frac{DB}{Dt} + Q_f \tag{10.13}$$

This set of conservation laws can constitute a closed system when it is supplemented by appropriate constitutive equations for the field variables such as polarization. The most general theory of constitutive equations determining the polarization, electric conduction current, heat flux, and Cauchy stress tensor has been developed by Eringen and Maugin.[10]

CHAPTER 11

UPDATE ON CONTROLLING INSTABILITY OF ELECTROSPUN NANOFIBERS (PART II)

SHIMA MAGHSOODLOU* and S. PORESKANDAR

Textile Engineering, University of Guilan, Rasht, Iran

Corresponding author. E-mail: sh.maghsoodlou@gmail.com

CONTENTS

ABSTRACT

Controlling the formation of electrospun polymer nanofibers are vital in the electrospinning process. Applying Ac potential on the resulting mats exhibits a significant reduction in the amount of fiber whipping, and a higher degree of fiber alignment. By utilizing this method, fiber radius can be easily controlled. In this chapter, a mathematical model will be investigated to explore the physics behind AC-electrospinning. A relationship between the radius of the jet and the axial distance from nozzle, and a scaling relation between fiber radius and the AC frequency will be obtained.

11.1 INTRODUCTION

Deposition of electrospun nanofibers on a flat collector is essentially random and disordered due to the chaotic and unstable motion of the electrospinning jet as it travels to the collector. Fiber diameter is an important characteristic for electrospinning, because of its direct influence on the properties of the produced webs.[1-3]

Depending on several solution parameters, different results can be obtained using the same polymer and electrospinning setup.[4] As a concept, successful electrospinning requires an understanding of the complex interaction of electrostatic fields, properties of polymer solutions, and component design and system geometry.[5] During this process, the jet has shown instability behavior (Fig. 11.1).[6]

These instabilities arise owing to the charge–charge repulsion between the excess charges present in the jet, which encourages the thinning and elongation of the jet. At high electric forces, the jet is dominated by bending and whipping instability, causing the jet to move around and produces wave in the jet. These instabilities vary and increase with distance, electrical field, and fiber diameter at different rates depending on the fluid parameters and performing conditions.[7]

Some applications need the parallel and oriented nanofibers with highly improved mechanical properties. By increasing the rotational speed of the collecting drum, introducing a potential across a gap or series of gaps in the collecting electrode, introducing an external lens element or a viscous liquid environment, or rapidly oscillating a grounded frame within the liquid jet, well-aligned electrospun fibers can be produced.[8-10]

Considering the chaotic nature of the electrospun jet motion, the buildup of electrical surface charges on the collector and the nanometer size of the electrospun fibers into account, the collector design can control the electrospun nanofibers architecture. By also controlling the geometrical shape and strength of the macroscopic electric field between the spinneret and the collector, it should be possible to control the jet path, to improve its stability and to achieve aligned nanofibers. In recent years, researchers have developed several collection mechanisms to control the deposition of the electrospun nanofibers and to obtain continuous fiber alignment by manipulating the dynamic motion of the collector or the electric field strength and geometrical shape or both (Fig. 11.2; Table 11.1).[8,11]

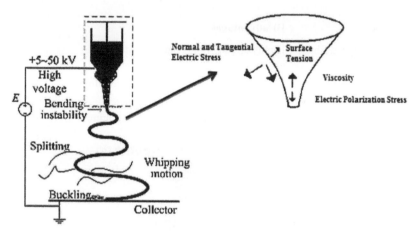

FIGURE 11.1 Jet formation instabilities.

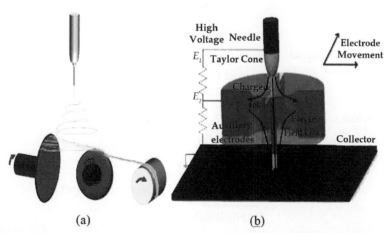

(a)   (b)

FIGURE 11.2 Fiber-deposition control by (a) application of different collectors and (b) secondary electric fields in near-field electrospinning.

TABLE 11.1   Nano-Aligning Mechanisms Based on Dynamic Mechanical Collection.

Collector		References
Rotating drum	Rotating target	[12–14]

Advantages	**Disadvantages**
Large area of aligned fibers may be fabricated	Highly aligned fibers may be difficult to achieve
	Fiber breakages may occur if rotating drum speed is too high
	Twist cannot be applied for yarn spinning

Collector		References
Scanned disk fabrication	←Needle tip Rotating disk having sharp edge	[15,16]

Advantages	**Disadvantages**
Rapid fabrication of aligned polymeric nanowires	Vaporization of solvent is not complete
	Thicker jet formation
	Twist cannot be applied to form a yarn

Collector		References
Coagulation bath collector	Non-solvent Rotating target Coagulation bath	[17,18]

TABLE 11.1   *(Continued)*

	Collector	References
Advantages	**Disadvantages**	
Nanofibers can be collected without sticking	Applicable to some polymers	
	Partial alignment of fibers	
	Growth of bead deposits	
	Twist cannot be applied to form a yarn	

Most electrospinning processes are carried out using high-voltage direct current (DC). However, high-voltage alternating current (AC) has also been shown to be able to create electrospun fibers.[19]

Many of the nanofiber research reported so far have focused on nanofibers made from DC potential. In DC-electrospinning, the fiber whipping instability makes it difficult to control the fiber location and the resulting microstructure of electrospun materials. To overcome these limitations, some new technologies can be applied in the electrospinning process.[20]

Using AC-electrospinning bring interesting results. It is not necessary to use a collector for nanofibrous deposition—the process is powered by electric gradient instead. Nanofibers are electrically neutral. Fibers join to form a linear structure by attraction of electric charges. Nanofiber layer structure is different from the one generated by using DC-electrospinning method. By manipulating the strength, frequency, position and geometric shape of the electric field, and the polarity of the charges, aligned and preordered nanofibers can be achieved.[21] An additional parameter when using AC high voltage for electrospinning is the frequency of the current. When the frequency is too high, the transfer of charges or mobility of the ion may not be fast enough to sufficiently charge the solution for electrospinning. For each material, the frequency range for AC-electrospinning needs to be determined although it is typically between 500 Hz and 1 kHz. When the frequency is too low, the electrospinning jet traveling between the tip and the collector may comprise mainly of a single polarity instead of periodic positive and negative segments.[22,23]

The periodicity of positive and negative segments on an AC-electrospinning jet has been shown to facilitate a more stable jet formation compared to DC. When collecting on a rotating drum, the more stable AC electrospun jet produces fiber with greater alignment compared to fibers from DC. However, the fiber diameter from AC-electrospinning is significantly larger compared to DC-electrospinning. This is probably due to suppression of bending instability in AC-electrospinning.[19]

Presence of positive and negative segments on an AC-electrospinning jet has also been known to induce self-bundling of the electrospinning jet. Using AC-electrospinning can be terminated to obtain spontaneous self-bundled yarn. In this case, the jet first spreads out from the spinneret tip and subsequently converges in mid-flight to form yarn. During the flight of the spinning jet, segments of the jet with positive charges will be attracted to segments with negative charges. The electrospinning jet immediately enters the bending instability when the solution emerges from the nozzle. Multiple jets erupting from the meniscus and higher order branching are also recorded. Increasing applied frequency can be found to increase the fiber thickness and likelihood of beads formation due to reduced stretching of the jet in the presence of differing charges.[24]

The investigations proved that the AC potential resulted in a significant reduction in the amount of fiber whipping and the resulting mats exhibit a higher degree of fiber alignment but contained more residual solvent. In AC-electrospinning, the jet is inherently unsteady due to the AC potential unlike DC ones, so all thermal, electrical, and hydrodynamic parameters are considered to be effective in the process (Fig. 11.3; Table 11.2).[20]

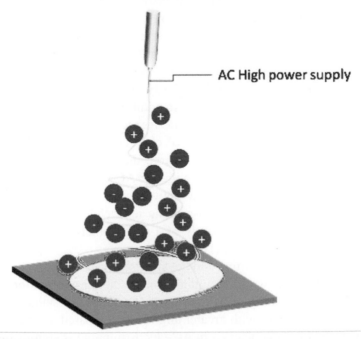

FIGURE 11.3 Illustration of AC-electrospinning.

TABLE 11.2   Nano-Aligning Mechanisms Based on Manipulation of the Electric Field.

Electrode		References
Auxiliary rings electrodes		[25,26]

Advantages	Disadvantages
Controlled fibers deposition area and location	Complicate assembly
Twist can be applied	Small deposited area

Auxiliary AC circular electrodes		[27,28]

Advantages	Disadvantages
Highly aligned fibers are easily obtained	Complicated set-up
Aligned fibers are easily transferable	Thicker layer of aligned fibers are not possible
	There is a limit in the length of the fibers
	Twist cannot be applied to spin a yarn

Auxiliary AC parallel electrodes		[21,29]

TABLE 11.2 *Continued*

Electrode		References
Advantages	**Disadvantages**	
Highly aligned fibers are easily obtained	Complicated set-up	
	Thicker layer of aligned fibers are not possible	
Aligned fibers are easily transferable		
No limit in the length of fibers produced		
Parallel auxiliary electrodes		[30,31]
Advantages	**Disadvantages**	
Simple set-up	Thicker layer of aligned fibers are not possible	
Highly aligned fibers are easily obtained	There is a limit in the length of the fibers produced	
Aligned fibers are easily transferable	Twist cannot be applied to spin a yarn	

By controlling the geometric profile and strength of the electric field and by using different dynamic collecting systems, greater alignment in the nanofibers assembly can be achieved. Researchers successfully fabricated fibrous assemblies by using of both rotating collectors and manipulation of the electric field profile.[32]

A suitable theoretical model of the electrospinning process is one that can show a strong–moderate–minor rating effects of these parameters on the nanofiber diameter.[33] The main focus in this chapter is investigated AC model to give a better understanding of unstable behavior of the electrospinning process. In the next part, more details were investigated about this model.

11.2 GOVERNING EQUATIONS OF AC MATHEMATICAL MODEL

The regulation of scale is an intriguing and enduring problem after the technology was invented by Formhals in 1934. Regulatory mechanisms for controlling the radius of electrospun fibers are clearly illustrated in the different states in Ref. [34]. Generally, the relationship between radius r of jet and the axial distance z from nozzle can be expressed as an allometric equation of the form $r \sim z^b$, the values of the scaling exponent (b) for the initial steady stage, instability stage, and terminal stage are, respectively, $-1/2$, $-1/4$, and 0, that is,

Kessick et al.[19] analyzed the role of AC potentials in electrospinning process. The AC potential was resulted in a significant decreasing in the amount of fiber whipping. Applying this potential on the resulting mats exhibited a higher degree of fiber alignment but noted to contain more residual solvent. Yet, theoretical modeling of the AC-electrospinning process remains a bottleneck, severely hampering further improvement in both quality and efficiency. This chapter establishes a mathematical model to explore the physics behind AC-electrospinning. In DC-electrospinning, the jet can be considered as a steady stream. However, in AC-electrospinning, the process is inherently unsteady due to the AC potential (Fig. 11.4).[35]

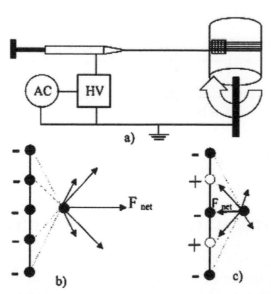

FIGURE 11.4 (a) Schematic diagram of the electrospinning apparatus. Force balance on a displaced segment of an electrically charged fiber in (b) DC-electrospinning and (c) AC-electrospinning.[8]

It consists of modified Maxwell's equations governing electrical field in a moving fluid, the modified Navier–Stokes equations governing fluid flow under the influence of electric field, and constitutive equations describing behavior of the fluid. The governing equations for an unsteady flow of an infinite viscous jet pulled from a capillary orifice and accelerated by an AC potential can be expressed as follows. The conservation of mass equation gives

$$r^2 \frac{\partial}{\partial t}(\rho) + \frac{\partial}{\partial z}(\rho r^2 u) = 0. \tag{11.1}$$

The conservation of charge becomes

$$\frac{\partial}{\partial t}(2\pi r \sigma) + \frac{\partial}{\partial z}(E\pi kr^2 E + 2\pi r \sigma u) = 0 \tag{11.2}$$

The current is composed of two parts: the Ohmic bulk conduction current: $Jc = \pi r^2 kE$, and the surface convection current: $Js = 2\pi r \sigma u$. The Navier–Stokes equations are defines as

$$\frac{\partial u}{\partial t} + u \frac{\partial u}{\partial z} = -\frac{1}{\rho}\frac{\partial p}{\partial z} + g + \frac{2\sigma E}{\rho r} + \frac{1}{r^2}\frac{\partial \tau}{\partial z} \tag{11.3}$$

where p is the internal pressure of the fluid expressed as

$$p = \left(\frac{1}{R_1} + \frac{1}{R_2}\right)V - \frac{\varepsilon - \bar{\varepsilon}}{8\pi}E^2 - \frac{2\pi}{\bar{\varepsilon}}\sigma^2 \tag{11.4}$$

Power-law constitutive equation can be described as the rheological behavior of many polymer fluids as below:

$$\tau = \mu_0 \frac{\partial u}{\partial z} + \sum_{n=1}^{m} a_n \left(\frac{\partial u}{\partial z}\right)^{2n+1} \tag{11.5}$$

11.3 ALLOMETRICAL SCALING LAWS IN AC-ELECTROSPINNING PROCESS

Electrospinning applies electrically generated motion to spin fibers; therefore, it is difficult to predict the size of the produced fibers, which mainly depends on the applied voltage. Therefore, the relationship between the radius of the jet and the axial distance from the nozzle became a subject of investigation. It can be described as an allometric equation by using the values of the scaling exponent for the initial steady stage, instability stage, and terminal stage. In addition, understanding the regulation of allometry in AC-electrospinning would have broad implications on furthering our knowledge of the process and on controlling the diameter of the electrospun fibers. Several authors have described experiments and searched for a ubiquitous scaling law in DC-electrospinning.[34,36,37] The stream of jet assume incompressible. Under such an assumption, the conservation of mass reduces

$$\pi r^2 u = Q \tag{11.6}$$

Equation 11.6 reveals that r and u are independent of time, so the current balance can be written in the form:

$$2r\frac{\partial}{\partial t}(\sigma) + \frac{\partial}{\partial z}\left(kr^2 E + 2r\sigma u\right) = 0 \tag{11.7}$$

Suppose the AC potential can be expressed in the form

$$E = \bar{E}\cos(\Omega t + \alpha) \tag{11.8}$$

We assume that it changes simultaneously with the AC potential:

$$\sigma = \bar{\sigma}\cos(\Omega t + \beta) \tag{11.9}$$

Substituting eq 11.8 and 11.9 into eq 11.7, and integrating the result from zero to T yields

$$\int_0^T \left(2r\frac{\partial \sigma}{\partial t} + \frac{\partial}{\partial z}\left(kr^2 E + 2r\sigma u\right)\right) dt = 0 \tag{11.10}$$

From eq 11.10, we have

$$2r\Omega\frac{\sqrt{2}}{2}\bar{\sigma} + \frac{\partial}{\partial z}\left(\frac{\sqrt{2}}{2}kr^2\,\bar{E} + 2r\frac{\sqrt{2}}{2}\bar{\sigma}u\right) = 0 \tag{11.11}$$

or

$$2r\Omega\bar{\sigma} + \frac{\partial}{\partial z}\left(kr^2\bar{E} + 2r\bar{\sigma}u\right) = 0 \tag{11.12}$$

Introducing a special functional Φ defined as

$$\frac{\partial \Phi}{\partial z} = 2r\Omega\bar{\sigma} \tag{11.13}$$

Equation 11.12 can be rewritten in the form

$$\frac{\partial}{\partial z}\left(\Phi + kr^2\bar{E} + 2r\bar{\sigma}u\right) = 0 \tag{11.14}$$

Therefore, the following useful equation is obtained

$$\Phi + kr^2\bar{E} + 2r\bar{\sigma}u = I \tag{11.15}$$

where I can be considered as equivalent current passing through the jet, which consists of three parts: (1) the Ohmic bulk conduction current: $J_c = \pi kr^2 E$; (2) the surface convection current: $J_s = 2\pi r \sigma u$; and (3) AC-induction surface current Φ. Fiber diameter is approximately proportional to jet length. The jet length is measured from the tip of the spinning drop to the onset of waves in the fiber. He, Wan, and Yu[38] obtained an allometric scaling law for fiber diameter before instability in DC-electrospinning, their theory can be readily extended to AC-electrospinning. In the absence of an electric field, a meniscus is formed at the exit of the capillary. The meniscus is pulled out into a cone when the electric force is applied. When the electric force surpasses a threshold value, the electric force exceeds the surface tension, and a fine charged jet is pulled out and is accelerated. When the jet is accelerated by the electrical force, the viscous resistance becomes higher and higher, and the jet becomes instability when the value of the viscous resistance almost reaches or surpasses that of the electrical force. Under such a condition, a slight perturbation by air might lead to oscillation. Experiment showed that the AC potential resulted in a significant reduction in the amount of fiber "whipping." Before fiber "whipping," where electrical force is dominant over other forces acting on the jet, the balance equation reduces to

$$\frac{d}{dz}\left(\frac{u^2}{2}\right) = \frac{2\sigma E}{\rho r} \tag{11.16}$$

Integrating eq 11.16 from zero to T results in

$$\int_0^T \frac{\partial}{\partial z}\left(\frac{1}{2}u^2\right)dt = \int_0^T \frac{2\sigma E}{\rho r}dt \tag{11.17}$$

Many experiment shows scaling relationship between r and z, which can be expressed as an allometric equation of the form

$$r \approx z^b \tag{11.18}$$

Allometrical method is widely applied in biology[38] and in engineering[34,36] as well. Assume that the volume flow rate (Q) and the maximal AC voltage (max E) keep unchanged during the electrospinning procedure, we have the following scaling relations: $Q \approx r^0$ and $\tilde{E} \approx r^0$.

From eq 11.6, So

$$u \approx r^{-2} \tag{11.19}$$

Rewrite eq 11.15 in the form:

$$\overline{E} = \frac{I}{kr^2} - \frac{2u\overline{\sigma}}{kr} - \frac{\varphi}{kr^2} \approx r^0 \tag{11.20}$$

the following scaling relationships are obtained

$$I \approx r^2 \tag{11.21}$$

$$\overline{\sigma} \approx r^3 \tag{11.22}$$

and

$$\Phi \approx r^2 \tag{11.23}$$

Substituting eqs 11.19 and 11.22 into eq 11.17 results in

$$\frac{\partial}{\partial z}(u^2) \approx \frac{2\overline{\sigma}E}{\rho r} \tag{11.24}$$

or

$$\frac{\partial}{\partial z}(r^{-4}) \approx r^2 \tag{11.25}$$

Therefore, the following scaling relationship for AC-electrospinning can be expressed as

$$r \approx z^{\frac{-1}{6}} \tag{11.26}$$

By substituting eqs 11.22 and 11.23 in eq 11.13, the result is

$$\frac{\partial}{\partial z}(r^2) \approx r^4\Omega \tag{11.27}$$

In view of eq 11.26, from eq 11.27, we have a scaling relationship between the fiber radius and AC frequency:

$$r \approx \Omega^{\frac{1}{4}} \tag{11.28}$$

The allometric scaling laws eq 11.26 and 11.28 might be useful in theoretical and experimental analyses.

11.4 CONCLUDING REMARKS

In this chapter, a brief review was applied for investigating AC mathematical model for controlling the unstable behavior of this process. The governing equations in this model were conservation of mass, conservation of charge and the Navier–Stokes equation. Also, the power-law constitutive equation was used for describing the rheological behavior of polymer fluids. Using these governing equations, the final model of AC-electrospinning was able to find the relationship between the radius of the jet and the axial distance from nozzle, and a scaling relation between fiber radius and the AC frequency. This model described a complex dynamic process from the theory, and it required less empirical or semi-empirical inputs.

KEYWORDS

- electrospinning process
- instability behavior
- AC-electrospinning model
- fiber whipping
- fiber alignment

REFERENCES

1. Huang, Z. M.; Zhang, Y. Z.; Kotaki, M.; Ramakrishna, S. A Review on Polymer Nanofibers by Electrospinning and their Applications in Nanocomposites. *Compos. Sci. Technol.* **2003,** *63,* 2223–2253.
2. Haghi, A. K. Electrospun Nanofiber Process Control. *Cellul. Chem. Technol.* **2010,** *44* (9), 343–352.

3. De, V, S.; Van, C, T.; Nelvig, A.; Hagström, B.; Westbroek, P.; De, C, K. The Effect of Temperature and Humidity on Electrospinning. *J. Mater. Sci.* **2009**, *44* (5), 1357–1362.
4. Sill, T. J.; Recum, H. A. Electrospinning: Applications in Drug Delivery and Tissue Engineering. *Biomaterials* **2008**, *29* (13), 1989–2006.
5. Lukáš, D.; Sarkar, A.; Martinová, L.; Vodsed'álková, K.; Lubasova, D.; Chaloupek, J.; Pokorný, P.; Mikeš, P.; Chvojka, J.; Komarek, M. Physical Principles of Electrospinning (Electrospinning as a Nano-Scale Technology of the Twenty-First Century). *Text. Progr.* **2009**, *41* (2), 59–140.
6. Yarin, A. L.; Koombhongse, S.; Reneker, D. H. Bending Instability in Electrospinning of Nanofibers. *J. Appl. Phys.* **2001**, *89* (5), 3018–3026.
7. Baji, A.; Mai, Y. W.; Wong, S. C.; Abtahi, M.; Chen, P. Electrospinning of Polymer Nanofibers: Effects on Oriented Morphology, Structures and Tensile Properties. *Compos. Sci. Technol.* **2010**, *70* (5), 703–718.
8. Sarkar, S.; Deevi, S.; Tepper, G. *Biased AC Electrospinning of Aligned Polymer Nanofibers. Macromol. Rapid Commun.* **2007**, *28* (9), 1034–1039.
9. Yang, F.; Murugan, R.; Ramakrishna, S.; Wang, X.; Ma, Y. X.; Wang, S. *Fabrication of Nano-Structured Porous PLLA Scaffold Intended for Nerve Tissue Engineering. Biomaterials* **2004**, *25* (10), 1891–1900.
10. Mo, X.; Weber. H. J. Electrospinning P (LLA-CL) Nanofiber: A Tubular Scaffold Fabrication with Circumferential Alignment. In *Macromolecular Symposia*. Wiley Online Library, 2004.
11. Kakade, M. V.; Givens, S.; Gardner, K.; Lee, K. H.; Chase, D. B.; Rabolt, J. F. Electric Field Induced Orientation of Polymer Chains in Macroscopically Aligned Electrospun Polymer Nanofibers. *J. Am. Chem. Soc.* **2007**, *129* (10), 2777–2782.
12. Bashur, C. A.; Dahlgren, L. A.; Goldstein, A. S. Effect of Fiber Diameter and Orientation on Fibroblast Morphology and Proliferation on Electrospun Poly(D,L-Lactic-*co*-Glycolic Acid) Meshes. *Biomaterials* **2006**, *27* (33), 5681–5688.
13. Deitzel, J. M.; Kleinmeyer, J. D.; Hirvonen, J. K.; Tan, N. B. Controlled Deposition of Electrospun Poly(Ethylene Oxide) Fibers. *Polymer* **2001**, *42* (19), 8163–8170.
14. Mathew, G.; Hong, J. P.; Rhee, J. M.; Leo, D. J.; Nah, C. Preparation and Anisotropic Mechanical Behavior of Highly-Oriented Electrospun Poly (Butylene Terephthalate) Fibers. *J. Appl. Polym. Sci.* **2006**, *101* (3), 2017–2021.
15. Kameoka, J.; Craighead, H. G. Fabrication of Oriented Polymeric Nanofibers on Planar Surfaces by Electrospinning. *Appl. Phys. Lett.* **2003**, *83* (2), 371–373.
16. Yang, F.; Murugan, R.; Wang, S.; Ramakrishna, S. Electrospinning of Nano/Micro Scale Poly(L-Lactic Acid) Aligned Fibers and their Potential in Neural Tissue Engineering. *Biomaterials* **2005**, *26* (15), 2603–2610.
17. Zhong, S.; Zhang, Y.; Lim, C. T. Fabrication of Large Pores in Electrospun Nanofibrous Scaffolds for Cellular Infiltration: A Review. *Tissue Eng., B: Rev.* **2011**, *18* (2), 77–87.
18. Meli, L.; Miao, J.; Dordick, J. S.; Linhardt, R. J. Electrospinning from Room Temperature Ionic Liquids for Biopolymer Fiber Formation. *Green Chem.* **2010**, *12* (11), 1883–1892.
19. Kessick, R.; Fenn, J.; Tepper, G. The Use of AC Potentials in Electrospraying and Electrospinning Processes. *Polymer* **2004**, *45* (9), 2981–2984.
20. Rafiei, S.; Maghsoodloo, S.; Saberi, M.; Lotfi, S.; Motaghitalab, V.; Noroozi, B.; Haghi, A. K. New Horizons in Modeling and Simulation of Electrospun Nanofibers: A Detailed Review. *Cellul. Chem. Technol.* **2014**, *48* (5–6), 401–424.

21. Kim, G. H. Electrospinning Process Using Field-Controllable Electrodes. *J. Polym. Sci., B: Polym. Phys.* **2006,** *44* (10), 1426–1433.
22. Sarkar, S.; Rixen, S. T.; Hamilton, G.; Seifalian, A. M. Achieving the Ideal Properties for Vascular Bypass Grafts Using a Tissue Engineered Approach: A Review. *Med. Biol. Eng. Comput.* **2007,** *45* (4), 327–336.
23. Agarwal, S.; Wendorff, J. H.; Greiner, A. Progress in the Field of Electrospinning for Tissue Engineering Applications. Adv. Mater. **2009,** *21* (32–33), 3343–3351.
24. Ali, U.; Zhou, Y.; Wang, X.; Lin, T. Electrospinning of Continuous Nanofiber Bundles and Twisted Nanofiber Yarns. In *Nanofibers—Production, Properties and Functional Applications*; Lin, T., Ed.; InTech: Rigeka, 2011; pp 153–174.
25. Park, S.; Park, K.; Yoon, H.; Son, J.; Min, T.; Kim, G. Apparatus for Preparing Electrospun Nanofibers: Designing an Electrospinning Process for Nanofiber Fabrication. *Polym. Int.* **2007,** *56* (11), 1361–1366.
26. Migliaresi, C.; Ruffo, G. A.; Volpato, F. Z.; Zeni, D. Advanced Electrospinning Setups and Special Fibre and Mesh Morphologies, In *Electrospinning for Advanced Biomedical Applications and Therapies. Nanomedicine*, 2012, pp 23–68.
27. Yin, G. B. Control of Jet Flows and their Effects on the Alignment of Electrospun Nanofibers. In *Advanced Materials Research*. Trans Tech Publ., 2012.
28. Long, Y. Z.; Sun, B.; Zhang, H. D.; Duvail, J. L.; Gu, C. Z.; Yin, H. L. Fabrication and Applications of Aligned Nanofibers by Electrospinning. *Nanotechnol. Res. J.* **2014,** *7* (2), 155–178.
29. Woo, L. Y.; Wansom, S.; Ozyurt, N.; Mu, B.; Shah, S. P.; Mason, T. O. Characterizing Fiber Dispersion in Cement Composites Using AC-Impedance Spectroscopy. *Cem. Concr. Compos.* **2005,** *27* (6), 627–636.
30. Arras, M. M. L.; Grasl, C.; Bergmeister, H.; Schima, H. Electrospinning of Aligned Fibers with Adjustable Orientation Using Auxiliary Electrodes. *Sci. Technol. Adv. Mater.* **2012,** *13* (3), 035008–035015.
31. Carnell, L. S.; Siochi, E. J.; Wincheski, R. A.; Holloway, N. M.; Clark, R. L. Aligned Mats from Electrospun Single Fibers. *Macromolecules* **2008,** *41* (14), 5345–5349.
32. Mandal, D.; Yoon, S.; Kim, K. J. Origin of Piezoelectricity in an Electrospun Poly(Vinylidene Fluoride-Trifluoroethylene) Nanofiber Web-Based Nanogenerator and Nano-Pressure Sensor. *Macromol. Rapid Commun.* **2011,** *32* (11), 831–837.
33. Fridrikh, S. V.; Yu, J. H.; Brenner, M. P.; Rutledge, G. C. Controlling the Fiber Diameter during Electrospinning. In *Physical Review Letters*; American Physical Society, 2003; pp 144502–144505.
34. He, J. H. Wan, Y. Q. Allometric Scaling and Instability in Electrospinning. *Int. J. Nonlin. Sci. Numer. Simul.* **2004,** *5* (3), 243–252.
35. Wan, Y. Q.; Guo, Q.; Pan, N. Thermo-Electro-Hydrodynamic Model for Electrospinning Process. *Int. J. Nonlin. Sci. Numer. Simul.* **2004,** *5* (1), 5–8.
36. He, J. H.; Wan, Y. Q. Allometric Scaling for Voltage and Current in Electrospinning. *Polymer* **2004,** *45* (19), 6731–6734.
37. Shin, Y. M.; Hohman, M. M.; Brenner, M. P.; Rutledge, G. C. Experimental Characterization of Electrospinning: The Electrically Forced Jet and Instabilities. *Polymer* **2001,** *42* (25), 09955–09967.
38. He, J. H.; Wu, Y.; Pang, N. A Mathematical Model for Preparation by AC-Electrospinning Process. *Int. J. Nonlin. Sci. Numer. Simul.* **2005,** *6* (3), 243–248.

CHAPTER 12

UPDATE ON THERMO-ELECTROHYDRODYNAMIC MODEL FOR ELECTROSPINNING PROCESS

SHIMA MAGHSOODLOU* and S. PORESKANDAR

Textile Engineering, University of Guilan, Rasht, Iran

*Corresponding author. E-mail: sh.maghsoodlou@gmail.com

CONTENTS

ABSTRACT

Vibration-electrospinning is utilized for producing finer nanofibers. Thus, in this chapter, a mathematical model, which is explained the physics behind this electrospinning process, will be discussed. On the other hand, the thermal factor is one of the critical factors for a polymer with high molten temperature and a thermo-electrohydrodynamics description of electrospinning is needed for better understanding of the process. So, in the following of this chapter, a thermo-electrohydrodynamic model of the vibration-electrospinning process which can be applied to numerical study will be investigated. This model can offer in-depth insight into the physical understanding of many complex phenomena which cannot be fully explained experimentally.

12.1 INTRODUCTION

Industrial application of electrospun nanofibers, due to their weakness in strength, has been limited. One of the critical reason for the weakness of electrospinning is the involvement of large volume solvent which causes numerous pores and structural defects when the solvent evaporates for fibers during and after electrospinning.[1,2]

Therefore, reduce the usage amount of solvent or the utilization of polymer solution with higher concentration/viscosity is theoretically a pathway to strengthen electrospun nanofibers. It is obvious that the viscosity of polymer solutions for electrospinning constricts to a very narrow range.[3,4]

On the other hand, for a polymer with high molten temperature, the thermal factor is critical for the process. So a rigorous thermo-electrohydrodynamics description of electrospinning is needed for better understanding of the process. For this cause, vibration-electrospinning, the novel strategy produces finer nanofibers than those obtained without vibration and can produce nanofibers which cannot be done by electrospinning without vibration.[5]

Application of vibration technology in polymer processes such as injection molding, extrusion, and compression molding/thermoforming proved that it works well in reduction of melt viscosity and enhancement of mechanical properties of polymer products. Theory analysis experimental data reveal many parameters, such as fiber diameter, threshold voltage, fiber strength, are related to the polymer viscosity. By vibration technology, finer nanofibers under lower applied voltage can be produced and the reduction in the applied voltage terminates to increase the fiber strength.[6,7]

The mechanism is deceptively simple: in the absence of an electric field, the fluid forms a drop at the exit of the capillary, and its size is determined by surface tension. When an electric field is present, it induces charges into the fluid.

These quickly relax to the fluid surface. The coupling of surface charge and the external electric field creates a tangential stress, resulting in the deformation of the droplet into a conical shape (Taylor cone).

Once the electric field exceeds the critical value needed to overcome the surface tension, a fluid jet ejects from the apex of the cone.[8]

In this chapter, a simple model and a thermo-electrohydrodynamic model of the vibration-electrospinning process which can be applied to numerical study are established.

12.2 VIBRATION TECHNOLOGY IN POLYMER PROCESSING

Vibration technology has been introduced into polymer processing for many years. Initially, it was only applied in research for polymer melt viscosity measurement. Subsequently, the principles of melt vibration has been offered into practical applications including injection molding, extrusion, and compression molding/thermoforming for reduction of viscosity to lowering processing temperature and pressure to the elimination of melt defects and weld lines, and enhancement of mechanical properties by modification of the amorphous and semicrystalline texture and orientational state (Fig. 12.1).[7,9,10]

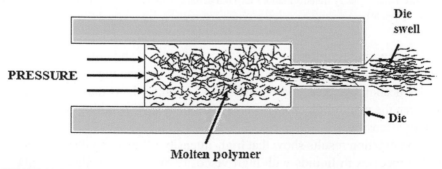

FIGURE 12.1   Die swelling of molten polymer.

In polymers, intramolecular bonds are due to primary valence bonds (covalent) while the intermolecular attractions usually are due to secondary bonding forces. The intermolecular forces are opposed by thermal agitation, which induces vibration, rotation, and translation of a molecular system. Atomic vibrations exist at all temperature levels. The stability of the molecular system depends on the vibration energy of the chemical bonds. In polymers, thermal degradation occurs when the energy of vibration exceeds the primary bonding between atoms, while the transitional phenomena associated with the crystalline melting point, the glass transition temperature, and the polymer deformations are related to rotation and vibration of molecular chains (Fig. 12.2).[11,12]

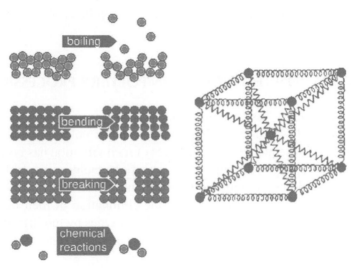

FIGURE 12.2 Energy transformations turn out at the atomic level cause of heat and vibration (cubic vibration)

When a vibrating force is applied a concentrated and entangled polymer solution or melt, the weak van der Waals' force connecting with macromolecules become weaker, and entanglement is relaxed, so that viscous force between the macromolecules decreases dramatically, resulting in the reduction of viscosity.

Investigation results show that high-intensity ultrasonic vibrations accelerate processes in liquids with high viscosity or density of disperse phase. There are typical examples of this application: polymerization and depolymerization of high molecular compounds, modification of epoxy oligomers, melt mixing, emulsions, dissolving of sludge, dispersion, and uniform spraying of solid particles in polymer materials and technical oils (Fig. 12.3).[13–15]

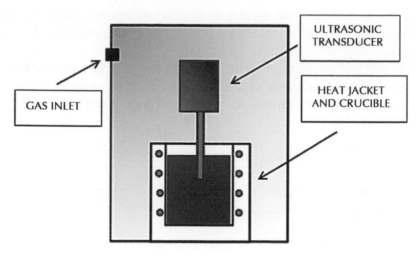

FIGURE 12.3   Ultrasonic vibration and heating for polymer melt or solution.

Viscosity decreases due to several processes taking place during the ultrasonic processing[16,17]:

- Destruction of macromolecules formed during the process caused by the action of van der Waals forces and capillary forces, the action of shock waves, liquid microjets, microflows, or in the vicinity of cavitations bubbles and friction flows formed by the deceleration of acoustic flows near the surface.
- Mixing of the coating material under the influence of acoustic flows. Particles of the coating material become uniform under the influence of sound waves.
- Depolymerization occurring due to the difference between the vibration speed of the solvent and that of the polymer.
- Dispersion of pigment particles as a result of cavitations.
- Heating that leads to a decrease in viscosity which enhances all the processes listed above.

An allometric scaling relation between the viscosity and the oscillating frequency, for low frequency, can be obtained in the form[8]:

$$\eta \approx \omega^{-0.4}. \tag{12.1}$$

When frequency surpasses a threshold value, high frequency will not affect very much on the reduction of viscosity. Though many experimental observations show that the oscillation in polymer solution leads to reduction of its viscosity, but no theory analysis, so far, has appeared in the open literature.

The polymer solutions or melts consist of a system of macromolecules entangled with each other and connected with the weak van der Waals' force.

By using potential model, the van der Waals' energy term can be expressed as[8,18]

$$U = 4k(A/x)^{12} - (B/x)^{6}. \tag{12.2}$$

When the macromolecules are subjected to an external periodic force applied, its equation can be expressed as

$$\rho\ddot{x} + \eta\dot{x} + 6B^{6}x^{-7} - 48kA^{12}x^{-13} = a\cos\omega t \tag{12.3}$$

This is a strongly nonlinear forced oscillation, to solve this equation, it can be referred to the homotopy perturbation method,[19,20] or modified Lindstedt–Poincare method.[21,22] After obtaining an approximate solution, the relationship between η, ω and a can be arrived at.

12.3   VIBRATION TECHNOLOGY FOR MOLDING PROCESS MODIFICATION

There are three categories of processes using vibration to modify the molding process and/or the properties of molded materials[9,23,24]:

1.   The common practical feature among the patents of the first category is their use of mechanical shaking/oscillation or ultrasonic vibration devices to homogenize and increase the density of the material molded, either in the liquid stage or in the solidifying stage, and either at a macroscopic or microscopic level (Fig. 12.4).

The vibration processes and equipment use for shaking or local micro-shaking in the case of ultrasonic vibration. In some patents of this first category, the initial states of the materials treated are granules or pellets and the vibration is applied to this state.

The result of applying vibration is to compact the granules and powders and to combine the effect of heat and mechanical shaking so as to avoid the trapping of bubbles between the granules and thereby obtain a more homogeneous product with better mechanical properties. The shaking can also be applied to the melt to decrease the number of bubbles in the melt.

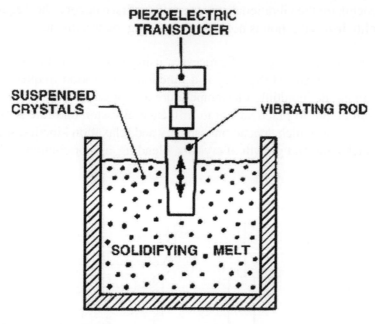

FIGURE 12.4   Vibration devices to homogenize and increase the melt density.

This process relates to an injection molding technique claimed to eliminate part defects and increase strength, particularly with fiber-reinforced materials, by use of multiple gates and injection pistons to oscillate the melt back and for thin the mold to achieve desired orientation effects across weld lines.[25,26]

Another application of melt oscillation to create orientation effects in melts containing fibers.

This process is also based on oscillation of the melt in the mold to improve the quality, and uses a molding machine with twin injection units, such as is used in multicomponent or two-color work, along with software modifications to control the movement of the injection units. After filling a two-gated mold, one injection screw advances while the other retracts, creating the oscillation of melt in the cavity. Substantial benefits of the technique are said to be enhanced fiber-reinforcement effect and reduction of defects from internal weld lines.[27,28] For processes using ultrasonic energy, the objective is to alter the kinetics of nucleation and growth of crystals in the melt to obtain more homogeneous solidified parts. Therefore, ultrasonic vibrations may be utilized for improving the strength and other physical characteristics of the solidified sample.[29]

In general, for the vibrational processes of the first category, the frequency and amplitude of vibration is not changing during the treatment.

2. The second category of processes using vibration is based on the fact that material rheology is a function of vibration frequency and amplitude in addition to temperature and pressure (Fig. 12.5). This can be put to practical use to influence diffusion and rate sensitive processes which depend on viscosity and relaxation kinetics, such as nucleation and growth of crystals, blending, and orientation.[30,31]

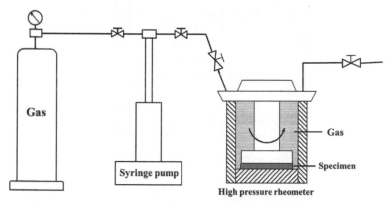

FIGURE 12.5 Controlling rheological parameters by vibration frequency and high pressure.

The second category provides a method and apparatuses for transforming the physical characteristics of a material by controlling the influence of rheological parameters. In particular, it provided a method for molding by vibration to control or modify the physical properties of the molded materials, notably their mechanical and optical properties. The process uses vibrational means in order to influence and/or tailor a change in state, either a transitional state (melting transition, glass transition) or a relaxation state, that is, the internal friction related to viscosity and orientation.[32,33]

This is based on the fact that material rheology is a function of vibration frequency and amplitude in addition to temperature and pressure.

3. In a third category, vibration is essentially used to generate heat locally by internal friction or to decrease surface stresses at the

wall interface between the melt and the barrel or the die to increase throughputs. The heat generated locally by pressure pulsation can be significant enough, in injection molding, as to avoid the premature freezing of the gate, resulting in a significant reduction of the shrinkage in the final part (Fig. 12.6).[34,35]

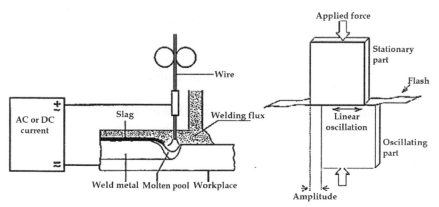

FIGURE 12.6   The nozzles of injection molding machines by vibrating of wall surfaces for extrusion.

For the third category, vibration of wall surfaces for extrusion and runners allows lower processing temperature and lower pressure. The throughput in annular dies can be increased by 30–50% by the effect of wall vibration.

The nozzles of injection molding machines by vibrating terminate in higher throughputs and parts with less shrinkage. The absence of shrinkage causes to the delay in freezing by friction heating under continuous compensating pressure.

The viscosity at the wall surface of extruders or runners is the result of shear frictional forces, which slow down the passage of the melt of the molten plastic. Throughputs can be increased by increasing pressure, which requires more powerful and more expensive equipment, or by increasing the temperature of the melt which reduces viscosity.[36,37]

12.4   APPLICATION OF VIBRATION TECHNOLOGY TO POLYMER ELECTROSPINNING

In electrospinning method, the nanofibers produce by using a very high voltage. The voltage is applied to polarize dielectrics, where the charges

$$I = I_s + I_v + I_{ps} + I_b + I_{pb} \tag{12.4}$$

$$I_s = 2\pi r\, \sigma u \tag{12.5}$$

$$I_v = 2\pi r\, \sigma v \cos \omega t \tag{12.6}$$

$$I_{ps} = 2\pi r \varepsilon_p E u \tag{12.7}$$

$$I_b = k\pi r^2 E \tag{12.8}$$

$$I_{pb} = \pi r^2 \varepsilon_p E \tag{12.9}$$

$$v = f\lambda \tag{12.10}$$

Thus, the above equations can be rewritten as

$$I = 2\pi r u \left(\sigma + \varepsilon_p E\right) + 2\pi r\, \sigma v \cos \omega t + \pi r^2 E\left(k + \varepsilon_p E\right). \tag{12.11}$$

12.6.2 LINEAR MOMENTUM EQUATION

The two extra energies generated with the currents due to the coupling of polarization with the electrostatic field play an important role in this system. Moreover, the heat generated by the applied vibration will significantly decrease the viscosity of the solution, thus accelerates the motion of the jet.[45,47]

$$\frac{\partial}{\partial z}\left(\frac{u^2}{2}\right) = -\frac{1}{\rho}\frac{\partial p}{\partial z} + \frac{1}{r^2}\frac{\partial \tau}{\partial z} + \frac{2\left(\sigma + \varepsilon_p E\right)E}{\rho r} + \frac{2\sigma \cos \omega t}{\rho r}\frac{\partial v}{\partial z} + \frac{k + \varepsilon_p E}{\rho}\frac{\partial E}{\partial z} \tag{12.12}$$

12.6.3 CONSERVATION OF ENERGY

As above-mentioned, the ultrasound energy is mainly transferred into two types of energy: kinetic energy (vibration induced by the ultrasonic vibration) and thermal energy[45]:

$$\frac{1}{\pi r^2 \rho}\frac{\partial U}{\partial t} = Cu - v^2 \omega \sin \omega t \cos \omega t \tag{12.13}$$

$$U = U_0 \cdot e^{-a(z/\lambda)} \tag{12.14}$$

$$\alpha = \frac{16\pi^2 f \eta}{3v^2 \rho}.$$ (12.15)

12.7   SIMPLIFIED MODEL FOR VIBRATION-ELECTROSPINNING

To make the model more applicable in practice, some reasonable omission is necessary to simplify the model.

The conservation of mass equation will be the same to the typical one since the application of ultrasonic vibration won't change the flow rate of the solution and the macroscopic axial velocity of the jet[48,49]:

$$\rho \pi r^2 u = Q.$$ (12.16)

Since polarization of polymer molecules is strongly enhanced due to the free of polymer molecular chains by the directly applied vibration and the increased temperature of the solution, the surface and bulk-charge-generated currents become negligible[45]:

$$I = 2\pi r \left(\varepsilon_p E u + \sigma v \cos \omega t \right) + \pi r^2 E^2 \varepsilon_p.$$ (12.17)

The forces dominate the motion of the electrospinning jet are mainly the viscous and electric forces, so it can be rewritten as[45]

$$\frac{\partial}{\partial z}\left(\frac{u^2}{2} \right) = \frac{1}{r^2}\frac{\partial \tau}{\partial z} + \frac{2\varepsilon_p E^2}{\rho r} + \frac{2\sigma \cos \omega t}{\rho r}\frac{\partial v}{\partial z} + \frac{\varepsilon_p E}{\rho}\frac{\partial E}{\partial z}.$$ (12.18)

The conservation of energy equation should be the same as the previous section.

12.8   GOVERNING EQUATIONS OF THERMO-ELECTROHYDRODYNAMIC

Spivak et al.[50] established a model of steady state jet in the electrospinning process. The equations in this model were introduced as follow:

1. Equation of mass balance gives

$$\nabla \cdot u = 0.$$ (12.19)

2. Linear momentum balance is

$$\rho(u \cdot \nabla)u = \nabla T^m + \nabla T^e. \tag{12.20}$$

3. Electric charge balance reads

$$\nabla \cdot J = 0. \tag{12.21}$$

The right-hand side of eq 12.20 is the sum of viscous and electric forces. This is a simple model without considering thermal effect. In this chapter, we consider the couple effects of thermal, electricity, and hydrodynamics. A complete set of balance laws governing the general thermo-electrohydrodynamics flows has been derived by Ko and Dulikravich[51] and other researchers.[52] It consists of modified Maxwell's equations governing electrical field in a moving fluid, the modified Navier–Stokes equations governing heat and fluid flow under the influence of electric field, and constitutive equations describing behavior of the fluid:

$$\frac{\partial q_e}{\partial t} + \nabla \cdot J = 0 \tag{12.22}$$

$$\rho \frac{Du}{Dt} = \nabla \cdot t + \rho \cdot f + q_e E + (\nabla E) \cdot P \tag{12.23}$$

$$\rho_c P \frac{DT}{Dt} = Q_h + \nabla \cdot q + J \cdot E + E \cdot \frac{DP}{Dt}. \tag{12.24}$$

This set of conservation laws can constitute a closed system when it is supplemented by appropriate constitutive equations for the field variables such as polarization. The most general theory of constitutive equations determining the polarization, electric conduction current, heat flux, and Cauchy stress tensor has been developed by experts.[52]

$$P = \varepsilon_p E \tag{12.25}$$

$$J = kE + \alpha u + \sigma_T \nabla T \tag{12.26}$$

$$q = k \nabla T + k_E E \tag{12.27}$$

$$t = -\tilde{p}I + \eta \left[\nabla v + (\nabla v)^t \right]. \tag{12.28}$$

Here, coefficients ε_p, μ_m, k, σ, σ_T, K, K_E, η are material properties and depend only on the temperature in the case of an incompressible.[51] Equation 12.10 is valid only for Newtonian flows.

12.9 MATHEMATICAL MODEL FOR ONE-DIMENSIONAL CASE

An unsteady flow of an infinite viscous jet pulled from a capillary orifice and accelerated by a constant external electric field is considered in this section.

1. The conservation of mass equation gives

$$\frac{\partial}{\partial t}\left(r^2\right)+\frac{\partial}{\partial z}\left(r^2 u\right)=0. \tag{12.29}$$

2. Conservation of charge reduces into

$$\frac{\partial}{\partial t}\left(2\pi r\left(\sigma+\varepsilon_p E\right)\right)+\frac{\partial}{\partial z}\left(2\pi r\left(\sigma+\varepsilon_p E\right)u\right)+\pi r^2 kE+\pi r^2\,\sigma_T\frac{\partial T}{\partial z}=0. \tag{12.30}$$

The current is composed of three parts: (1) the Ohmic bulk conduction current: $J_c=\pi r^2 kE$; (2) surface convection current: $J_s=2\pi r\sigma u$; and (3) current caused by temperature gradient: $J_T=\pi r^2\sigma T\partial T/\partial z$.

3. The Navier–Stokes equations becomes

$$\frac{\partial u}{\partial t}+u\frac{\partial u}{\partial z}=-\frac{1}{\rho}\frac{\partial p}{\partial z}+g+\frac{2\sigma E}{\rho r}+\frac{1}{r^2}\frac{\partial \tau}{\partial z}+\frac{1}{r^2}\varepsilon_p E\frac{\partial \sigma}{\partial z} \tag{12.31}$$

$$\rho_c p\left(\frac{\partial T}{\partial t}+u\frac{\partial T}{\partial z}\right)=Q+\frac{\partial}{\partial z}\left(k\frac{\partial T}{\partial z}+k_E E\right)+\left(2\pi r\sigma u+\pi r^2 kE+\pi r^2\sigma_T\frac{\partial T}{\partial z}\right)$$
$$E+\varepsilon_p E\left(\frac{\partial E}{\partial t}+u\frac{\partial E}{\partial z}\right) \tag{12.32}$$

where p is the internal pressure of the fluid expressed as

$$P=k\gamma-\frac{\varepsilon-\overline{\varepsilon}}{8\pi}E^2-\frac{2\pi}{\overline{\varepsilon}}\sigma^2 \tag{12.33}$$

$$K=\frac{1}{R_1}+\frac{1}{R_2}. \tag{12.34}$$

Rheologic behavior of many polymer fluids can be described by power-law constitutive equation in the form:

$$\tau = \mu_0 \frac{\partial u}{\partial z} + \sum_{n=1}^{m} a_n \left(\frac{\partial u}{\partial z} \right)^{2n+1}. \tag{12.35}$$

In addition to conducting bodies, there are also dielectrics. In dielectrics, the charges are not completely free to move, but the positive and negative charges that compose the body may be displaced in relation to one another when a field is applied. The body is said to be polarized. The polarization is given in terms of a dipole moment per unit volume p, called the polarization vector. The bound charge or polarization charge in the dielectric is given by

$$q_p = -\nabla \cdot p. \tag{12.36}$$

In an isotropic linear dielectric case, the polarization is assumed to be proportional to the field that causes it, thus

$$p = \varepsilon_p E \tag{12.37}$$

12.10   BRATU EQUATION AND BIFURCATION IN THE PROCESS

In this section, the steady state jet ignoring the thermal effort is considered. In case electrically generated force is dominant, the momentum equation becomes

$$u \frac{\partial u}{\partial z} = \frac{2 \sigma E}{\rho r}. \tag{12.38}$$

From the charge balance equation:

$$2r \sigma u + r^2 k E = I. \tag{12.39}$$

Equation 12.40 can be expressed in the form

$$u \frac{\partial u}{\partial z} = \frac{E \left(I - r^2 k E \right)}{\rho r^2 u}. \tag{12.40}$$

Introducing a new variable, u, defined as

$$u = e^{-\nu/6}. \tag{12.41}$$

Substituting 12.41 into 12.40 results in

$$\frac{\partial v}{\partial z} = -\frac{\sigma E\left(I - r^2 kE\right)}{\rho r^2} e^{v/z} = 0. \tag{12.42}$$

Differentiating 12.42 with respect to z, and assuming $\partial r/\partial z \approx 0$, yields

$$\frac{\partial^2 v}{\partial z^2} = -\frac{3E\left(I - r^2 kE\right)}{\rho r^2} e^{v/z} \frac{\partial v}{\partial z}. \tag{12.43}$$

In view of eq 12.42, eqs 12.44 and 12.45 becomes the well-known Bratu equation[51]

$$\frac{\partial^2 v}{\partial z^2} + \lambda e^v = 0 \tag{12.44}$$

$$\lambda = \frac{18E^2\left(I - r^2 kE\right)^2}{\rho^2 r^4}. \tag{12.45}$$

Equation 12.44 comes originally from a simplification of the solid fuel ignition model in thermal combustion theory.[51] There are two solutions to eq 12.44 for values $0 < \lambda < \lambda_c$, and no solutions for $\lambda > \lambda_c$. In case $\lambda = \lambda_c$, there is only one solution. By the semi-inverse method,[53] various variational principles can be easily obtained for electrospinning.

12.11   CONCLUDING REMARKS

The vibration-electrospinning is utilized for the strategy produces finer nanofibers than those obtained without vibration and can produce nanofibers which cannot be done by electrospinning without vibration. Application of vibration technology in polymer processes proved that it works well in reduction of melt viscosity and enhancement of mechanical properties of polymer products. By vibration technology, finer nanofibers under lower applied voltage can be produced and the reduction in the applied voltage terminates to increase the fiber strength. In addition, theoretical models of the spinning process remains a bottleneck severely hampering further improvement in both quality and efficiency. Therefore, a thermo-electrohydrodynamic model of the vibration-electrospinning process which considers the couple effects of thermal field, electric field, vibration force was investigated in this chapter. Because of the difficult usage of this model for investigating numerical analysis, a one-dimensional thermo-electrohydrodynamic model can be derived to numerical study. The model can offer in-depth insight into the physical understanding of many complex phenomena which cannot be fully explained experimentally. Also, an allometric scaling relation between the viscosity and the oscillating frequency can be obtained.

KEYWORDS

- vibration technology
- thermo-electrohydrodynamic model
- electrospinning
- high molten temperature
- thermal factor

REFERENCES

1. Persano, L.; Camposeo, A.; Tekmen, C.; Pisignano, D. Industrial Upscaling of Electro-spinning and Applications of Polymer Nanofibers: A Review. *Macromol. Mater. Eng.* **2013,** *298* (5), 504–520.
2. Subbiah, T.; Bhat, G. S.; Tock, R. W.; Parameswaran, S.; Ramkumar, S. S. Electrospin-ning of Nanofibers. *J. Appl. Polym. Sci.* **2005,** *96* (2), 557–569.
3. Weitz, R. T.; Harnau, L.; Rauschenbach, S.; Burghard, M.; Kern, K. Polymer Nanofi-bers via Nozzle-Free Centrifugal Spinning. *Nano Lett.* **2008,** *8* (4), 1187–1191.
4. Zong, X.; Kim, K.; Fang, D.; Ran, S.; Hsiao, B. S.; Chu, B. Structure and Process Relationship of Electrospun Bioabsorbable Nanofiber Membranes. *Polymer* **2002,** *43* (16), 4403–4412.
5. Xu, L.; Wang, L.; Faraz, N. A Thermo-electro-hydrodynamic Model for Vibration-elec-trospinning Process. *Therm. Sci.* **2011,** *15* (Suppl. 1), 131–135.
6. He, J. H.; Wan, Y. Q.; Xu, L. Nano-Effects, Quantum-Like Properties in Electrospun Nanofibers. *Chaos, Solit. Fract.* **2007,** *33* (1), 26–37.
7. Wan, Y. Q.; He, J. H.; Wu, Y.; Yu, J. Y. Vibrorheological Effect on Electrospun Polyac-rylonitrile (PAN) Nanofibers. *Mater. Lett.* **2006,** *60* (27), 3296–3300.
8. He, J. H.; Wan, Y. Q.; Yu, J. Y. Application of Vibration Technology to Polymer Electro-spinning. *Int. J. Nonlin. Sci. Numer. Simul.* **2004,** *5* (3), 253–262.
9. Ibar, J. P. Control of Polymer Properties by Melt Vibration Technology: A Review. *Polym. Eng. Sci.* **1998,** *38* (1), 1–20.
10. Taub, A. I.; Krajewski, P. E.; Luo, A. A.; Owens, J. N. The Evolution of Technology for Materials Processing Over the Last 50 Years: The Automotive Example. *J. Miner., Met. Mater. Soc.* **2007,** *59* (2), 48–57.

11. Bershtein, V. A.; Egorov, V. M.; Egorova, L. M.; Ryzhov, V. A. The Role of Thermal Analysis in Revealing the Common Molecular Nature of Transitions in Polymers. *Thermochim. Acta* **1994**, *238*, 41–73.

12. Wojtecki, R. J.; Meador, M. A.; Rowan, S. J. Using the Dynamic Bond to Access Macroscopically Responsive Structurally Dynamic Polymers. *Nat. Mater.* **2011**, *10* (1), 14–27.

13. He, J. H.; Liu, Y.; Xu, L. Apparatus for Preparing Electrospun Nanofibres: A Comparative Review. *Mater. Sci. Technol.* **2010**, *26* (11), 1275–1287.

14. Mezger, T. G. *The Rheology Handbook: For Users of Rotational and Oscillatory Rheometers*; Vincentz Network GmbH & Co KG: Hannover, 2011; p 423.

15. Brunsveld, L.; Folmer, B. J. B.; Meijer, E. W.; Sijbesma, R. P. Supramolecular Polymers. *Chem. Rev.* **2001**, *101* (12), 4071–4098.

16. Livanskiy, N. A.; Prikhodko, M. V.; Sundukov, K. S.; Fatyukhin, S. D. Research on the Influence of Ultrasonic Vibrations on Paint Coating Properties. *Trans. FAMENA* **2016**, *40* (1), 129–138.

17. Sinisterra, J. V. Application of Ultrasound to Biotechnology: An Overview. *Ultrasonics* **1992**, *30* (3), 180–185.

18. Lim, T. C. The Relationship Between Lennard-Jones (12-6) and Morse Potential Functions. *Zeitsch. Naturforsch. A* **2003**, *58* (11), 615–617.

19. He, J. H. Homotopy Perturbation Method: A New Nonlinear Analytical Technique. *Appl. Math. Comput.* **2003**, *135* (1), 73–79.

20. He, J. H. Homotopy Perturbation Method for Solving Boundary Value Problems. *Phys. Lett. A* **2006**, *350* (1), 87–88.

21. He, J. H. Modified Lindstedt–Poincare Methods for Some Strongly Non-Linear Oscillations: Part I: Expansion of a Constant. *Int. J. Nonlin. Mech.* **2002**, *37* (2), 309–314.

22. Liu, H. M. Approximate Period of Nonlinear Oscillators with Discontinuities by Modified Lindstedt–Poincare Method. *Chaos, Solit. Fract.* **2005**, *23* (2), 577–579.

23. Qu, J.; He, G.; He, H.; Yu, G.; Liu, G. Effect of the Vibration Shear Flow Field in Capillary Dynamic Rheometer on the Crystallization Behavior of Polypropylene. *Eur. Polym. J.* **2004**, *40* (8), 1849–1855.

24. Ibar, J. P. Viscosity Control for Molten Plastics Prior to Molding, Google Patents, 1999.

25. Barker, G. C. Computer Simulations of Granular Materials. In *Granular Matter*; Springer: Berlin, 1994; pp 35–83.

26. Lu, P.; Ding, B. Applications of Electrospun Fibers. *Rec. Pat. Nanotechnol.* **2008**, *2* (3), 169–182..

27. Khan, S. A.; Prud'Homme, R. K. Melt Rheology of Filled Thermoplastics. *Rev. Chem. Eng.* **1987**, *4* (3–4), 205–272.

28. Tsori, Y.; Tournilhac, F.; Leibler, L. Orienting Ion-Containing Block Copolymers Using AC Electric Fields. *Macromolecules* **2003**, *36* (15), 5873–5877.

29. Jevtic, M.; Mitric, M.; Skapin, S.; Jancar, B.; Ignjatovic, N.; Uskokovic, D. Crystal Structure of Hydroxyapatite Nanorods Synthesized by Sonochemical Homogeneous Precipitation. *Cryst. Growth Des.* **2008**, *8* (7), 2217–2222.

30. Yan, Z.; Shen, K. Z.; Zhang, J.; Chen, L. M.; Zhou, C. Effect of Vibration on Rheology of Polymer Melt. *J. Appl. Polym. Sci.* **2002**, *85* (8), 1587–1592.

31. Zeng, G. S.; Qu, J. P. Rheological Behavior of a Polymer Melt Under the Impact of a Vibration Force Field. *J. Appl. Polym. Sci.* **2007**, *106* (2), 1152–1159.

32. Ibar, J. P. Method and Apparatus for Transforming the Physical Characteristics of a Material by Controlling the Influence of Rheological Parameters. Google Patents, 1984.

33. Ibar, J. P. Apparatus for Controlling Gas Assisted Injection Molding to Produce Hollow and Non-Hollow Plastic Parts and Modify their Physical Characteristics. Google Patents, 1998.

34. Ibar, J. P. Method and Apparatus to Control Viscosity of Molten Plastics Prior to a Molding Operation. Google Patents, 2001.

35. Chen, X.; Qu, J. Extrusion Characteristics of Round-Section Dies with VFF. *Polym.–Plast. Technol. Eng.* **2008,** *47* (2), 203–208.

36. Ibar, J. P. Processing Polymer Melts Under Rheo-Fluidification Flow Conditions, Part 2: Simple Flow Simulations. *J. Macromol. Sci., B* **2013,** *52* (3), 442–461.

37. Stevens, M. J.; Covas, J. A. Practical Extrusion Processes and their Requirements. In *Extruder Principles and Operation*; Springer: Berlin, 1995; pp 4–26.

38. He, J. H.; Xu, L.; Wu, Y.; Liu, Y. Mathematical Models for Continuous Electrospun Nanofibers and Electrospun Nanoporous Microspheres. *Polym. Int.* **2007,** *56* (11), 1323–1329.

39. Moghadam, M. S.; Dong, Y.; Davies, I. J. Recent Progress in Electrospun Nanofibers: Reinforcement Effect and Mechanical Performance. *J. Polym. Sci., B: Polym. Phys.* **2015,** *53* (17), 1171–1212.

40. Tian, W.; Yung, K. L.; Xu, Y.; Huang, L.; Kong, J.; Xie, Y. Enhanced Nanoflow Behaviors of Polymer Melts Using Dispersed Nanoparticles and Ultrasonic Vibration. *Nanoscale* **2011,** *3* (10), 4094–4100.

41. He, J. Huan.; Liu, Y.; Xu, L.; Yu, J. Y.; Sun, G. BioMimic Fabrication of Electrospun Nanofibers with High-Throughput. *Chaos, Solit. Fract.* **2008,** *37* (3), 643–651.

42. De. V, S.; Van. C, T.; Nelvig, A.; Hagström, B.; Westbroek, P.; De, C. K. The Effect of Temperature and Humidity on Electrospinning. *J. Mater. Sci.* **2009,** *44* (5), 1357–1362.

43. Chronakis, I. S. Novel Nanocomposites and Nanoceramics based on Polymer Nanofibers Using Electrospinning Process—A Review. *J. Mater. Process. Technol.* **2005,** *167* (2), 283–293.

44. Chandrasekar, R.; Zhang, L.; Howe, J. Y.; Hedin, N. E.; Zhang, Y.; Fong, H. Fabrication and Characterization of Electrospun Titania Nanofibers. *J. Mater. Sci.* **2009,** *44* (5), 1198–1205.

45. Wan, Y. Q.; Qiang, J.; Yang, L. N.; Cao, Q. Q.; Wang, M. Z. Vibration and Heat Effect on Electrospinning Modeling. *Advanced Materials Research*. Trans Tech Publ., 2014.

46. He, J. H.; Wan, Y. Q. Allometric Scaling for Voltage and Current in Electrospinning. *Polymer* **2004,** *45* (19), 6731–6734.

47. Xu, L. A Mathematical Model for Electrospinning Process under Coupled Field Forces. *Chaos, Solit. Fract.* **2009,** *42* (3), 1463–1465.

48. Greenfeld, I.; Fezzaa, K.; Rafailovich, M. H.; Zussman, E. Fast X-Ray Phase-Contrast Imaging of Electrospinning Polymer Jets: Measurements of Radius, Velocity, and Concentration. *Macromolecules* **2012,** *45* (8), 3616–3626.

49. Saville, D. A. Electrohydrodynamics: The Taylor–Melcher Leaky Dielectric Model. *Annu. Rev. Fluid Mech.* **1997,** *29* (1), 27–64.

50. Spivak, A. F.; Dzenis, Y. A. Asymptotic Decay of Radius of a Weakly Conductive Viscous Jet in an External Electric Field. *Appl. Phys. Lett.* **1998,** *73* (21), 3067–3069.

51. Ko, H. J.; Dulikravich, G. S. Non-Reflective Boundary Conditions for a Consistent Model of Axisymmetric Electro-Magneto-Hydrodynamic Flows. *Int. J. Nonlin. Sci. Numer. Simul.* **2000,** *1* (4), 247–256.
52. Wan, Y. Q.; Guo, Q.; Pan, N. Thermo-Electro-Hydrodynamic Model for Electrospinning Process. *Int. J. Nonlin. Sci. Numer. Simul.* **2004,** *5* (1), 5–8.
53. He, J. H. Variational Theory for Linear Magneto-Electro-Elasticity. *Int. J. Nonlin. Sci. Numer. Simul.* **2001,** *2* (4), 309–316.

CHAPTER 13

UPDATE ON NUMERICAL ANALYSIS AND METHODS FOR SOLVING EQUATIONS

SHIMA MAGHSOODLOU* and S. PORESKANDAR

Textile Engineering, University of Guilan, Rasht, Iran

*Corresponding author. E-mail: sh.maghsoodlou@gmail.com

CONTENTS

ABSTRACT

Numerical analysis is a rapidly growing field, with new techniques being developed constantly. The branch of mathematics concerned with finding accurate approximations to the solutions of problems whose exact solution is either impossible or infeasible to determine. In addition to the approximate solution, a realistic bound is needed for the error associated with the approximate solution. Typically, a mathematical model for a particular problem, generally consisting of mathematical equations with constraint conditions, is constructed by specialists in the area concerned with the problem. Numerical analysis is concerned with devising methods for approximating the solution to the model and analyzing the results for stability, speed of implementation, and appropriateness to the situation. The aim of this chapter is studying the most important numerical methods for solving equations.

13.1 INTRODUCTION

Numerical analysis is a rapidly growing field, with new techniques being developed constantly. The branch of mathematics concerned with finding accurate approximations to the solutions of problems whose exact solution is either impossible to determine.[1] A mathematical model for a particular problem, generally consisting of mathematical equations with constraint conditions, is constructed by specialists in the area concerned with the problem. Numerical analysis is utilized for approximating the solution to the model. Numerical analysis is not only the design of numerical methods but also their analysis. Three central concepts in this analysis are[2]

- **convergence:** whether the method approximates the solution,
- **order:** how well it approximates the solution, and
- **stability:** whether errors are damped out.

Some definitions about numerical analysis were summarized in the flow chart (Fig. 13.1).

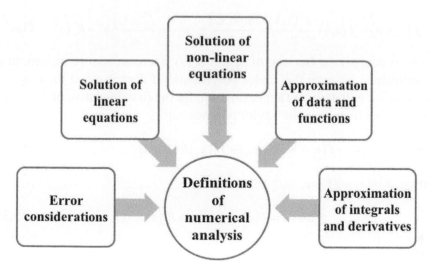

FIGURE 13.1 Some definitions of numerical analysis.

13.2 NUMERICAL METHODS

Numerical methods for ordinary differential equations (ODEs) are methods used to find numerical approximations to the solutions of ODEs. Their use is also known as "numerical integration," although this term is sometimes taken to mean the computation of integrals. Many differential equations cannot be solved using symbolic computation. For practical purposes, however, such as in engineering, a numeric approximation to the solution is often sufficient. The algorithms studied here can be used to compute such an approximation. An alternative method is to use techniques from calculus to obtain a series expansion of the solution. ODEs occur in many scientific disciplines, for instance in physics, chemistry, biology, and economics. In addition, some methods in numerical partial differential equations convert the partial differential equation into an ODE, which must then be solved. In the next part, more details were investigated about these methods.[3,4]

13.2.1 DERIVATION FROM TAYLOR'S POLYNOMIAL

First, assuming the function whose derivatives are to be approximated is properly behaved, by Taylor's theorem, we can create a Taylor series expansion[5]:

$$f(x_0 + h) = f(x_0) + \frac{f'(x_0)}{1!}h + \frac{f^{(2)}(x_0)}{2!}h^2 + \cdots + \frac{f^{(n)}(x_0)}{n!}h^n + R_n(x) \quad (13.1)$$

where $n!$ denotes the factorial of n, and $R_n(x)$ is a remainder term, denoting the difference between the Taylor polynomial of degree n and the original function. We will derive an approximation for the first derivative of the function f by first truncating the Taylor polynomial[5]:

$$f(x_0 + h) = f(x_0) + f'(x_0)h + R_1(x) \quad (13.2)$$

Setting, $x_0 = a$, we have,

$$f(a + h) = f(a) + f'(a)h + R_1(x) \quad (13.3)$$

Dividing across by h gives

$$\frac{f(a+h)}{h} = \frac{f(a)}{h} + f'(a) + \frac{R_1(x)}{h} \quad (13.4)$$

Solving for $f'(a)$:

$$f'(a) = \frac{f(a+h) - f(a)}{h} - \frac{R_1(x)}{h} \quad (13.5)$$

Assuming that $R_1(x)$ is sufficiently small, the approximation of the first derivative of f is

$$f'(a) \approx \frac{f(a+h) - f(a)}{h} \quad (13.6)$$

$$f(x) = \frac{1}{5 + 4\cos x} \quad (13.7)$$

Example: solve above equation with Taylor polynomial by Matlab program
　Matlab scripts:

```
≫ syms x
≫ f = 1/(5 + 4×cos(x));
≫ t = Taylor(f, 8)
t = 1/9 + 2/81×x² + 5/1458×x⁴ + 49/131,220×x⁶
```

$$(13.8)$$

13.2.2 EULER METHOD

A very simple finite-difference method (FDM) for the numerical solution of an ODE is Euler method. The method is named after Leonhard Euler who described it in his book *Institutionum calculi Integral is* in 1768. In mathematics and computational science, this method is a first-order numerical procedure for solving ODEs with a given initial value. It is the most basic explicit method for numerical integration of ODEs and is the simplest Runge–Kutta method.[6,7]

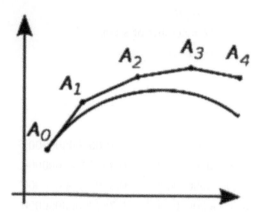

FIGURE 13.2 Illustration of the Euler method. The unknown curve is in blue and its polygonal approximation is in red.

As shown in Figure 13.2, from any point on a curve, you can find an approximation of a nearby point on the curve by moving a short distance along a line tangent to the curve.

We replace the derivative y' by the finite difference approximation[6]:

$$y'(t) \approx \frac{y(t+h) - y(t)}{h} \tag{13.9}$$

which when re-arranged yields the following formula:

$$y(t+h) \approx y(t) + hy'(t) \tag{13.10}$$

Finally:

$$y(t+h) \approx y(t) + hf\left(t, y(t)\right) \tag{13.11}$$

This formula is usually applied in the following way. We choose a step size h, and we construct the sequence $t_0, t_1 = t_0 + h, t_2 = t_0 + 2h, \ldots$. We denote by y_n a numerical estimate of the exact solution $y(t_n)$. We compute these estimates by the following recursive scheme:

$$y_{n+1} = y_n + hf(t_n, y_n) \tag{13.12}$$

This is the *Euler method* (or *forward Euler method*).

$$\begin{aligned}
E &= \text{Euler}(f, a, b, ya, M) \\
y_0 &= ya \\
M &= \text{number of steps} \\
x &= a - b
\end{aligned} \tag{13.13}$$

$y' = f(x, y) = -2x - y, y(0) = -1$
$\gg f = \text{inline}("-2 \times x - y," " " x," " " y");$
$\gg E = \text{Euler}(f, 0, 0.4, -1, 4)$

$$
\begin{array}{ll}
 & 0 \qquad\qquad\qquad -0.10000000000000 \\
 & 0.10000000000000 \quad -0.90000000000000 \\
E = & 0.20000000000000 \quad -0.83000000000000 \\
 & 0.30000000000000 \quad -0.78700000000000 \\
 & 0.40000000000000 \quad -0.76830000000000
\end{array} \tag{13.14}
$$

13.2.3 RUNGE–KUTTA METHOD

A one-step method for numerically solving the Cauchy problem for a system of ODEs is Rung–Kutta method. The principal idea of the Runge–Kutta method was proposed around 1900 by the German mathematicians C. Runge and M. W. Kutta.[8] Let an initial value problem be specified as follows.[9]

$$\begin{aligned}
\dot{y} &= f(t, y) \\
y(t_0) &= y_0
\end{aligned} \tag{13.15}$$

y is an unknown function of time t; Also, $\dot{y}$, the rate at which y changes, is a function of t and of y itself. At the initial time t_0, the corresponding y-value is determined y_0. The function f and the data t_0, y_0 are given. A step-size $h > 0$ was chosen and defined as given below[9]:

$$y_{n+1} = y_n + \frac{1}{6}\left[k_1 + 2k_2 + 2k_3 + k_4\right]$$

$$t_{n+1} = t_n + h$$

(13.16)

For $m = 0, 1, 2, 3, \ldots$, using

$$k_1 = f(t_n, y_n)$$

$$k_2 = f\left(t_n + \frac{1}{2}h, y_n + \frac{1}{2}k_1\right)$$

$$k_1 = f\left(t_n + \frac{1}{2}h, y_n + \frac{1}{2}k_2\right)$$

$$k_1 = f(t_n + h, y_n + k_3)$$

(13.17)

Example: Solve this equation with Rung–Kutta method by Matlab program

$$\begin{cases} y_1' = y_1 + y_2, y_1(0) = 0 \\ y_2' = -y_1 + y_2, y_2(0) = 1 \end{cases} \quad x_s = 0.1$$

(13.18)

The function should be defined as an m-file like Figure 13.3.

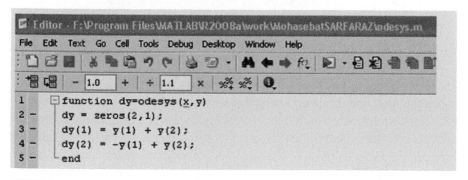

```
function dy=odesys(x,y)
    dy = zeros(2,1);
    dy(1) = y(1) + y(2);
    dy(2) = -y(1) + y(2);
end
```

FIGURE 13.3 The function written as an m-file.

Then:

$$\gg \text{options} = \text{odeset}\left(\text{"Reltol,""0.0001,""Abstol,"}[0.0001, 0.0001]\right);$$

$$\gg [X, Y] = \text{ode45}\left(@\,\text{odesys}, [0, 0.1], [0, 1], \text{options}\right)$$

(13.19)

13.2.4 FINITE-DIFFERENCE METHOD

In mathematics, FDMs are numerical methods for solving differential equations by approximating them with difference equations, in which finite differences approximate the derivatives. FDMs are thus discretization methods. Today, FDMs are the dominant approach to numerical solutions of partial differential equations (Fig. 13.4).[5,6]

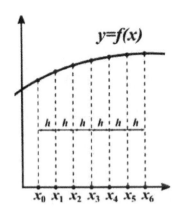

FIGURE 13.4 The finite-difference method relies on discretizing a function on a grid.

13.2.4.1 ACCURACY AND ORDER

The error in a method's solution is defined as the difference between the approximation and the exact analytical solution. The two sources of error in FDMs are round-off error, the loss of precision due to computer rounding of decimal quantities, and truncation error or discretization error, the difference between the exact solution of the original differential equation and the exact quantity assuming perfect arithmetic (i.e., assuming no round-off).

To use a FDM to approximate the solution to a problem, one must first discretize the problem's domain. This is usually done by dividing the domain into a uniform grid. Note that this means that FDMs produce sets of discrete numerical approximations to the derivative, often in a "time-stepping" manner.[10]

Solve the nonlinear boundary value problem:

$$y'' = (1/8) \times (32 + 2x^3 - yy'), \quad \text{for } 1 < x < 3,$$
$$\text{where } y(1) = 17 \text{ and } y(3) = 43/3 \tag{13.20}$$

Step 1:
Create the function f as a separate m-file and save it in the current working directory.

$$\text{function } f = f(x, y, yp)$$
$$f = (1/8) \times (32 + 2 \times x^3 - y \times yp); \%\text{Note that } yp = y' \tag{13.21}$$

Step 2:
In the command window, type

$$\gg Y = \text{nonlinear BVP\_FDM } (1,3,17,43/3); \tag{13.22}$$

Note that $Y(:,1)$ represents x and $Y(:,2)$ is vector $y(x)$.
The solution is then plotted in a new figure.
If the exact solution is given, plot it for comparison.

$$\gg y_{\text{exact}} = (Y(:,1)).^2+16./Y(:,1); \text{plot}(Y(:,1), y_{\text{exact}}, "c") \tag{13.23}$$

13.2.5 B-SPLINE QUASI-INTERPOLATION ODES SOLVER

Consider partition $\pi = (x_0, x_1, \ldots, x_n)$ of interval $[a,b]$ and suppose that $X_n = (x_j)_{j=-d}^{n+d}$ subject to $x_{-d} = x_{-d+1} = \ldots = x_{-1} = a, x_n = x_{n+1} = \ldots = x_{n+d} = b$. According to recurrence relation of B-spline, the jth B-spline of degree d for the knot sequence X_n is denoted by $B_{Xn,j,d}$ or B_j and can be defined as eqs 13.24 and 13.25.[11,12]

$$B_{X_{n,j,d}}(r) = \omega_{X_{n,j,d}} B_{X_{n,j,d-1}}(r) + \left(1 - \omega_{X_{n,j+1,d}}\right) B_{X_{n,j+1,d-1}}(r) \tag{13.24}$$

$$\omega_{X_{n,j,d}}(r) = \frac{r - x_j}{x_{j+d-1} - x_j}, B_{X_{n,j,d}}(r) = \begin{cases} 1, & \to x_j \leq r < x_{j+1} \\ 0, & \to \text{otherwise} \end{cases} \tag{13.25}$$

With these notations, the support of $B_{Xn,j,d}$ is $\text{Supp}(B_{Xn,j,d}) = [x_{j-d-1}, x_j]$ and the sets $= (B_1, B_2, \ldots, B_{n+d})$ forms a basis over the region $a \leq x \leq b$. The univariate B-spline quasi-interpolants can be defined as operators of the form in eq 13.26.[11]

$$Q_d f = \sum_{j=1}^{n+d} \mu_j(f) B_j \tag{13.26}$$

For $f \in C^{d+1}(I)$, we have $|f - Q_d f\|_\infty = O(h^4)$. Let $f_i = f(x_i), j = 0,1,\ldots,n$, the coefficient functional for cubic QI $(d = 3)$ are, respectively, in eq 13.27.[11]

$$\mu_1(f) = f_0$$

$$\mu_2(f) = \frac{1}{18}(7f_0 + 18f_1 - 9f_2 + 2f_3)$$

$$\mu_j(f) = \frac{1}{6}(-f_{j-3} + 8f_{j-2} - f_{j-1}), \quad 3 \le j \le n+1 \qquad (13.27)$$

$$\mu_{n+2}(f) = \frac{1}{18}(2f_{n-3} - 9f_{n-2} + 18f_{n-1} + 7f_n)$$

$$\mu_{n+3}(f) = f_n$$

Now, let f be a given function of x, y, where y is a function of x and y' is the first derivative with respect to x. Consider the first order of ODE as follows as eq 13.29, where the initial condition y_0 is a given number and the answer of y is unique.

$$y'(x) = f(x, y), \quad y(0) = y_0 \qquad (13.28)$$

Using cubic QI as an approximation of $F(x) := f(x, y(x))$ in eq 13.29 and integrating in the interval $[x_i, x_{i+1}]$, $i = 0, 1, \dots, n - 1$, we have eq 13.29.

$$\int_{x_i}^{x_{i+1}} y'(x)\, dx = \int_{x_i}^{x_{i+1}} \sum_{j=1}^{n+3} \mu_j(F) B_j(x)\, dx \qquad (13.29)$$

and from B-spline properties, we have eq 13.30, where $y_i = y(x_i)$.

$$y_{i+1} - y_i = \sum_{k=1}^{4} \mu_{i+k}(F) \int_{x_i}^{x_{i+1}} B_{i+k}(x)\, dx \qquad (13.30)$$

From eq 13.29, the following ODEs solver will be achieved, where $F_i = F(x_i)$, $i = 0, 1, \dots, n - 1$. The nonlinear system eq 13.31 is formed of n nonlinear equations in n unknowns $y_1, y_2, \dots, y_n$ that can be solved using trust-region-dogleg method.[11,12]

$$y_1 = y_0 + h\left(\frac{3}{8}F_0 + \frac{19}{24}F_1 - \frac{5}{24}F_2 + \frac{1}{24}F_3\right)$$

$$y_2 = y_1 + h\left(-\frac{7}{144}F_0 + \frac{41}{72}F_1 + \frac{1}{2}F_2 - \frac{1}{72}F_3 - \frac{1}{144}F_4\right)$$

$$y_{i+1} = y_i + h\left(-\frac{1}{144}F_{i-2} - \frac{1}{48}F_{i-1} + \frac{19}{36}F_i + \frac{19}{36}F_{i+1} - \frac{1}{48}F_{i+2} - \frac{1}{144}F_{i+3}\right) \quad i = 2, 3, \dots, n-3 \qquad (13.31)$$

$$y_n = y_{n-1} + h\left(\frac{1}{24}F_{n-3} - \frac{5}{24}F_{n-2} + \frac{19}{24}F_{n-1} + \frac{3}{8}F_n\right)$$

Note that since we have eq 13.32.

$$\left\| \int_{x_i}^{x_{i+1}} \left((F)(x) - Q_3 F(x) \right) dx \right\| \le h \left\| F - Q_3 F \right\| \tag{13.32}$$

Then from eq 13.26, the truncation error of present method will be $O(h^5)$.

13.3 CONCLUDING REMARKS

Numerical analysis with new techniques is being developed constantly. To the approximate solution, a realistic bound is needed for the error associated with the approximate solution. Numerical analysis is concerned with devising methods for approximating the solution to the model and analyzing the results for stability, speed of implementation, and appropriateness to the situation. In this chapter, the most important numerical methods, which can be used for electrospinning simulation, such as Euler method were reviewed for solving equations.

KEYWORDS

- **numerical analysis**
- **accurate approximations**
- **Euler method**
- **Runge–Kutta method**
- **finite-difference method**
- **B-spline method**

REFERENCES

1. Stoer, J.; Bulirsch, R. *Introduction to Numerical Analysis*, vol. 12. Springer Science & Business Media: Dordrecht, 2013.
2. Epperson, J. F. *An Introduction to Numerical Methods and Analysis*. John Wiley & Sons: Hoboken, NJ, 2013.
3. Chapra, S. C.; Canale, R. P. *Numerical Methods for Engineers*, vol. 2. McGraw-Hill: New York, 1998.
4. Marchuk, G. I.; Ruzicka, J. *Methods of Numerical Mathematics*, vol. 2. Springer-Verlag: New York, 1975.

5. Smith, G. D. *Numerical Solution of Partial Differential Equations: Finite Difference Methods.*: Oxford University Press: Oxford, 1985.
6. Reddy, J. N. *An Introduction to the Finite Element Method*, vol. 2. McGraw-Hill: New York, 1993.
7. James, M. L.; Smith, G. M.; Wolford, J. C. *Applied Numerical Methods for Digital Computation*, vol. 2. Harper & Row: New York, 1985.
8. Butcher, J. Runge–Kutta Methods. *Scholarpedia* **2007,** *2* (9), 3147.
9. Butcher, J. C. *The Numerical Analysis of Ordinary Differential Equations: Runge–Kutta and General Linear Methods.* Wiley-Interscience: Hoboken, NJ, 1987.
10. Brian, P. L. T. A Finite-Difference Method of High-Order Accuracy for the Solution of Three-Dimensional Transient Heat Conduction Problems. *AIChE J.* **1961,** *7* (3), 367–370.
11. Aminikhah, H. Alavi, J. B-spline Collocation and Quasi-Interpolation Methods for Boundary Layer Flow and Convection Heat Transfer over a Flat Plate. *Calcolo* **2016,** *54*, 1–19.
12. Hon, Y. C.; Wu, Z. A Quasi-Interpolation Method for Solving Stiff Ordinary Differential Equations. *Int. J. Numer. Methods Eng.* **2000,** *48* (8), 1187–1197.

PART II
Special Topics

CHAPTER 14

THE MULTILEVEL MODELING OF THE NANOCOMPOSITE COATING PROCESSES BY ELECTROCODEPOSITION METHOD

A. V. VAKHRUSHEV[1,2] and E. K. MOLCHANOV[1]

[1]Institute of Mechanics, Ural Division, Russian Academy of Sciences, Izhevsk, Russia

[2]Kalashnikov Izhevsk State Technical University, Izhevsk, Russia

*Corresponding author. E-mail: vakhrushev-a@yandex.ru

CONTENTS

ABSTRACT

This chapter reviews the recent experimental results from literature on electrocodeposition of nanoparticles in a metallic matrix. The mathematical model for hydrodynamic simulation of electrocodeposition of copper and alumina composite coating on rotating cylinder electrode is presented. The model can describe the hydrodynamics, particles size, current density, and concentration of particles. It is found that the unsteady diffusion layer is formed close to the rotating electrode surface and the electrokinetics forces are the driving forces of the process. The good correlations with experimental data are received.

14.1 INTRODUCTION AND THE PROBLEM STATEMENT

The composite materials represent the hetero-phase system which is consisted of two or more components, which can be classified as reinforcing elements and as a binder matrix. The properties of the composite material are determined by the ratio of the parameters of reinforcing elements and binder matrix, as well as of technology their production. As a result of combining of utility of matrix and reinforcing elements it is formed the new complex properties of the composite, which not only reflects the original properties of its components but also includes properties that do not had the isolated components.

The composite coating with improved and unique operational characteristics, such as wear resistance, cracking resistance, antifriction properties, corrosion resistance, radiation resistance, and high adhesion to the substrate can be produced by this technology.

There are many kinds of traditional techniques for formation of surface layers with improved physicochemical properties. The most widely used are the surface hardening, surface strengthening, and various methods of chemical and thermal machining, for example, the carburizing, nitriding, boriding, etc. In recent time, the methods influencing on the workpiece surface by beams of particles (ions, atoms, clusters) or high-energy quantum (ion-plasma surface treatment, laser machining) are used extensively. The methods of gas-phase deposition of composite coatings at atmosphere pressure or at vacuum environment have a significantly development. Also, the thermal sprayed coating methods receive a powerful development in connection with practical application of technique of plasma and detonation spraying of powdered materials.

A separate new branch of knowledge is the process of composite electro-chemical coatings formation.[1-5]

Metal matrix composite electrochemical coatings (MMEC) are prepared from the suspensions, representing electrolyte solutions with additives of certain quantity of a superfine powder. The particles are adsorbed onto cathode surface in combination with metal ions during electrocodeposition (ECD) process and the metal matrix composite coating is formed. MMEC consists of galvanic metal (dispersion phase) and particles (dispersed phase).

There are the following steps of the ECD process: (1) the particles in suspension obtain a surface charge; (2) the charged particles and metal ions are transported through the liquid by the application of an electric field (electrophoresis), convection, and diffusion; (3) the particles and metal ions are adsorbed onto the electrode surface; and (4) the particles adhere to the electrode surface through van der Waals forces, chemical bonding, or other forces and, simultaneously, adsorbed metal ions are reduced to metal atoms. Metal matrix are encompassed the adsorbed particles and thus the MMEC is formed.

The ECD process is schematically displayed in Figure 14.1.

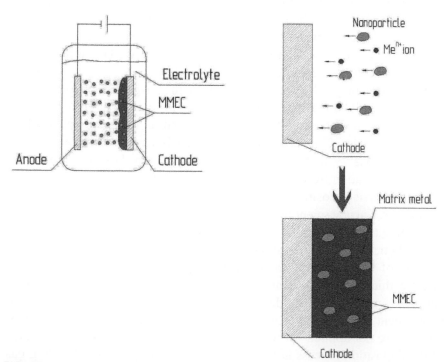

FIGURE 14.1   The ECD process.

Initially, all electrolyte inclusions are considered as impurities that degraded the quality of the electroplated metal. Efforts were made to prevent incorporation of undesirable particles by enclosing soluble anodes in a bag to prevent dissolved anode material from being codeposited with the plated metal at the cathode or by periodic filtration of electrolyte. Filters were used to remove from the plating solution unwanted suspended particles which could cause dull, rough, and poorly adherent deposits.[6]

The first application of MMEC dates back to the beginning of 20th century. The nickel matrix with sand particles composite was utilized as antislip coatings on ship stairs.[7] Also in 1928, Fink and Prince[8] investigated the possibility of using ECD process to produce self-lubricating copper–graphite coatings on part of car engines. Despite this the systematical scientific investigation of the ECD process occurred only in the early 1960s, immediately after the electrophoretic deposition method has an industrial application for coating metallic substrates by charged particles.[9,10] In the electrophoretic deposition process, the suspended charged particles moves toward and deposit onto the substrate surfaces as the result of an applied electric field. The codeposition process was developed with the intent of increasing the versatility of the electrophoretic deposition process by combining it with electroplating.

The particles of synthetic diamond, ceramic, or organic materials can be used as dispersed phase.[4] Also, it can be used the particles of metals that can't be electrodeposited from aqueous electrolyte. The used particles usually have a dimensions smaller than 100 μm.

Also, the nanoparticles can be used as the dispersed phase in ECD process. It is well known that the nanoparticles have dimensions in the range of 1–100 nm, also these particles are identified as nanomaterial or ultradispersed particles. Presently, the nanomaterials can be manufactured by various methods[11–13] including the sol–gel deposition, laser ablation, sputtering, condensation from an inert atmosphere, thermal vacuum deposition, flame-hydrolysis deposition, pyrolysis, high-energy milling, or electrochemical process. Current scientific research have pointed out that ECD process is an effective technique for the nanocomposite coatings production.

The benefits of EPD process amongst other coating methods are the lowering of waste material in contrast with spraying or dipping technologies, the thickness and chemical composition homogeneity of composite coating even if the covered detail have a complex shape, the low-level of environment pollution.[1] Furthermore, the high operation temperature or pressure are not applied in ECD process. Also, it is not required to carry out the deposition process in vacuum or in an atmosphere of shielding gas at this method. The ECD process realized at relatively low temperatures (in vicinity of ambient temperature), which is reduced to a minimum the undesirable chemical reactions or interdiffusion.[5] The thickness of coating can be accurately controlled by adjusting of applied current density. Nevertheless, there are a number of difficulties, for example, the agglomeration or sedimentation of nanoparticles and the nonuniformity on the nanoparticles concentration in the electrolyte bath.

The metals as Cu, Ni, Co, Cr, Zn, Ag, Fe, Au, As, and their alloys[14-17] can be employed as metal matrix of composite coating. The particles of oxides Al_2O_3,[1,18-22] ZrO_2,[23-25] TiO_2,[26-28] SiO_2,[20,29,30] and Cr_2O_3,[31] carbides SiC,[32-36] WC,[37] TiC,[38] various allotropic form C[39-41] nitrides Si_3N_4,[42] polymers polystyrene,[43] PTFE,[44,45] and a number of various materials can be used as dispersed phase. Usually, the particles with diameters from 4 to 800 nm are used in ECD process.

The concentration of suspended particles in electrolyte bath usually ranges from 2 up to 200 g/l, which is producing composites coating with 1 ± 10 vol% of embedded particles.[1-5] It has been noted[46] that the concentration of embedded particles can be reach up to 50 vol% when the horizontal cathode was used. The ECD process is applied to manufacture both hard-magnetic materials[47-49] and soft-magnetic materials for sensor implementation.[50-52] Furthermore, it was found that the magnetically hard particles of barium ferrites significantly increases the coercivity of the composite coating based of Ni or Ni–alloy matrix.

The MMEC have applications as lubrication, abrasion, and wear-resistant coatings, as coating for high hardness tools,[47] as protection against oxidation and hot corrosion,[53] and as a dispersion-strengthened alloys.[54] The ECD process has been applied to produce cathodes with high surface area, which are used in water electrolysis.[55] The ECD process is also used in such industry areas as oil and gas production, construction, electricity generation, airspace industry, and car manufacturing. The MMEC of nickel matrix and particles of aluminum oxides have applied in various field of utilization as coating with high friction and corrosion resistances.[56] The Cu–SiC MMEC has a more appreciable grain refinement and more than 61% higher microhardness than pure copper coating.[57]

On the structure, morphology and the properties of the MMEC are affected by next electrodeposition parameters of ECD process like a electrolysis conditions (bath composition, type of electrolyte agitation, additives of surfactant, pH, and temperature of electrolyte), particle properties (type, size, shape, surface charge, concentration and dispersion in the bath), interaction between particles and the electrolyte ions, and the character and velocity of fluid motion and applied current. The nanoparticles surface charge is an considerable parameter of the ECD process. It is found[20] that the negatively charged particles are more easily embedded MMEC in comparison with positively charged particles. Also, the agitation of electrolyte during ECD process is a considerable parameter that influences on the uniformity of nanoparticles concentration in the electrolyte bath and on the mass transfer of nanoparticles and electrolyte ions to cathode surface.

Also, the MMEC properties are greatly influenced by the current density and voltage parameters. As an example, the deposition with alternating current[58] result in deposition of MMEC with enhanced characteristic like wear and corrosion resistance, good surface roughness and low grain size in contrast with the direct current electrocodeposition. Also, the particles content in the MMEC[1,5] is significantly increased by using alternating current electrocodeposition. The particles content in the MMEC is a fundamental characteristic which defines the MMEC properties such as wear resistance, high temperature corrosion, oxidation resistance, the coefficient of friction. The nanoparticles distribution uniformity in the MMEC is another important factor.

Also there are range of kinds of electrochemical cells used in experimental work, like rotating cylindrical[60,61] or disk[59] electrodes, impinging jet electrode (IJE),[58] or parallel plate electrodes. The IJE can be used for high speed and selective deposition. The rotating cylinder electrode (RCE) can be used for different electrochemical processes[62] studied the deposition of alloys, metal-ion recovery, corrosion and effluent treatment, and in Hull cell studies. RCE is also exceedingly well appropriate for studies of mass transport in the turbulent flow.

14.2   PROCESS MECHANISM AND MATHEMATICAL MODELS

Historically, the ECD process as an independent processing procedure was proposed in 1962 by Whiters.[63] Also, he made a first attempt to explain the mechanism of ECD process. He suggested that the particles suspended in the electrolyte solution may have a positive charge. Consequently, such particles are attracted to the negatively charged electrode (cathode) by electrophoretic forces. According to Whinters, this action is the main force of the ECD process. However, at that time, Whinters model has not received a development because it was thought that the influence of electrophoretic forces on the mass transfer of substances in a strong electrolyte solution can be ignored.

In 1964, Martin and Williams[64] suggested that the particles moves throughout electrolyte volume to the cathode surface due to the convective flow caused by agitation of the electrolyte. And, then particles, located close to cathode surface, are mechanically entrapped by growing coating. They have made assumption that the particles weight content in the composite coating depends on the contact duration of particles with cathode surface and on the intensity of metal matrix growth.

In 1967, Brands and Goldthorpe[65] rejected the theory of mechanical entrapment of particles proposed earlier by Martin and Williams. They suggested that van der Waals attraction force acts on the particles which were adsorbed on the electrode surface. This force keeps the particles on the surface during the time necessary for covering particles by metal matrix. Consequently, the authors considered all of particles adsorbed on the surface like a particles which will be covered by a layer of growing metal matrix and does not take into account the influence of the desorption process of particles from the electrode surface.

In 1968, Saifullin and Khalilova[66] present a mathematical eq 14.1, which allows to determine the weight content of adsorbed particles in a metal matrix. However, the model has not received a further development because it was based on the theory of mechanical entrapment of particles.

$$a_m = \frac{0.1C_{\text{part}}}{\rho_m + 0.001C_{\text{part}}\left(1 - \rho_m/\rho_p\right)}, \tag{14.1}$$

where a_m is the particle weight content in the MMCC, C_{part} is the real particle concentration in electrolytic solution (g/l), ρ_m is density of deposited metal (g/cm^3), and ρ_p is the particle density (g/cm^3).

In 1972, Buzzard and Boden[67] suggested that the mechanism of ECD process can be divided into a series of sequential steps. On the first step, the particles mass transfer to cathode surface occurs by the convection of electrolyte. After that the particles collides with the electrode surface and absorbs on it. Then, the adsorbed particles are covered by the growing metal matrix. However, it is necessary some time to cover the adsorbed particles by the growing metal matrix. If the particle located at the electrode surface less time than required to cover by the growing matrix, then this particle desorb from the electrode surface to the electrolyte. The authors proposed eq 14.2 to determine the weight content of the particles in the MMEC. However, it was established by the authors of model that the proposed equation poorly reflects the real physical mechanism of the process.

$$\text{wt\%} = \frac{(4/3)\,\pi \cdot r_p^3 \rho_p}{(4/3)\,\pi \cdot r_p^3 \rho_p + 4\pi \cdot \rho_p^2 e_m i \cdot t}, \tag{14.2}$$

where wt% is the weight content of particles in the MMEC, r_p is the particle radius, e_m is the electrochemical equivalent of deposited metal, i is the current density, t is the process time.

In 1972, Guglielmi[68] proposed the mathematical model, which was a big step forward on the way to an explanation of the mechanism of the ECD process. It was the first mathematical model, which was experimentally verified. Two successive steps were laid in the basis of this model. These steps were called weak and strong adsorption. The Guglielmi mechanism of MMEC formation is schematically shown in Figure 14.2.

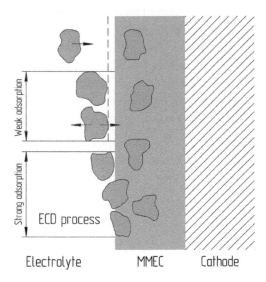

FIGURE 14.2   Guglielmi mechanism of ECD process.

During the first step, there is only weak physical adsorption of the particles that have reached the electrode surface. On the surface of such particles, there is a thin layer of adsorbed electrolyte ions which inhibits chemical reactions between the particles and the cathode surface. Consequently, at this step, the interaction between them have a purely physical nature. Guglielmi used the Langmuir adsorption isotherm[6] to determine the dimensionless fraction of the cathode surface occupied by the weakly adsorbed particles, σ, (14.3). Langmuir adsorption isotherm relates the concentration of particles on the electrode surface with the concentration of particles in the electrolyte volume, c_p, reaction rate constant of particle adsorption process, k, which is reflected the intensity of the interaction between the particle and the electrode, and the dimensionless fraction of free cathode surface $(1 - \theta)$, where θ is the dimensionless fraction of the cathode surface occupied by the strongly adsorbed particles.

$$\sigma = \frac{kc_p}{1 + kc_p}(1 - \theta), \tag{14.3}$$

The second step has an electrochemical nature. During this step, the external electric field is applied to electrodes of electrochemical cell. The ions of the electric double layer (EDL) around the particles are reduced on the cathode by the influence of electric field and metal matrix is growing around weakly adsorbed particles.

Guglielmi thought that the number of particles in the MMEC or their weight contentment is determined by σ and that this relationship is linear. Guglielmi suggested that on the analogy with Tafel equation, the volume flow of particles strongly adsorbed on the cathode surface, V_p, is related to the cathode overpotential, η, by the following equation:

$$V_p = \sigma v_0 e^{\beta \eta}, \tag{14.4}$$

where v_0 and β are the empirical coefficient, which are depend on the specific conditions of the experiment.

Guglielmi derived eq 14.5 which is tie together volume fraction of embedded particles, α, and concentration of particles in the electrolyte with electrode overpotential.

$$\frac{\alpha}{1 - \alpha} = \frac{zF \rho_m V_0}{M_m \cdot i_0} \cdot e^{(B-A)\eta} \cdot \frac{k \cdot c_{p,b}}{1 + k \cdot c_{p,b}}, \tag{14.5}$$

where F is Faraday constant, ρ_m and M_m are, respectively, density and molal mass of deposited metal, $c_{p,b}$ is the concentration of particles in the electrolyte volume, i_0 is the exchange current density, A is the Tafel constant of the reduction metal ions reaction, and z is the valence of metal ion.

The constants V_0, B, and k are dependent on the specific conditions of the experiment, that is, on the properties of the deposited metal and suspended particles and must be determined from the experimental data of ECD process.

Guglielmi model was experimentally verified on various electrochemical systems like Ni–SiC, Ni–TiO$_2$,[68] Cu–Al$_2$O$_3$,[69,70] Cr–C,[71] Ag–Al$_2$O$_3$,[72]

Cu–SiC,[73] and Cu–P.[74] However, this model wasn't able to predict the content of particles in the MMEC for Cr–Al$_2$O$_3$ system.[75]

Unfortunately in this model, the determination accuracy of the volume fraction of embedded particles is mainly determined by the correct assignment of empirical coefficients (V_0, B, k), since such system parameters as fluid dynamics, type, size and material of particles, particles surface functionalization method, fractional composition, pH, temperature, and chemical composition of the electrolyte were not taken into account in the model.

In 1974, Kariapper and Foster[76] based on experimental data suggested that the ions adsorbed on the surface of the particles have a dual role in the process. On the one hand, they define the zeta-potential of particles and thus effect on particles electrophoretic mobility, and on the other hand, they physically links particles with cathode surface by ion reduction at the cathode. The authors also found that intensity of electrolyte mixing has a significant influence on the weight content of particles in the MMEC. The authors proposed a mathematical model, which includes a number of empirical coefficients. This coefficients are determined based on the specific conditions of the electrochemical deposition and therefore cannot be calculated before carrying out the experiment. Kariapper and Foster divided the process mechanism on the two consecutive steps as well as Guglielmi. The first step is the physical adsorption; the second step has an electrochemical nature. On this step, the ions adsorbed on the particle surface are reduced to metal atom (Fig. 14.3).

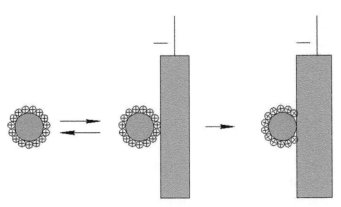

FIGURE 14.3   Kariapper and Foster mechanism of ECD process.

The authors proposed a mathematical eq 14.6 for determination of particles volume fraction in MMEC which takes into account the effect of the electrolyte–fluid dynamics.

$$\frac{dV_p}{dt} = \frac{N \cdot h \cdot C_v}{1 + h \cdot C_v}, \qquad (14.6)$$

where V_p is the volume fraction of particles in MMEC, N is the number of particles collisions with the cathode surface per unit time, C_v is the initial concentration of particles in the electrolyte solution, and h is the empirical coefficient, which is determined by the following equation:

$$h = h^* \left(q \ddot{A} E + L \cdot i^2 - a \cdot b \right), \qquad (14.7)$$

where h^* is the model constant, q is surface charge density of particle, ΔE is the electric-field gradient, i is the current density, L is the specific force of particles adsorption at the electrode surface, a is the coefficient associated with the shape, size, and density of particles, b is the coefficient which depend on the type and intensity of electrolyte mixing.

In 1983, Celis, Roos, and Buelens[77] proposed the mathematical model that allows determining the amount of the embedded particles for electrochemical systems with a rotating disk electrode. Their model took into account the particle transport mechanism from the bulk electrolyte to the electrode surface. They mechanism of the ECD process is suggested to consist of five consecutive steps which are shown schematically in Figure 14.4.

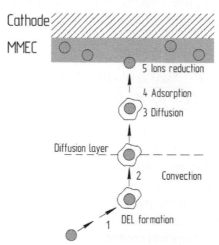

FIGURE 14.4   The scheme of the ECD process by Celisa.

On the first step, electrolyte ions are adsorbed onto surfaces of particles suspended in the electrolyte. Therefore, the EDL is formed around each particle. On the second step, the particles are transferred to Nernst diffusion layer due to the electrolyte convective flows caused by electrolyte mixing. Further, on the third step, the particles reach the electrode surface through diffusion mass transfer. In the sequel, the MMEC formations, as in Guglielmi model, occur at two successive steps. On the fourth step, the particles surrounded by double electrical layer (DEL) of electrolyte ions are weakly adsorbed on the electrode surface. On the last fifth stage, the weakly adsorbed particles are covered by a layer of growing metal matrix due to reduction of the metal ions located near at the cathode surface and adsorbed at surface of particles. The MMEC formation occurs when the amount of electrolyte ions adsorbed on the particle surface is decreasing when the particle moves to the electrode surface. The authors suggested that particle will be codeposited at the growing metal matrix if it situated on the surface of the cathode during some quite definite time. If particle resides at electrode surface for a time smaller than required then it desorbs from electrode surface and is not covered by growing matrix. Accounting for a residence time, the particle is a distinctive feature of the Celis et al. model from Guglielmi model, in which is considered that all particles that reached the surface of the electrode are adsorbed on it and covered by growing metal layer. At the same time, authors of the model suggested that the magnitude of the flux of particles reached the cathode surface is directly proportional to the flow of electroactive ions. Based on the stochastic approach, Celis and others derived the equation of mass fraction of particles embedded in MMEC.

$$w_{p,\text{th}} = \frac{4\pi \cdot r^3 \rho_P \cdot N_{\text{ion}} \left(c_{p,b}/c_{i,b} \right) \left(i_{\text{tr}}/i \right)^\alpha \cdot H \cdot P_{(k/K,i)}}{\left(3M_m i/zF \right) + 4\pi \cdot r^3 \rho_P \cdot N_{\text{ion}} \left(c_{p,b}/c_{i,b} \right) \left(i_{\text{tr}}/i \right)^\alpha \cdot H \cdot P_{(k/K,i)}} \cdot 100, \quad (14.8)$$

where $c_{p,b}$ and $c_{i,b}$ are, respectively, the concentration of particles and ions in the bulk of electrolyte solution, N_{ion} is the total amount of ions adsorbed on the particle, $P_{(k/K,i)}$ is the probability of particle deposition, which is determined as a function of residence time of the particles at the electrode surface, r is the particle radius, ρ_p is the particle density, z is the charge number of the metal ions, and M_m is the molar mass of the metal of matrix.

The dimensionless empirical coefficient H is used in the model to account the hydrodynamics effects on the process. H is equal to 1 for laminar flow and decreases to 0 at a strong turbulent motion. As in the preceding model, their model not able to predict the particle content in the MMEC without direct experimentation and determination of constants used in the model.

Moreover, it is necessary to accurately define the five different coefficients (K, H, k, i_{tr}, and α) for adjustment with results of experimental data.

The experiments[78] of particles ECD on a rotating disc electrode (RDE) were carried out in 1987. Based on the experimental observations on the influence of current density, electrode rotation rate, and electrolyte chemical composition on the particles codeposition process on an RDE. Celis et al. suggested that particles are covered by metal matrix when a certain amount of metal ions adsorbed on the surface were reduced on the cathode. This explains the transition between steps of weak and strong particles adsorption. The authors introduced the term of probability $P_{(k/K,i)}$ that at least k out of K adsorbed ions are reduced at current density i.

$$P_{(k/K,i)} = \sum_{j=k}^{K} \frac{K!}{j!(K-j)}(1-p_i)^{K-j} p_i^j \tag{14.9}$$

where p_i is the probability of reducing of one metal ion at current density i.

Then, a flux of codeposited particles can be determined according to the expression:

$$N_p = P_{(k/K,i)} H V_P N_p^n \tag{14.10}$$

where N_p^n is the flux of particles reaching the electrode surface at per unit time, V_p is the volume of one particle, H is the empirical coefficient that determines the hydrodynamics of electrolyte.

The authors of model assumed that N_p^n is directly proportional to the flux of electroactive ions. Value of N_p^n is determined according to the equation:

$$N_p^n = N_i \frac{c_p}{c_i}\left(\frac{i_{tr}}{i}\right)^{\alpha} \tag{14.11}$$

where c_i and c_p are respectively bulk concentration of ions and particles in the electrolyte, N_i is the flux density of electroactive ions, which is directly proportional to the current density i according to Faraday's law, i_{tr} is the current density which is corresponded to the transition between the kinetically controlled and diffusely controlled deposition regimes, α is the coefficient of proportionality. The values of α and i_{tr} are determined from the conditions of the experiment. Celis et al. find that above equation is true only in the region of kinetically controlled electrodeposition.

The authors of the model for the first time ever have experimentally realized ECD process on RDE. The Celis et al. model was a big step forward in understanding the mechanism of the ECD process because it took into account the particle transport process in the electrolyte volume. However, determining the number of particles colliding with electrode surface is the most vulnerable point in this model. The model also assumes that the transfer of particles to the electrode surface is proportional to the ion transport. These assumptions oversimplify the mechanism of particles behavior near the electrode surface.

In 1987, Valdes suggested the model,[79] which takes into account the mass transfer of the particles in the electrolyte by diffusion, convection, and migration. Valdes suggested that the intensity of particles adsorption process related to the intensity of reduction of electrolyte ions adsorbed on the particle surfaces (Fig. 14.5).

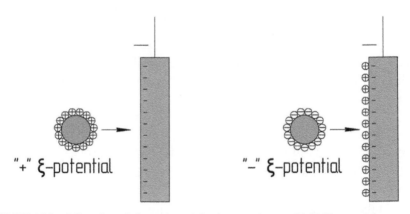

FIGURE 14.5   Migration of charged particles in accordance with Valdes model.

Valdes used the equation based on the Butler–Volmer expression[6] to describe the flow of particles transferred to cathode as boundary conditions on it for the equation of mass transport of particles in the electrolyte. The flow of particles has the following form:

$$V_p = k_0 c_{ion}^s \left[e^{((-\alpha_T nF/RT)\eta)} - e^{(((1-\alpha_T)nF/RT)\eta)} \right] \tag{14.12}$$

where k_0 is the rate constant of electrochemical reaction, c_{ion}^s is the concentration of electrolyte ions adsorbed on the particle surface, α_T is the transmission constant of electrochemical reaction, η is the overpotential on the cathode.

However, simulation results significantly diverged from the experimental data because the peak in the particles content in the MMEC according to the Valdes model should be located at deposited current density close to limiting diffusion values. While according to the experiment data, the peak of particles content is located at low-deposition current density. Valdes for the first time ever takes into account the particle mass transfer by solving the convective diffusion equation, but the flow of particles to cathode as in the preceding models was proportional to the flow of electrolyte ions.

In 1988, Guo[73] proposed the mathematical model that allowed to describe the particles mass transfer by using the similarity criterions, which are dimensionless values. The Sherwood similarity criterion was upgraded by Guo to describe the mass transfer process of particles.

$$Sh' = Co\,\mathrm{Re}^c\,Dm^d\,Sx^e Gq^f \qquad (14.13)$$

where Re is the Reynolds number of hydrodynamic system, Dm is dimensionless value characterized parameters of the particle EDL, and Co, c, d, e, and f are the empirical coefficients which are used to fit the model to the experimental data. More specifically, Co takes into account the van der Waals interactions. Sx is characterized by particle concentration in the bulk of electrolyte and intensity of adsorption process. Gq characterizes the covering process of adsorbed particles by layer of growing metal and is defined as the ratio of the particle diameter to the thickness of the metal layer which is deposited during residence time.

Valdes carried out a model debugging on the experimental ECD process of Cu–SiC in electrochemical cell with parallel plate electrodes. However, in the sequel, this model wasn't used because it utilized the large amount of empirical coefficients and it is unable to establish the optimal deposition conditions and determining the relationship between the concentration of particles in the MMEC and applied current density.

In 1991, Eng upgraded the Valdesa model for a description of the ECD process on a rotating cylindrical electrode (RCE). RCE has many advantages, including a more uniform distribution of the electric field in comparison with an RDE. The critical Reynolds number for system with RCE is much less than for RDE. Consequently, the hydrodynamic of RCE cell has often a turbulent character.

In the Eng model, the influence of the hydrodynamics was taken into account only by the magnitude of the diffusion layer near the surface of the cathode, which is determined in accordance with the empirical expression[80]:\

$$\delta = 1.61 \cdot D^{1/3} \cdot \Omega^{-1/2}\, v^{1/6} \qquad (14.14)$$

where D is the diffusion coefficient, Ω is the angular velocity of rotating electrode, and v is kinematic viscosity of electrolyte.

To test their theoretical assumptions, Eng conducted the experimental study of the ECD process of monodisperse polystyrene particles with copper. It was found that the amount of deposited particles increases with increasing of applied current density and, consequently, the electrophoretic forces have significant influences on content of particles in the MMEC. The largest amount of codeposited particles was at a rotation speed of electrode 615 rpm and was decreased at increasing or decreasing of rotation speed. The influence of current density was correctly defined by Eng model. However, the influence of hydrodynamics on the MMEC composition the model could not explain properly.

In 1992, Fransaer proposed mathematical model[81] to describe the ECD process on an RDE of non-Brownian particles, that is, particles with dispersion is less than 10^6 m^{-1}. This model is based on a trajectory analysis of particles, including convection and gravity, electrophoretic, Coulomb, and inertial forces. The codeposition process was separated into two steps: the reduction of metal ions and the coincident deposition of particles.

The rate of metal deposition was obtained from the diffusion convection equation for a rotating disk electrode. Fransaer used the Butler–Volmer equation as boundary conditions for the deposition of metal cations.

The mass transport of particles in the vicinity of the electrode surface was obtained from the second law of motion. The flow of particles directed to the surface of the electrode was determined by defining the limits of admissible trajectories of particle motion. Using these limits can separate the particles whose trajectories reach the electrode surface, from particles whose trajectories pass by it.

However, this model was not accurate near the electrode surface, because it was used assumption that all particles at some distance from the electrode surface are instantly and irreversibly covered by a layer of growing metal matrix.

It is found by author of model that the speed at which particles reaches the electrode surface is almost entirely determined by the hydrodynamic conditions of the electrolyte. Therefore, the author suggests that the effect of electrokinetic forces on the mass transport of particles in the model can be neglected.

The probability of particles codeposition depends on the balance of forces acting on a particle near the electrode surface (Fig. 14.6).

Fransaer thought that if the interaction force between the particle and the surface of the cathode, $F_{adhesion}$, is more than the total shear force, F_{shift}, then particle codeposited into metal matrix otherwise particle leave the cathode surface.

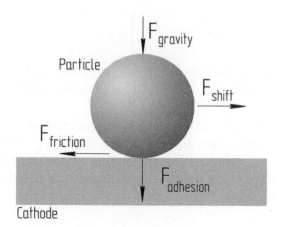

FIGURE 14.6   The forces acting on a particle near the electrode surface.

In 1993, Hwang and Hwang[82] proposed the more universal model than Celis model. They also suggested that the particle codeposition process is caused by reduction of electrolyte ions adsorbed on the particles. The intensity of which is determined by the parameters of the convective diffusion. Authors have focused on the study of reduction mechanism of ions H^+ and Co^{2+} because they investigated the ECD process of Co–SiC system.

The authors found that in investigated diapason of current density, there are three different ranges: at low current density, only H^+ ions are reduced; at intermediate current densities, the reduction of Co^{2+} ions is observed while the reduction rate of H^+ ions has reached its limiting value; and at high current density, the reduction rate is at its limiting value for both ions.

As in Guglielmi model the deposition rate of metal matrix is defined as

$$V_M = \frac{M_M}{\rho_M nF} i \Gamma_M (1 - \theta),\tag{14.15}$$

where Γ_M is the current efficiency. At low current density, the particle codeposition rate V_p is determined by the reduction of the adsorbed ions H^+ and can be deduced with next eq 14.16:

$$V_P = k_1 c_{H^+}^s \cdot \sigma \cdot e^{\beta_1 \eta},\tag{14.16}$$

where $c_{H^+}^s$ is the concentration of adsorbed H^+ ions on the particles surface, which decreases with increasing the H^+ reduction. The equation for determining $c_{H^+}^s$ is represented below:

$$c_{H^+}^s = \left(1 - \frac{V_p}{V_{P,H^+}}\right) c_{H^+}^b, \tag{14.17}$$

where V_{p,H^+} is the maximum rate of the particle codeposition process due to reduction of the ions H^+ and $c_{H^+}^b$ is the concentration of H^+ ions in the bulk of electrolyte.

In the intermediate current density range, the rate of particle codeposition process caused by H^+ reduction is at its limit value, V_{p,H^+}, while the contribution of metal ions reduction is similar to that of effect of H^+ ions at the low current densities. Consequently, the equation for determining V_p in this current density range is given by

$$V_p = V_{p,H^+} + k_2 \left(1 - \frac{V_p}{V_{p,M}}\right) c_M^b \cdot \sigma \cdot e^{\beta_2 \eta} \tag{14.18}$$

where $V_{p,M}$ is the limiting particle codeposition rate caused by metal ions reduction and c_M^b is the concentration of metal ions in the bulk of electrolyte.

Finally, in the high current density range, the particle codeposition rate is entirely determined by diffusion process and is independent of the current density and the adsorbed ions concentration.

$$V_p = k_3 \sigma \tag{14.19}$$

In eqs 14.16, 14.18, 14.19, k_1, k_2, k_3 are constants of particle deposition rate, B_1 and B_2 are the model constants.

The volume fraction of the embedded particles can be calculated using eqs 14.15 and 14.16, 14.18, or 14.19 depending on the current density range. In Hwang and Hwang model, the mass transfer of particles and electrolyte ions are described by diffusion–convection equations and the calculated area is limited by the thickness of diffusion layer, within which the influence of convection is not considered.

In 2000, Vereecken proposed the mathematical model[83] to describe the ECD process of nanoparticles on an RDE. The movement of the particles and the residence time with the electrode surface are considered in this model. The author investigated the mass transport of nanoparticles in the ECD process within the diffusion layer with convective diffusion equation. Alumina Al_2O_3 nanoparticles with dimension of 50 and 300 nm were used as the dispersed phase, while metallic nikel was used as dispersion phase. The experimental process was carried out in the sulfamate electrolytic solution.

As in previous works, Vereecken realize the mathematical simulation only within the Nernst diffusion layer.

He found that for 300 nm particles, the spatial orientation of cathode has a significant impact on the particle weight content, for example, at horizontal position of the cathode the particle weight content in composite coating was up to 40% higher than at vertical cathode position. While for 50-nm particles, the spatial orientation of cathode is not influence on ECD process. The gravity has a little effect on the deposition of 50-nm nanoparticles because they have a low sedimentation rate and buoyancy force compensates the gravity.

According to Fick's first law and Nernst diffusion layer theory, Vereecken used the following expression to define volume content of particles in the MMEC:

$$\frac{x_V}{1-x_V} = \frac{4\pi \cdot r^3 zFN_A}{3V_{m,M}}\left(1.554D_p^{2/3}\,\upsilon^{-1/6}\right)\left(c_{p,b}-c_{p,s}\right)\frac{\omega^{1/2}}{i}, \qquad (14.20)$$

where $c_{p,b},\ c_{p,s}$ are respectively the particle concentration in bulk of electrolyte and adsorbed onto cathode surface, υ is the kinematic viscosity of electrolyte, $V_{m,M}$ is the molar volume of metal matrix, r is the particle radius, ω is the angular velocity of electrode, and D_p is the particle diffusion coefficient in solution.

Vereecken suggested that the particles are transported to the cathode surface by convection and diffusion. The influence of convection is taken into account by the thickness of the diffusion layer. In its turn, the particle codeposition process depends on the residence time and on the intensity of the metal matrix growing process. In the author's work, it was noted that the quantity of the volume content of the nanoparticles in MMEC defined by eq 14.20 corresponds to the experimental data only in the range of limiting current density.

In 2006, Huerta[84] proposed a mathematical model that describes the deposition of copper from sulfuric acid electrolyte with and without presence of chlorine anions and additives of polyethylene glycol nanoparticles. Author considered the adsorption process of polyethylene glycol nanoparticles and the reduction process of copper ions on the electrode as the multistep reactions during which intermediate compounds are formed. To describe this process, author used the diffusion–convection equation and the equation of conservation of mass, as well as electroneutrality condition of the electrolyte. However, author performed the mathematical modeling within computational domain only at three times greater than thickness of the diffusion layer. Simulation was made in a one-dimensional scale. The influence of hydrodynamics on the mass transport of nanoparticles and electrolyte ion within computational domain was not accounted.

In 2007, Lee and Talbot,[85] on the basis of models of Celis and Vereecken, developed a mathematical model describing the ECD process of alumina Al_2O_3 nanoparticles with copper matrix on an RCE. Authors have changed the basic model algorithm of Vereecken to describe the ECD process at RCE.

However, as in the previous models, the computational domain was a narrow region near the cathode surface with the thickness of diffusion layer. Therefore, the study of the mass transfer of electrolyte ions and nanoparticles was produced within their diffusion layers without accounting of convection.

In 2013, Eroglu published a paper[86] on the modeling of ECD process of MMEC of Ni–SiC on an RCE. Eroglu proposed a new mechanism for description of nanoparticles adsorption. But as in previous models, the effect of electrolyte hydrodynamics on mass transfer of nanoparticles and ions was taken into account only by the thickness of diffusion layer.

14.3 EXPERIMENTAL PARAMETERS

The number of particles in the MMEC is a key factor which largely determines the properties of the composite coating. For the production of the MMEC with specified physical and chemical properties, it is necessary to define the main relation between the parameters of the ECD process and the properties of the MMEC. The determination of such empirical relationships, in addition to its purely practical significance, will allow developing the general production methods of the MMEC with required properties and the more accurately mathematical model.

It was found that there are a number of parameters that directly or indirectly effect on the properties of the MMEC. These parameters may be classified into three main categories (Table 14.1).

However, in practice, it is difficult to reliably determine the influence of any single parameter on the particles content in the MMEC, since the influence of various parameters are often interconnected with each other. Also, unfortunately, there are no experimental studies dedicated to determinate the impact on the ECD process, a series of above-mentioned parameters.

It is indisputable that the intensity and the way of mixing of the electrolyte, its viscosity, the concentration of particles in the bath, particle density, and other parameters significantly affect the mechanism of mass transfer of nanoparticles and ions in the volume of an electrochemical cell and, respectively, on the effectiveness of the ECD process and properties of the resulting MMEC. In turn, the interactions between the particles and the electrode are depended on the properties of the contacting surfaces, which are mainly determined by the material of particles and metal matrix, the chemical composition of the bath, pH index of the electrolyte and others parameters.

TABLE 14.1 The System Parameters.

Electrolyte parameters	The particle parameters	Parameters of electrochemical deposition
Chemical composition	Particle material	Concentration of substances in the electrolyte
pH value;	Particle size	
Pollutants or additives	Shape of the particles	Current density
Life time of electrolyte		Type of electrolyte mixing, as well as its intensity
		Temperature of the electrolyte

14.3.1   ELECTROLYTE PARAMETERS

The choice of the electrolyte composition, concentration, and pH level is largely determined by the requirements for the quality of the MMEC and cannot be changed arbitrarily. However, various additives in the electrolyte allow to greatly increase the content of particles in the MMEC, but also, in turn, they may cause the deterioration quality of the metallic matrix and adversely effects on the properties of the MMEC.

14.3.1.1   THE CHEMICAL COMPOSITION OF THE ELECTROLYTE

At the same experimental conditions of the ECD process, there is a difference in the content of particles in the MMEC for different types of electrolyte. For example, the MMEC of Cu–γ-Al$_2$O$_3$ is successfully deposited only from cyanide electrolyte, while under the same conditions and without the addition of any surfactant in the electrolyte solution from the sulfuric acid electrolyte the γ-Al$_2$O$_3$ codeposition was not occured.[65,87] In works[71,88] was found that particles contents in the MMEC increases with increasing of concentration of metal salt in the electrolyte. This is confirmed in Ref. [89], wherein the content of SiO$_2$ particles in the MMEC based on the Zn–Fe alloys increases with increasing concentration of the Fe metal salt in the electrolyte.

Kariapper and Foster[76] have found that the amount of metal ions adsorbed on the particle increases with increasing of the concentration of metal ions in the electrolyte, which means that the zeta-potential of the particle is changed. For example, in Ref. [90], it was found that the zeta-potential of SiC particles increases with increasing of electrolyte concentration. Consequently, electrokinetic attraction of particles to the cathode and particles weight content in the MMEC increases too.

The brighteners and wetting agents are added to the electrolyte to improve the characteristics of metal plating. Some of these additives act as surfactants and may largely effect on the particles content in the MMEC. Greco and Baldauf[91] found that wetting agents can increase the content of particles in the MMEC. However, in Ref. [65], it was determined that the particles content decreases with addition of wetting agents.

14.3.1.2   pH OF ELECTROLYTE

In Refs. [38,45,65,88,92–95], the effect of the concentration of H^+ ions in the electrolyte solution on the content of the particle in the MMEC was investigated. It was found that at pH below 2.3, the content of particles in the MMEC is sharply reduced, while the particles content is not practically changed with increasing of pH greater than 3. Except the ECD process of TiC particles in a metal matrix Ni^{37} where particle content continuously decreases with increasing of pH level. The changing of the electrolyte pH level with addition of particles into electrolyte solution indicates that the H^+ ions are also adsorbed onto particles surface along with other electrolyte ions and, consequently, effect on the mass transfer and adsorption process of particles.

The reduction of the particle content in the MMEC with decreasing of electrolyte pH level is explained by increasing of H^+ ions concentration in the electrolyte solution with decreasing of pH level. Therefore, this leads to increasing of the H^+ ions concentration onto particle surface, while the concentration of metal ions on the particle surface accordingly decreases. The content of particles is decreased with decreasing of H^+ ions concentration onto the particle surface decrease since the metal ions adsorbed onto the particle surface are reduced on the cathode and physically connected the particle with the electrode surface. In confirmation of this, it was found[96] that concentration of the metal ions adsorbed on the particle surface and the current efficiency increases with increasing of the pH indicator.

14.3.1.3 ADDITIVES

It is assumed that the additives increase the content of particles in the MMEC by increasing the amount of ions adsorbed on the particles. However, the relationship between the amount of adsorbed ions and the particles content is still a thoroughly unstudied. Additives also influence on the deposition process of the metallic matrix because they form a complex compound with metal ions (NH_4^+) or adsorbed on the surface of the cathode (Tl^+, amides). A small addition of monovalent cations such as Tl^+, Ce^+, Rb^+, NH_4^+, and amides such as ethylenediamine and tetraethylenepentamine, alanine, and ethylenediaminetetraacetic acid increases the particles content and improves the uniformity of the particle distribution in the MMEC.[76,87,92]

Another class of additives that improved the ECD process is the surfactants.[8,90,97–99] Surfactants are usually used to stabilize the suspension to prevent the agglomeration of the particles in the electrolyte solution. In strong electrolytes, the DEL around the particle is considerably compressed by the action of surrounding electrolyte ions. Consequently, the electrokinetic repulsive forces between the particles become a much smaller than the attracting forces of dispersive interactions and, therefore, the particles is agglomerated. In turn, the adsorption of surfactant molecules and electrolyte ions onto the particle surface increase the magnitude of the repulsive electrokinetic forces and thus prevent the particles agglomeration process.

It is believed that the positive effect of the using of cationic surfactants explained that the surfactant cations adsorbed on the particle surface impart to the particle a positive charge and, therefore, this particle is further attracted to the negatively charged cathode due to the action of electrokinetic forces. However, it should be noted that the influence of cationic surfactants on the parameters of the ECD process was not thoroughly studied as Hu[99] reported that the zeta-potential of particles become more positive with the increasing of the cationic surfactant concentration, but at the same time the particles content was not changed.

14.3.1.4 ELECTROLYTE LIFE TIME

The electrolyte life time is one of the main factors determining the possibility of production the ECD process in industrial scale. Narayan and Chattopadhyay[75] have found that the use of the electrolyte with a life time up to 18 days was not affected on the Al_2O_3 particles content in the matrix metal

Cr. In other investigation, it was found that the electrolyte with a life time of 18–50 days reduces the content of SiC particles in the MMEC from 1% to 0.3% by weight. Perhaps, this is explained by the changing in time of adsorption–desorption processes between the particles and the electrolyte ions or by the agglomeration and sedimentation of particles. Actually, the aging effect of electrolyte is currently poorly understood.

14.3.2 PARTICLE PARAMETERS

Particle parameters are the least controllable process parameters. The choice of the particles material is previously limited by the requirements properties of MMEC. Furthermore, the selected particle material limit particle shape and dimensions.

14.3.2.1 PARTICLE MATERIAL

The accurately determined influence of the particle material on ECD process is not always possible, because, in many cases, the changing of the material, also alter the shape and size of the particles. The density of the particles and the type of functionalization of their surface also influence on the ECD process of ECD particles. Greco and Baldauf[91] noticed that the number of particles of titanium dioxide (TiO_2) in Ni matrix with the same conditions of the ECD process is more than three times greater than particles of electro-corundum Al_2O_3.

Also, the allotropic form of particles SiC[97] and Al_2O_3[100] leads to different its weight content in the MMEC. However, it has been shown in experimental work that by changing the method of surface functionalization, it is possible to overcome the impossibility of introduction of α-Al_2O_3 particles in the MMEC based on Cr[65] and γ-Al_2O_3 in the MMEC based on Cu.[100] More specifically, the α-Al_2O_3 particles were previously grinded and the particles γ-Al_2O_3 were annealed. The BN particles are strongly aggregate in electrolyte solutions because they has a hydrophobic properties, and consequently the particle content in the MMEC is small. The incorporation of aggregated particles is inhibited by a reduction of their concentration in the bulk of electrolyte, due to surfacing of the particles with a strong foaming on top of electrolyte surface or particles sedimentation. The addition of surface active agents (surfactants) reduces the aggregation of BN particles and allows increasing particles volume content from 1% to 15% in Ni matrix.

Also, in spite of the fact that only chemically inert materials are used in ECD as the dispersed phase, in the bulk of electrolyte, interactions occurring between the particles, ions, and electrodes are always present.

In particular, the adsorption of the electrolyte ions occurs on the surface of particles.[101] The amount of adsorption on the particle surface ions and the initial functionalization of the particle surface determine the electrical charge of particle. This charge forms the EDL around the particle in the electrolyte solution. In the electrolytes, EDL plays a decisive role in the interactions between the particles and between the particles and the electrodes. Thus, according to the theory of stability of colloidal systems (DLVO theory),[102,103] the particles in the electrolyte solution reacts with each other due to the forces of attraction and repulsion.

For this reason, Tomaszewski[92] suggested that the positively charged particles codeposit considerably more easily, because they are attracted to the negatively charged cathode. Also Tomaszewski noted that in the electrolyte-containing sodium sulfate (Na_2SO_4), negatively charged SiO_2 particles are adsorbed onto the cathode surface in much smaller quantities than the positively charged Al_2O_3 particles. Lee and Wan[87] explain this by a variety of sign of particle zeta-potential. They found that in dilute copper sulfate bath the α-Al_2O_3 particles have a positive zeta-potential, while the γ-Al_2O_3 particles has a negative zeta-potential. Consequently, the content of α-Al_2O_3 particles in the MMEC achieved a considerable value, while the MMEC with γ-Al_2O_3 particles was not obtained.

The mass transport of particles in the electrolyte solution is also affected by forces gravity and buoyancy, which depend on the density and particle size and density of the electrolyte. Consequently, in addition to the surface properties of the particles, their bulk properties also influence the ECD process. Unfortunately, at the present time, the scientific literature of studies on the effect of particle density on their content in the MMEC are not represented. However, it was found[104] that the orientation of the cathode surface relative to the direction of the forces of gravity and buoyancy significantly effects the particles content in the MMEC. For example, under the same conditions the content of particles whose density is greater than the density of the electrolyte reaches its maximum at the horizontally arranged cathode, as gravity additionally effects the deposition of the particles onto the horizontal surface of the cathode.

The magnetic properties of the particles also influence on the content of particles in the MMEC. Tacken in his work[105] pointed out that the previously magnetized particles Ni particles retain a some residual magnetization in the electrolyte. Such magnetic particles are attracted to a ferromagnetic cathode surface (steel) due to the magnetic field, whereby the content of previously magnetized particles Ni particles in MMEC is three times greater than the unmagnetized.

The conductive particles behave different in ECD process from isolating particles. Conductive particles tend to agglomerate on the surface of the cathode during the deposition;[106] therefore, the distribution of such particles in the volume of MMEC is nonuniform.

14.3.2.2   PARTICLES SIZE

For system of Ni–Al$_2$O$_3$,[93] Ag–Al$_2$O$_3$,[107] and Ni–V$_2$O$_5$,[93] increase of particles content was found if particles size was increased. In Ref. [108], it was determined that the content of particles P in Cu-based MMEC linearly increases with increasing average particle size. However, the reverse pattern has been detected in the experimental studies of ECD Ni–SiC,[97] Ni–Al$_2$O$_3$,[65] and Ni–Cr.[106] Perhaps, this contradiction is due to the fact that the authors, in their works, using different units of measurement in determining the values of particle content in the MMEC.

In Ref. [109], Bozzini has established the relationship between particle size distribution and particles content of the MMEC. He determined that the volume content of particles in the MMEC decreases with increasing particle size. Thus, it was found that particles with smaller size are deposited in the MMEC more readily than particles with a large size. In particular, the large particles adsorbed on the cathode surface occupied a larger surface area than particles with small sizes[88,109] and consequently the free surface available for deposition of metal or other particles are proportionally reduced. Unfortunately, the experiments that take into account the effect of the distribution of the particles sizes in the electrolytic bath in ECD process were not presented in the literature at the present time.

14.3.2.3   SHAPE OF THE PARTICLES

Effect of particle shape has been studied in Ref. [110], where it was found that the amount of Al$_2$O$_3$ particles varies depending on the particle shape in the MMEC based on Ni metal matrix, as follows: for particles in the form of fibers are observed minimum content, for particles having a spherical shape are observed the maximum content, while for particles having a shape other surface are observed average content. However, as in the case of particle sizes, these results were obtained with the same weight concentration of particles in the electrolyte volume, but not to the same molar concentration of particles in the electrolyte bulk. Since the particles in the form of fibers are large, compared with spherical particles, and hence their molar concentration in the electrolyte volume at the same weight concentration have to be less. This may explain the fact that the number of particles in the form of fibers is less in the MMEC.

Particle shape also determines the specific surface area of the particles, the amount of electrolyte ions adsorbed on the particle surface, and hence the force with which particles interacts with cathode surface. In Ref. [111], it has been found that the weight content of Al_2O_3 particles increases with decreasing the specific surface area of particles.

14.3.3 PARAMETERS OF ELECTROCHEMICAL DEPOSITION

Such of electrochemical deposition parameters as the concentration of particles in the electrolyte, the current density, the process of mixing the electrolyte and its temperature can be changed during the ECD process over a wide range. So, the concentration of particles in the electrolyte is the most obvious parameter that effects the content of particles in the MMEC.

14.3.3.1 THE CONCENTRATION OF PARTICLES IN THE ELECTROLYTE

It was found that the volume content of Al_2O_3 particles in the matrix metal of Ni increases with increasing particle concentration in the electrolyte volume. This correlation was also confirmed for a wide range of the investigated systems in further studies.[38,43,45,68,71,75,88,91,93,106,108,109,113] However, in some studies,[45,93] it was found that particles content reduced at very high concentration of particles in the electrolyte. This can be connected with the agglomeration of the particles and their subsequent sedimentation or with the decreasing of electrolyte conductivity due to the adsorption of electrolyte ions onto the particles surface.

14.3.3.2 CURRENT DENSITY

The ECD process occurs at applying of the external electric field, which characterized by the values of cathode and anode potentials and current density.

The current density is the one of the most thoroughly studied parameters. From experimental studies,[68,71,76,91,93,108] it can be concluded that particles content continuously decreases with increasing current density. However, the one or two maximum peaks in the plot of the particle content from applied current density were observed in some experimental works.[3,43,45,75,76]

The height of the peak and the corresponding current density are directly dependent on the intensity of electrolyte mixing,[97,113] on the concentration of particles in the electrolyte[45,75,113] and on the particle material.[97] Also, it has been found[114] that the using of an alternating pulsed deposition can significantly increase the content of γ-Al_2O_3 nanosized particles in a metal matrix of copper. The highest content of such nanoparticles in the MMEC is achieved when the thickness of metal matrix layer to be deposited per one cycle is comparable to the dimension of the nanoparticles. Using an alternating pulsed deposition is allowed to receive MMEC with Al_2O_3 particles content four times larger compared to the deposition with constant current.

14.3.3.3 MIXING OF THE ELECTROLYTE

The main purpose of electrolyte mixing during ECD process is to prevent the agglomeration and sedimentation or floating up of the particles. Also, the electrolyte mixing improves the convective mass transfer of particles and electrolyte ions. The electrolyte mixing is achieved by shaking, sparging, recycling, magnetic mixing of the electrolyte, as well as through the using of rotating electrode. The increase of the electrolyte mixing intensity differently influences on the ECD process. On the one hand, it increases the amount of particles in the MMEC by means of the increase the particle flow[91] traveling from the electrolyte bulk to the electrode surface, and, on the other hand, it decreases the particle content at the excessive mixing intensity, because the particles are carried away from the electrode surface by convective flow of liquid.

14.3.3.4 ULTRASONIC TREATMENT

The problem of agglomeration of nanoparticles is still not completely solved during the ECD process. There are a variety of chemical and physical methods for preventing the agglomeration of nanoparticles in the ECD process. Currently, in addition to aforementioned prevention methods of nanoparticles agglomeration in the ECD process, the technique of ultra-sonic treatment of the initial suspension of nanoparticles before the ECD process[115,116] as well as, during the ECD process has actively been used.[117,118] In the experimental work,[118] it was found that the effect of ultrasonic treatment with the specific power of 225 W/cm^2 during the formation of the MMEC of Cu–α-Al$_2$O$_3$ allows obtaining a more uniform distribution of particles in a galvanic matrix of copper.

More specifically, the MMEC of Cu–α-Al$_2$O$_3$ obtained by using of ultra-sonic treatment have a coating hardness more than 28% greater than untreatable bath. In Ref. [118], it was found that ultrasonic treatment reduced the grain size of a pure metal coating.

14.3.3.5 TEMPERATURE

The ECD process is usually carried out at temperatures typical for the galvanic deposition process. The electrolyte temperature has a different effect depending on parameters of ECD process. A continuous decrease of the Ni particles content in a Zn matrix was observed[105] with increasing of electrolyte temperature. At the same time, the inverse relations was found for system of Ni–TiC.[37] For systems of Ni–PTFE,[40] Ni–V$_2$O$_5$,[93] Cr–graphite,[71] and Ni–BN,[40] the maximum of particles content was observed at 50°C.

The temperature of the electrolyte effects the condition of the DEL around the particles, on the viscosity and density of electrolyte, on the rate of chemical reactions occurring in the electrolyte bulk and onto electrode surface. In the experimental work,[75] it was found that the decrease of the content of Al$_2$O$_3$ particles in a metal matrix Cr with increasing of electrolyte temperature is connected with the lowering amounts of ions adsorbed on the surface of the particles.

14.4 HYDRODYNAMIC MULTILEVEL MATHEMATICAL MODEL OF ELECTROCHEMICAL DEPOSITION

This work is based on the fundamental works of Stojak and Talbot[60,61] where the process of elctrocodeposition of the Cu–Al$_2$O$_3$ MMEC on an RCE was experimentally investigated. Three electrode system, which consists of an RCE, a concentric stationary anode, and a saturated calomel reference electrode placed in a Luggin capillary is used in the fundamental works as the electrochemical cell. The ECD process was realized from water electrolyte solution of 0.1 M $CuSO_4$ and 1.2 M H_2SO_4 with various concentrations of nanoparticles: 39, 120, and 158 g/l and at different electrode rotational rates: 500, 1000, and 1500 rpm. Figure 14.7 depicted the geometric parameters of using electrochemical cell. This article summarizes the results of the previous work of the authors.[119,120]

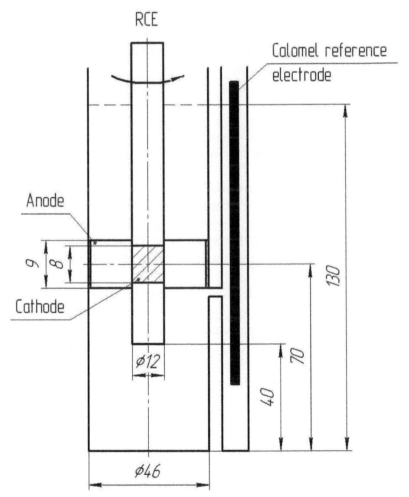

FIGURE 14.7   The electrochemical cell configuration.

The 2D-axisymmetric space (Fig. 14.8) was used as the computational domain in this simulation.

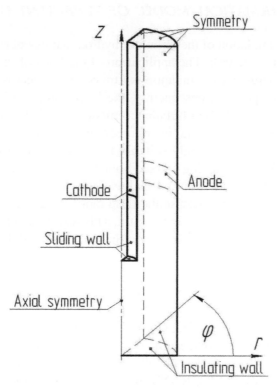

FIGURE 14.8 The 2D-axisymmetric computational space.

It is thought that the Reynolds number characterize the hydrodynamics of fluid. The Reynolds number of hydrodynamic system with RCE depends on rotational speed of RCE, electrochemical cell configuration, and electrolyte kinematic viscosity and is defined by the following equation:

$$Re = \frac{\Omega \cdot r_i \cdot (r_o - r_i)}{v},$$

(14.21)

where $r_i = 6$ mm is the inner cylinder radius, Ω is RCE rotating velocity (rad/s), $v = 1.155$ mm²/s is the kinematic viscosity of electrolyte, $r_o = 24$ mm is the of the outer cylinder radius. The Reynolds numbers are equal to 4896, 9792, and 14,688 for, respectively, rotational rates of 500, 1000, and 1500 rpm. It is found[121] that critical Reynolds number for this type of hydrodynamic system is equal to 200. Therefore, it is necessary to account in simulation the influences of electrolyte turbulent flow on the mass transfer of nanoparticles and electrolyte ions at all rotating rates of RCE.

14.4.1 MATHEMATICAL MODEL OF TURBULENT FLOW

Mathematical simulation of the turbulent hydrodynamics of rotating fluid is the not easily soluble task. The application of methods of direct numerical simulation and large-eddy simulation is limited by the necessity to use a lot of computational power. Consequently, the k–epsilon (k–ε) Reynolds averaged Navier–Stokes (RANS) turbulence model is widely used to simulate the turbulent motion of the fluid in engineering applications.[122]

However, this model is only suitable to simulate the turbulent flow at some distance from the moving or fixed boundaries and is not suitable for the modeling of rotating fluid.[123] The special wall function is used to describe the flow close to the surfaces because the fluid motion near the borders is very different from the fluid motion in the bulk. This wall function is represented the analytical equation which depend on the grid mesh and flow parameters.

Consequently, it is assumed that the computational domain is displaced from some distance of walls. However, it is necessary to account the concentrations of electrolyte ions at close proximity of surface of electrode for definition of cathode concentration polarization. Therefore, the k–ε RANS model is not suited for modeling of ECD process. In this work, we used the low Reynolds number k–ε turbulence model (low Re k–ε RANS), wherein the wall function isn't used, to describe the turbulent fluid flow. There are many various types of low Re k–ε RANS model. The wide used models are the Chang–Hsieh–Chen,[124] the Abe–Kondoh–Nagano (k–ε AKN),[125] the Yang–Shih (k–ε YS),[126] and the Launder–Sharma (k–ε LS).[127] According to the review paper,[128] where the foregoing models were compared, the k–ε AKN model is used in this work to simulate the hydrodynamics of electrolyte.

RANS equation is represented below:

$$\rho\frac{\partial \mathbf{U}}{\partial t} + \rho(\mathbf{U}\cdot\nabla)\mathbf{U} = \nabla\left[-P+\left(\mu+\mu_T\right)\left(\nabla\mathbf{U}+(\nabla\mathbf{U})^{\mathrm{T}}\right)-\frac{2}{3}\rho k\mathbf{I}\right]+\mathbf{F} \quad (14.22)$$

where P is the averaged pressure, ρ is the density of electrolyte, $\mathbf{U}$ is the averaged velocity field, μ is the dynamic viscosity of electrolyte, which is depended only on the electrolyte physical properties, μ_T is the eddy viscosity (eq 14.26), which is taken into account the influences of fluctuations of velocity field, ε is the turbulent dissipation rate, k is the turbulent kinetic energy, and F is the volume forces vector.

The equation of continuity is represented below:

$$\nabla\mathbf{U} = 0. \quad (14.23)$$

The equations for turbulent dissipation rate, ε, turbulent kinetic energy, k, and eddy viscosity are represented in the following.

$$\rho\frac{\partial \varepsilon}{\partial t}+\rho\mathbf{U}\cdot\varepsilon=\nabla\cdot\left[\left(\mu+\frac{\mu_T}{\sigma_\varepsilon}\right)\nabla\varepsilon\right]+C_{\varepsilon 1}\frac{\varepsilon}{k}P_k-f_\varepsilon C_{\varepsilon 2}\rho\frac{\varepsilon^2}{k}, \tag{14.24}$$

$$\rho\frac{\partial k}{\partial t}+\rho(\mathbf{U}\cdot\nabla)k=\nabla\cdot\left[\left(\mu+\frac{\mu_T}{\sigma_k}\right)\nabla k\right]+P_k-\rho\varepsilon, \tag{14.25}$$

$$\mu_T=\rho f_\mu C_\mu\frac{k^2}{\varepsilon}, \tag{14.26}$$

where f_μ, f_ε are the damping functions (14.27 and 14.28), P_k is the production term of turbulent kinetic energy (14.29) and σ_ε, σ_k, C_μ, $C_{\varepsilon 1}$, $C_{\varepsilon 2}$ are the constant of the model (14.38–14.42).

$$f_\mu=\left(1-e^{-l^*/14}\right)^2\cdot\left(1+\frac{5}{R_t^{3/4}}e^{-(R_t/200)^2}\right), \tag{14.27}$$

$$f_\varepsilon=\left(1-e^{-l^*/3.1}\right)^2\cdot\left(1-0.3e^{-(R_t/6.5)^2}\right), \tag{14.28}$$

$$P_k=\frac{\mu_T}{2}\left|\nabla\mathbf{U}+(\nabla\mathbf{U})^T\right|^2, \tag{14.29}$$

Next constants used in this model:

$$\sigma_\varepsilon=1.5, \tag{14.30}$$

$$\sigma_k=1.4, \tag{14.31}$$

$$C_\mu=0.09, \tag{14.32}$$

$$C_{\varepsilon 1}=1.5, \tag{14.33}$$

$$C_{\varepsilon 2}=1.9, \tag{14.34}$$

In eqs 14.27 and 14.28, R_t is the turbulent Reynolds number (14.35), l^* is the dimensionless number characterizing the distance to the closest wall (14.36),

$$R_t=\frac{\rho k^2}{\mu\varepsilon}, \tag{14.35}$$

$$l^* = \frac{(\rho u_\varepsilon l_w)}{\mu}$$ (14.36)

where l_w is the distance to the immediate wall (14.37), u_ε is the Kolmogorov velocity scale (14.38).

$$l_w = \frac{1}{G} - \frac{l_{ref}}{2},$$ (14.37)

$$u_\varepsilon = \left(\frac{\mu \varepsilon}{\rho}\right)^{1/4},$$ (14.38)

where G is the solution of modified eikonal equation.[129] The characteristic length, l_{ref}, depends on the geometry of the model and usually equal to a half of smallest length of rectangle, which is confined the geometry of the model. In this work, l_{ref} is equal to a half of external cylinder radius.

The modified eikonal equation is represented below:

$$\nabla G \cdot \nabla G + \sigma_w G (\nabla \cdot \nabla G) = (1 + 2\sigma_w) G^4,$$ (14.39)

where σ_w is nondimensional constant varies from 0 to 0.5. In this work, σ_w equals 0.1.

Equation 14.40 is the boundary conditions for eq 14.39 on sliding and insulating wall boundaries.

$$G = \frac{2}{l_{ref}},$$ (14.40)

while the homogeneous Neumann condition (14.41) is boundary conditions for eq 14.39 on the other boundaries

$$\nabla G \cdot \mathbf{n} = 0.$$ (14.41)

Equation 14.42 is the initial condition for eq 14.39.

$$G = \frac{2}{l_{ref}}.$$ (14.42)

Equation 14.39 with boundary (14.40 and 14.41) and initial (14.42) conditions have to be solved before starting simulation.

14.4.2 MATHEMATICAL MODEL FOR MASS TRANSFER NANOPARTICLES AND ELECTROLYTE IONS

The concentrations of ions and nanoparticles near the surface of the cathode influences on the kinetics of electrode processes. In well-mixed electrolyte, the concentrations of ions and nanoparticles in the initial moment of the process are equal to the concentrations in the bulk solution. The concentrations of ions and nanoparticles close to the cathode surface changes when the ECD process is started. The diffusion mass transfer layer is formed near the cathode surface. The thickness of this layer depends on the parameters of electrolyte hydrodynamics and on the initial concentration distribution in the bulk of solution, on the particles and ions diffusion coefficients and on the viscosity of electrolyte. It is thought that the concentrations of electrochemical ions and nanoparticles are not changed outside of its diffusion layer and equal to its bulk values in a well-mixed electrolyte solution.

There are three mechanisms of mass transfer in the bulk of electrolyte: convection, diffusion, and migration. The hydrodynamic of fluid define the convective mass transfer. The changes in fluid density induce the natural convection of electrolyte, whereas the forced convection is created by mixing of electrolyte bath. On the mass transfer by diffusion is influenced the concentration gradient of the substance. Consequently, the diffusion of electrolyte ions and nanoparticle can occur either to the electrode surface or from it. The third mechanism of mass transfer of substance matter is a migration or electrophoresis. Migration is an electrokinetic phenomenon, which acts only on the ions and charged particles in contradistinction from diffusion and convection, which act on the all substance matter. It is assumed that the total substance flow N_Σ can be divided into three flows: diffusion N_d, migration N_m, and convection N_c.

$$N_\Sigma = N_d + N_m + N_c. \tag{14.43}$$

The diffusion–convection equations are used to realize the mathematical simulation of mass transfer. The mathematical simulation is carried out throughout the bulk of the electrolyte. The tertiary current distribution[130] is used as the boundary conditions for electrode processes. The tertiary current distribution is considered the ohmic resistance drop in electrolyte solution, the overpotential on the electrode, and ion activity near electrode surface.

The mass conservation law is used to define eq 14.44 of mass transfer of electrolyte ions and suspended nanoparticles.

$$\frac{\partial c_i}{\partial t} + \nabla \cdot \mathbf{N}_i = 0, \tag{14.44}$$

where subscript i denote Cu^{2+} ions, while subscript p denote a nanoparticles. $\mathbf{N}_i$ is the flux density of ion or nanoparticle. The Nernst–Planck eq 14.45 is used to define $\mathbf{N}_i$.

$$\mathbf{N}_i = -D_i \nabla c_i - z_i um_i F c_i \nabla \phi_l + c_i \mathbf{u}, \tag{14.45}$$

where c_i is the volume concentration, D_i is the diffusion coefficient, um_i is the electrophoretic mobility, z_i is the charge number, φ_l is the electrolyte potential, $\mathbf{u}$ is the vector of electrolyte velocity, and F is the Faraday constant.

The diffusion coefficient of electrolyte ion was determined in the experimental work.[61] Einstein's eq 14.46 was used to define the diffusion coefficient of nanoparticles.

$$D_p = \frac{k \cdot T}{6\pi \cdot \mu \cdot r_p}, \tag{14.46}$$

where T is the electrolyte temperature, k is Boltzmann constant and r_p is the nanoparticles radius.

Equation 14.47 is used to define the electrophoretic mobilities of Cu^{2+} ions and nanoparticles.

$$m_i = \frac{D_i}{R \cdot T}, \tag{14.47}$$

where R is the absolute gas constant.

The Maxwell equations are reduced to Poisson's equation for investigated electrochemical system. Poisson's equation is used to define the electrolyte potential, φ_l, and to close the system of equations.

$$-\nabla \cdot \left(\mathring{a} \nabla \phi_l \right) - F \sum_i z_i c_i = 0, \tag{14.48}$$

where ε_0 is the vacuum permittivity, ε_l is the medium-specific relative permittivity, ε is the permittivity, $\mathbf{I}$ is the unit tensor.

The current density depends on the mass flux of electroactive ion (14.50) because the electroactive ions are the charge carriers in the oxidation–reduction reaction.

$$\mathring{a} = \varepsilon_l \varepsilon_0 \mathbf{I}, \tag{14.49}$$

$$\mathbf{i}_l = F \cdot z_i \cdot \mathbf{N}_i, \tag{14.50}$$

The next equation describes the electrochemical reaction occurring on electrode surfaces.

$$Cu^{2+} + 2e^- \Leftrightarrow Cu_m^0, \tag{14.51}$$

where Cu_m^0 denote the metal copper electrodeposited on the cathode, while Cu^{2+} denote the Cu^{2+} electrolyte ions and $\Delta\varphi_{eq}$ is the standard electrode potential of reaction 14.51.

The Butler–Volmer theory[6] is used to describe the rates of cathode and anode reactions by the following:

$$i_n = i_0 \left(\frac{c_i}{c_i^b} \right) \cdot \left[\exp\left(\frac{\alpha_a F}{RT} \eta_s^a \right) - \exp\left(\frac{-\alpha_c F}{RT} \eta_s^c \right) \right], \tag{14.52}$$

where c_i^b is the bulk concentration of Cu^{2+} ions; η_s^c, η_s^a are, respectively, the overpotentials on cathode and anode electrodes, α_c, α_a, are, respectively, the transport coefficients of cathodic and anodic reactions.

The driving force for electrochemical reactions is the electrode overpotential, which can be defined by eq 14.53.

$$\eta_s^k = V_s^k - \varphi_l^k, \tag{14.53}$$

where superscript k denote the electrode (c – cathode, a – anode;), φ_l^k is the potential of electrolyte near the electrode surface, V_s^k is the electrode potential.

The difference between the electric potentials of anode and cathode electrodes is the voltage of electrochemical cell, U.

$$U = V_s^a - V_s^c. \tag{14.54}$$

Equation 14.55 describes the processes adsorption or desorption nanoparticles on cathode surface.

$$P \underset{r_p^d}{\overset{r_p^a}{\Leftrightarrow}} P_a, \tag{14.55}$$

where P_a denotes the adsorbed nanoparticles, while P denotes the suspended nanoparticles and r_p^d, r_p^a are, respectively, the rates of desorption and adsorption processes. It is thought in this work that the desorption process of nanoparticles from the cathode surface is not occurred. The process of nanoparticles adsorption is described by the adsorption isotherm:

$$r_p^a = k_p^a c_p RT, \qquad (14.56)$$

where k_p^a is nanoparticles adsorption coefficient.

The next equations are used as the initial conditions for system of equations.

$$c_i = c_i^0, \qquad (14.57)$$

$$c_p = c_p^0, \qquad (14.58)$$

$$U = f(t), \qquad (14.59)$$

$$\mathbf{u} = 0, \qquad (14.60)$$

$$p = 0 \qquad (14.61)$$

$$k_0 = \left(\frac{\mu}{\rho(0.1 \cdot l_{\text{ref}})} \right)^2, \qquad (14.62)$$

$$\varepsilon_0 = \frac{C_\mu \cdot k_0^{3/2}}{0.1 \cdot l_{\text{ref}}}. \qquad (14.63)$$

The boundary conditions on electrode surfaces:

$$\mathbf{N}_i = \frac{1}{F \cdot z_i} \cdot \mathbf{i}_l, \qquad (14.64)$$

$$\mathbf{N}_p = r_p^a \cdot \mathbf{n}. \qquad (14.65)$$

The boundary conditions on the insulating walls:

$$\mathbf{N}_i = \mathbf{N}_p = 0, \qquad (14.66)$$

$$V_s = 0, \qquad (14.67)$$

$$\mathbf{u} \cdot \mathbf{n} = 0, \qquad (14.68)$$

$$k = 0, \qquad (14.69)$$

$$\varepsilon = \frac{2\mu k}{\rho l_w^2}. \qquad (14.70)$$

The boundary conditions on the sliding wall:

$$u_r = 0, \tag{14.71}$$

$$u_\phi = \omega \cdot r, \tag{14.72}$$

$$u_z = 0, \tag{14.73}$$

$$k = 0, \tag{14.74}$$

$$\varepsilon = \frac{2\mu k}{\rho l_w^2}. \tag{14.75}$$

The next boundary conditions are used on the symmetry boundaries:

$$\mathbf{u} \cdot \mathbf{n} = 0, \tag{14.76}$$

$$\left(-p\mathbf{I} + \mu \left(\nabla \mathbf{u} + \left(\nabla \mathbf{u} \right)^T \right) \right) \cdot \mathbf{n} = 0, \tag{14.77}$$

$$\nabla k \cdot \mathbf{n} = 0, \tag{14.78}$$

$$\nabla \varepsilon \cdot \mathbf{n} = 0, \tag{14.79}$$

The finite element method[131] is used in this work to solve the received system of equations.

14.5   THE RESULTS OF MATHEMATICAL SIMULATION

The hydrodynamic simulation of rotating turbulent flow arising in RCE system was performed to compare the modeling results with the well-known experimental data.[132] The standard k–ε RANS turbulence model and low Re k–ε RANS described in this paper were used to define the more suitable models.

The test system configuration is depicted at Figure 14.9. The system consists of rotating inner and stationary outer concentricity cylinders ($r_i =$ 52.5 mm, $r_o = 59.46$ mm, a $L = 208.8$ mm). The inner cylinder is rotating with the rotation speed Ω. The simulations were performed with a 2D-axisymmetric computational domain during 100 s at slowly increasing Reynolds number from 0 to 2000.

It is found by Taylor[133] in 1923 that laminar Couette flow (Fig. 14.10a) is being unstable when the Reynolds number of system has reached the critical values ($Re = 120$) and the flow changes to the new form. This new flow named the toroidal Taylor vortex flow (Fig. 14.10b). The concentric volume between cylinders is filled by the vortices. The order of vortices disposition relative to the rotation axis is not been changing through time. The Taylor vortices flow is stable in a range of Reynolds number from 120 to 175. After that the flow is changing to the new form again. At second critical Reynold number, the Taylor vortices transforms to the new flow. At this new form, the previously stationary vortices start to migrate relative to the rotation axis. This new flow is named as the wave-vortex flow (Fig. 14.10c).

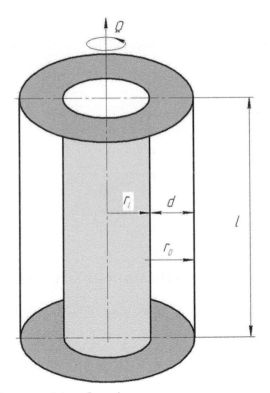

FIGURE 14.9   The test model configuration.

The mathematical modeling with low Re k–ε RANS model shows that laminar Couette flow is stable at Reynolds number, Re, from 0 to 200–250 (Fig. 14.11a). The Taylor vortex flow is stable at Re from 250 to 1000 (Fig. 14.11b). Then, when the Re number exceed this value, the wave-vortex flow is formed (Fig. 14.11c). The wave-vortex flow was stable to the simulation ending.

a) Couette flow   b) Taylor vortex   c) wave-vortex

FIGURE 14.10   The different types of flow.

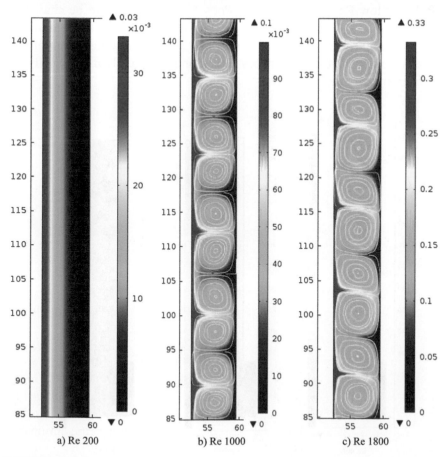

a) Re 200   b) Re 1000   c) Re 1800

FIGURE 14.11   Velocity flow profile (m/s).

It is found that the standard k–ε RANS turbulence model was not appropriated for mathematical simulation of turbulent rotating flow. The k–ε model was not able to simulate the previously mentioned flow changing. Thus, the mathematical modeling with k–ε model shows that only one vortex appears when the laminar Couette flow ceases to be stable. The results of simulation with the standard k–ε model are evidently diverging with the experiment.

To investigate the electrolyte hydrodynamics at different Reynolds numbers the mathematical modeling of electrolyte turbulent flow in RCE cell was done. The mass transfer of electrolyte ion and nanoparticles suspended in electrolyte was not considered at this modeling. At this modeling, the smooth increasing of rotational velocity was from 0 to 2000 rpm during 200 s.

It was discovered that at all examined rotational velocity of RCE the electrolyte hydrodynamics has a turbulent nature. The laminar Couette flow is stable at low Reynolds numbers ($Re < 400$); the flow can be characterized as laminar Couette flow. The flow changing happens at Re numbers from 400 to 900. But, the Taylor vortices do not appear at this Re numbers, whereas a number of small vortices appear and chaotically move relative to the rotation axis of RCE.

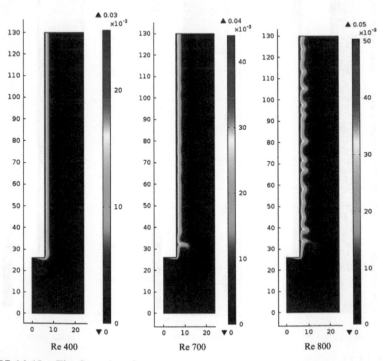

FIGURE 14.12   The flow changing.

Simulation of copper electrodeposition with account of turbulent electrolyte flow was done at all studied rotational velocities (500, 1000, 1500 rpm). The parameters of mathematical modeling corresponded the experimental conditions,[60,61] while the simulation of electrocodeposition process of Cu–Al$_2$O$_3$ was done at three different nanoparticles concentrations (39, 120, and 158 g/l) only at 1500 rpm. The computational time at all mathematical modeling was limited by 105 s. The voltage of electrochemical cell changes from the voltage of open circuit (+0.037 mV vs. SCE) to −0.5 V with a sweep rate of 5 mV/s during this time (Fig. 14.12), while the rotational velocity of RCE varied according to Figure 14.13.

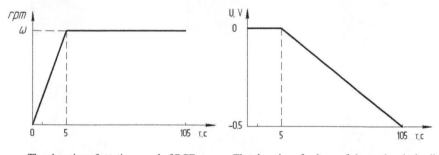

The changing of rotation speed of RCE   The changing of voltage of electrochemical cell

FIGURE 14.13   The conditions of mathematical simulation.

The mathematical simulations of Cu electrodeposition confirmed that the hydrodynamics of electrolyte has a complex turbulent character at all studied rotation speed (Fig. 14.14).

It is discovered that for all studied rotational speeds the concentration of Cu^{2+} ions close to surface of cathode is reduced to zero values when the cell voltage between the electrodes reaches −0.4 V relating to CRE. Therefore, at this cell voltage, the applied current density is limited by the strong concentration polarization. Therefore, the diffusion mass transfer of Cu^{2+} ions to the cathode surface becomes the time-dependent step which determines the intensity of reaction of Cu^{2+} ions reduction at the cathode.

The rate of redox reaction achieves a limiting value because the diffusion mass transfer is limited. It should be noted that the oscillation of current density arise at all rotating speed of RCE when the diffusion mass transfer achieved a limiting value (Figs. 14.15–17). The authors of this work think that these oscillations can be explained by the turbulent flow near the cathode surface. The substitution by each other of zones with low or high speed

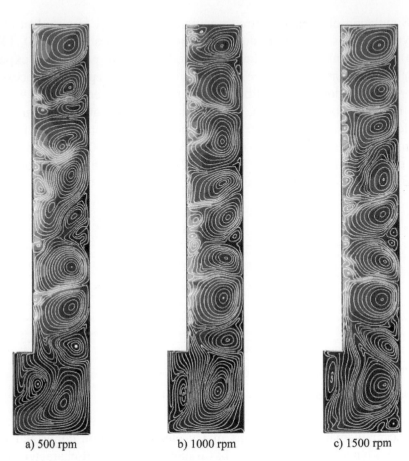

a) 500 rpm   b) 1000 rpm   c) 1500 rpm

FIGURE 14.14   The turbulent flow of electrolyte.

of electrolyte near the cathode surface leads to the appearance of current density oscillations (Fig. 14.18). More specifically, the stagnant zones with small concentration of electrolyte ions are being appeared in some time point near to surface of cathode. Consequently, the diffusion layer with larger thickness is arisen too (Fig. 14.19). This means that the time required to transfer the ions to the surface of cathode by diffusion increases, while the current density temporary decreases. However, the convective mass transfer of Cu^{2+} ions to the cathode surface will increase when a zone with a relative high velocity of electrolyte reaches the electrode surface. As a result of this, the time required to transfer the ions to the cathode surface by diffusion and the diffusion layer thickness decreases. Therefore, the current density is temporarily increasing. The detailed results of mathematical simulation of electrodeposition process with account of electrolyte hydrodynamics are presented in the paper.[120]

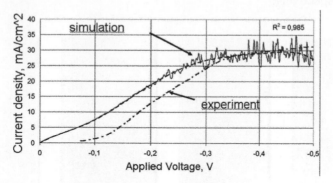

FIGURE 14.15   Oscillation of current density at 500 rpm of RCE.

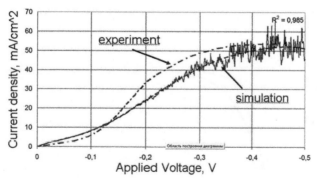

FIGURE 14.16   Oscillation of current density at 1000 rpm of RCE.

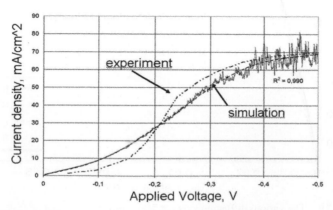

FIGURE 14.17   Oscillation of current density at 1500 rpm of RCE.

It is found for rotational velocity of RCE 500, 1000, and 1500 rpm that diffusion layer thicknesses varies from 30 to 60, from 20 to 50, from 15 to 40 μm, respectively.

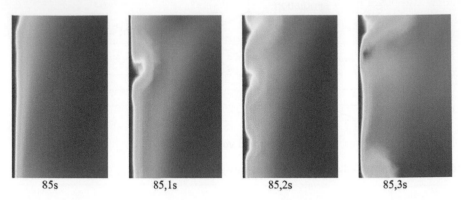

85s	85,1s	85,2s	85,3s

FIGURE 14.18   The substitution by each other of zones with low or high speed of electrolyte near the cathode surface.

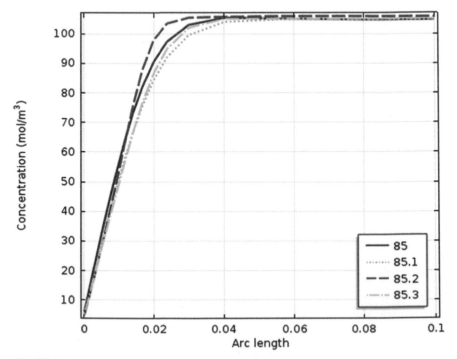

FIGURE 14.19   The various diffusion layers at 1000 rpm.

It was found by Eisenberg, Tobias, and Wilke[134] in 1954 that limiting diffusion current density is confirmed by the next empirical equation:

$$i_L = 0.0791 \frac{nFDC}{2r_i} Sc^{0.356} Re^{0.7},$$ (14.80)

where C, D, and n, are, respectively, the bulk concentration of electroactive ion, the diffusion coefficient of and the ions charge number, F is the Faraday constant.

Table 14.2 represents the values of limiting diffusion current density defined by eq 14.80 for the investigated system configurations. Eisenberg, Tobias, and Wilke deduced the equation of the thickness of the diffusion layer based on the Brunner's correlation:

$$\delta = 15,562 \cdot D^{0.356} \cdot v^{0.344} \cdot \Omega^{-0.7} \cdot r_i^{-0.4},$$ (14.81)

According to eq 14.81, the thickness of the diffusion layer is only governed by configuration of the electrochemical cell, electrolyte kinematic viscosity, electrode rotational velocity, and diffusion coefficient, while it is not time-depended value.

Results of mathematical simulation confirm that the diffusion layer is unsteady. Consequently, the theory of constant diffusion layer[134] is not fully sufficed to the RCE cell with turbulent hydrodynamics of electrolyte. It is necessary to realize the mathematical simulation of coupled process with taking into account the electrolyte hydrodynamics to improve the knowledge of the electrochemical process mechanism.

Commonly, the effect of electrokinetic forces on mass transfer of electroactive ions are not accounted into mathematical simulations of electrochemical processes like a cathodic metal reduction because these processes are realized in the strong electrolyte solution with high conductivity. But the electrokinetic forces have a considerable effect on mass transfer of nanoparticles and, consequently, their influence on the mass transfer of nanoparticles is necessary to account in model. The nanoparticles suspended in electrolyte solution have a DEL, the structure of which is significantly changed as function of properties of electrolyte and nanoparticles.

The particles zeta-potential characterizes the structure of DEL. The value of particles zeta-potential can be positive, negative, or equal zero. The nanoparticles with negative zeta-potential are repulsed from negatively charged cathode, while the nanoparticles with positive zeta-potential are additionally attracted to it during the electrocodeposition process. The mass

TABLE 14.2 Parameters of Process.

RCE rotating velocity (rpm)	Limited current density (mA/cm²)					Thickness of diffusion layer (μm)		Re
	Experiment[60,61]	Simulation		Calculation (eq 14.80)		Calculation (eq 14.81)	Simulation	
		Value	Error (%)	Value	Error (%)			
500	31	29	6.45	39.3	26.9	32,369	30–60	4896
1000	50	51.6	3.2	63.9	27.8	20,069	20–50	9792
1500	67	67.1	0.15	84.9	6.7	14,977	15–40	14,688

transfer of nanoparticles with zero value of zeta-potential does not affect the electrokinetic forces.

The analysis of the experimental results[60,61] revealed that the nanoparticles weight content is slowly reduced with increasing of current density. As can be deduced from the results of experimental work[60,61] that the increase of applied current density slowly reduces the weight content of nanoparticles in MMEC (Fig. 14.20).

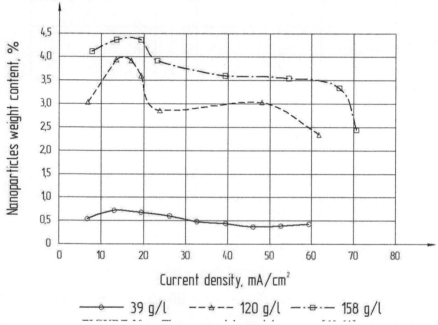

FIGURE 14.20 The nanoparticles weight content.[60,61]

However, the nanoparticles flow to the surface of cathode have to increase with increasing of current density, because the flow of ions which reduced on the cathode increase with the increase of current density according to the Faraday's law. Therefore, it is necessary to proportionally increase the flow of nanoparticles arriving in MMEC for keeping the weight content on a constant level with the increasing of current density. The adducted dependences of nanoparticles flow as function of applied current density and initial concentrations of nanoparticles are presented on Figure 14.21.

Figure 14.21 revealed that the flow of nanoparticles to the cathode linearly increases with increase of applied current density. In one's turn, the increase of applied current density is caused by the increase of cell voltage.

This means that the effect of electrokinetic forces on nanoparticles mass transfer have to be accounted in the mathematical modeling and that the zeta-potential of nanoparticles, which used in the basic experimental work,[60,61] should be a positive, because the flow linearly increases with increasing of the voltage of electrochemical cell.

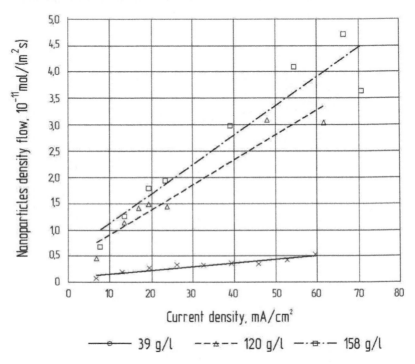

FIGURE 14.21   The adducted nanoparticles density flow.

In the work,[134] it was found that the zeta-potential of nanoparticles is the determinant factor, which defines the weight content of nanoparticles in MMEC. It also found that the γ-Al_2O_3 and α-Al_2O_3 nanoparticles have, respectively, negative and positive zeta-potential in same experimentation conditions (Fig. 14.22). Consequently, it was discovered that only Cu–α-Al_2O_3 MMEC are possible to deposit, while the composite coating of Cu–γ-Al_2O_3 was not received at the same experimental condition.

In the fundamental experimental work,[60,61] it was pointed out that it is obtained the composite coating Cu–γ-Al_2O_3 for the first time ever. However, the zeta-potential of γ-Al_2O_3 nanoparticles should have to be a negative for electrolyte concentration used in the work[60,61] based on the experimental data.[135] The zeta-potential can be varied by addition of surfactants to the electrolyte solution, but surfactants which could change the nanoparticles zeta-potential sign in experiment[60,61] were not used. Therefore, the codeposition of γ-Al$_2$O$_3$ nanoparticles in copper matrix must be restricted by the influences of electrokinetic forces and the MMEC formation could not occur.

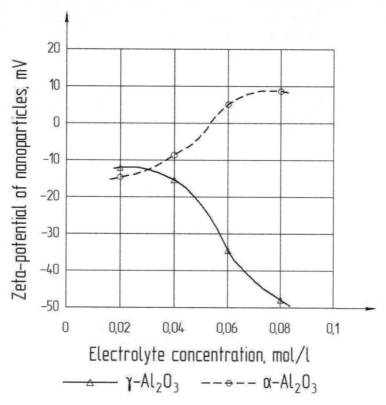

FIGURE 14.22   Zeta-potential of α- and γ-Al$_2$O$_3$ nanoparticles.

However, the authors of Refs. [60,61] established that used nanoparticles γ-Al$_2$O$_3$ were a powder mixture of γ-Al$_2$O$_3$ nanoparticles with the low concentration of α-Al$_2$O$_3$ nanoparticles. The real ratio between different phases has not been defined in the basic work. However, the weight content of each phase in powder mixture can be approximately defined from the divergence between the densities of the powder mixture of nanoparticles determined in Ref. [61] and reference density of γ- and α-Al$_2$O$_3$ material. The approximate ratio between α-Al$_2$O$_3$ and γ-Al$_2$O$_3$ materials was estimated, but the calculation is not given in this chapter. It was found that 88.3 wt% of powder mixture is the γ-Al$_2$O$_3$ nanoparticles, while the content of α-Al$_2$O$_3$ nanoparticles comprise only 11.7%. Based on the this, it was assumed that the γ-Al$_2$O$_3$ nanoparticles does not participate in the ECD process and the only mass transfer of α-Al$_2$O$_3$ was accounted in mathematical simulation.

The resulting dependencies of weight content on applied current density are depicted in Figure 14.23. The good agreement with experimental data[60,61] has been obtained. The experimental data are depicted by points on Figure 14.23.

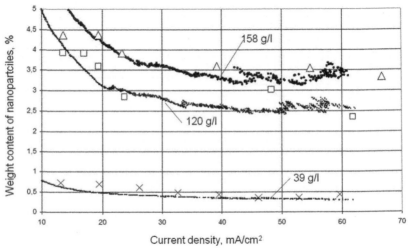

FIGURE 14.23   The weight content of nanoparticle in MMEC.

14.6   CONCLUSIONS

The EPD process is the advanced method for using in practice the nanotechnology to creation new materials and coatings with unique working characteristics. The mathematical modeling of EPD process is the significant and needful step on the road of explanation of composite coatings formation mechanism. A new mathematical model of electrocodeposition of nanocomposite coatings on the RCE, which is taking into account the electrolyte turbulent flow, is developed.

The results of mathematical modeling of ECD on RCE with consideration of electrolyte turbulent flow are presented. A good correlation with the published experimental date has been found. For the first time ever, it is found that near the RCE surface, the unsteady diffusion layer is formed by the reason of electrolyte turbulent flow. Also, it is found that when the current density achieves their limit value, the nonstationary diffusion layer with the variable thickness is formed near the electrode surface by the reasons of electrolyte turbulent flow.

The results of simulation were showed that the electrophoretic forces are considerable and principal effected on nanoparticles content in composite coatings. Consequently, it is necessary not only takes the influence into account but also utilized the influence on EPD process by indirect operation on sign and/or value of nanoparticles zeta-potential, which is achieved by choose a respective surfactant and electrolyte.

The intensification of EPD process is occurred by changing parameters of hydrodynamic. The experimental determination optimal parameters of hydrodynamic are complex and time-consuming process. Consequently, it is necessary to modeling hydrodynamic effect on studied process before the start to produce it at real conditions.

Next step is to develop the mathematical modeling of the ECD process using the methods of mesodynamics and molecular dynamics in the electrolyte turbulent flow, to expand the understanding of the ECD process.

KEYWORDS

- nanoparticles
- electrocodeposition
- coatings
- composite materials
- diffusion layer
- current density

REFERENCES

1. Stojak, J. L.; Fransaer, J.; Talbot, J. B. Review of Electrocodeposition. In *Advances in Electrochemical Science and Engineering*; Alkire, R. C., Kolb, D. M., Eds.; Wiley-VCH Verlag: Weinheim, 2002; vol. 9, p 457.
2. Bicelli, L. P.; Bozzini, B.; Mele, C. A Review of Nanostructural Aspects of Metal Electrodeposition. *J. Electrochem. Soc.* **2008**, *3*, 356–408.

3. Hovestad, A.; Janssen, L. J. J. Electrochemical Codeposition of Inert Particles in a Metallic Matrix. *J. Appl. Electrochem.* **1995**, *25*, 519–527.

4. Hovestad, A.; Janssen, L. J. J. Electroplating of Metal Matrix Composites by Codeposition of Suspended Particles. In *Modern Aspects of Electrochemistry*; Conway, B. E., Ed.; Kluwer Academic/Plenum Publishers: New York, 2005; vol. 38; p 564.

5. Gomes, A.; Pereira, I.; Fernández, B.; Pereir, R. Electrodeposition of Metal Matrix Nanocomposites: Improvement of the Chemical Characterization Techniques. In *Advances in Nanocomposites—Synthesis, Characterization and Industrial Applications*; Boreddy, R., Ed.; InTech: Rijeka, 2011; p. 538.

6. Bockris, J. O.; Reddy, A.; *Modern Electrochemistry. V.2A. Fundamentals of Electrodics*; Kluwer Academic Publishers: New York, 2002; 817 p.

7. Hiott, C. B. Le revetement Zinc–Nickel Slotoloy. *Galvano* **1964,** *333*, 709–718.

8. Fink, C. G.; Prince, J. D. The Codeposition of Copper and Graphite. *Trans. Am. Electrochem. Soc.* **1928,** *54*, 34–39.

9. Furusawa, K.; Matsumura, H. Colloidal nanoparticles: Electrokinetic characterization. In *Dekker Encyclopedia of Nanoscience and Nanotechnology*; Nova Science Publishers, second ed., 2009; pp 773–785.

10. Boccaccini, A. R.; Zhitomirsky, I. Application of Electrophoretic and Electrolytic Deposition Techniques in Ceramics Processing. *Curr. Opin. Solid State Mater. Sci.* **2002,** *6*, 251–260.

11. Shaw, L. L. Processing Nanostructured Materials: An Overview. *J. Miner., Met. Mater. Soc.* **2000,** *52*, 41–45.

12. Suryanarayana, C.; Froes, F. H. Nanocrystalline Metals: A Review. *Phys. Chem. Powder Met. Prod. Process* **1989**, *9*, 79–96.

13. Koch, C. C.; Nanostructured Materials for Structural Applications: Promise and Progress. *Process Fabr. Adv. Mater. VII, Proc. Symp.*, 7th, 1998; pp 497–507.

14. Low, C. T.; Willis, R. G.; Walsh, F. C. Electrodeposition of Composite Coatings Containing Nanoparticles in a Metal Deposit. *Surf. Coat. Technol.* **2006**, *201*, 371–383.

15. Tulio, P. C.; Carlos, I. A. Effect of SiC and Al_2O_3 Particles on the Electrodeposition of Zn, Co and ZnCo: II. Electrodeposition in the Presence of SiC and Al_2O_3 and Production of ZnCo–SiC and ZnCo–Al_2O_3 Coatings. *J. Appl. Electrochem.* **2009**, *39*, 1305–1311.

16. Muller, C.; Sarret, M.; Benballa, M. ZnNi–SiC Composites Obtained from an Alkaline Bath. *Surf. Coat. Technol.* **2002,** *162*, 49–53.

17. Tian, B. R.; Cheng, Y. F. Electrolytic Deposition of Ni–Co–Al_2O_3 Composite Coating on Pipe Steel for Corrosion/Erosion Resistance in Oil Sand Slurry. *Electrochim. Acta* **2007,** *53*, 511–517.

18. Bund, A.; Thiemig, D. Influence of Bath Composition and pH on the Electrocodeposition of Alumina Nanoparticles and Nickel. *Surf. Coat. Technol.* **2007,** *201*, 7092–7099.

19. Bund, A.; Thiemig, D. Influence of Bath Composition and pH on the Electrocodeposition of Alumina Nanoparticles and Copper. *J. Appl. Electrochem.* **2007,** *37*, 345–351.

20. Vidrich, G.; Castagnet, J.-F.; Ferkel, H. Dispersion Behaviour of Al_2O_3 and SiO_2 Nanoparticles in Nickel Sulfamate Plating Baths of Different Compositions. *J. Electrochem. Soc.* **2005,** *152*, 294–298.

21. Bahrololoom, M. E.; Sani, R. The Influence of Pulse Plating Parameters on the Hardness and Wear Resistance of Nickel–Alumina Composite Coatings. *Surf. Coat. Technol.* **2005,** *192*, 154–163.

22. Jung, A.; Natter, H.; Hempelmann, R. Nanocrystalline Alumina Dispersed in Nanocrystalline Nickel: Enhanced Mechanical Properties. *J. Mater. Sci.* **2009,** *44*, 2725–2735.

23. Gay, P.-A.; Berçot, P.; Pagetti, J. Electrodeposition and Characterisation of Ag–ZrO$_2$ Electroplated Coatings. *Surf. Coat. Technol.* **2001**, *140*, 147–154.
24. Wang, W.; Hou, F. Y.; Wang, H.; Guo, H. T. Fabrication and Characterization of Ni–ZrO$_2$ Composite Nano-coatings by Pulse Electrodeposition. *Scr. Mater.* **2005**, *53*, 613–618.
25. Hou, F.; Wang, W.; Guo, H. Effect of the Dispersibility of ZrO$_2$ Nanoparticles in Ni–ZrO$_2$ Electroplated Nanocomposite Coatings on the Mechanical Properties of Nanocomposite Coatings. *Appl. Surf. Sci.* **2006**, *252*, 3812–3817.
26. Gomes, A.; Pereira, S.; Mendonça, M. I. Zn–TiO$_2$ Composite Films Prepared by Pulsed Electrodeposition. *J. Solid State Electrochem.* **2005**, *9*, 190–196.
27. Punith Kumar, M. K.; Venkatesha, T. V. Fabrication of Zinc-nano-TiO$_2$ Composite Films: Electrochemical Corrosion Studies. *J. Chem. Pharm.* **2013**, *5*, 253–261.
28. Fustes, J.; Gomes, A.; Silva Pereira, M. I. Electrodeposition of Zn–TiO$_2$ Nanocomposite Films—Effect of Bath Composition. *J. Solid State Electrochem.* **2008**, *12*, 1435–1443.
29. Terzieva, V.; Fransaer, J.; Celis, J.-P. Codeposition of Hydrophilic and Hydrophobic Silica with Copper from Acid Copper Sulfate Baths. *J. Electrochem. Soc.* **2000**, *147*, 198–202.
30. Bin-shi, X.; Hai-dou, W.; Shi-yun, D.; Bin, J.; Wei-yi, T. Electrodepositing Nickel Silica Nano-Composites Coatings. *Electrochem. Commun.* **2005**, *7*, 572–575.
31. Srivastava, M.; Balaraju, J. N.; Ravishankar, B.; Rajam, K. S. Improvement in the Properties of Nickel by Nano-Cr$_2$O$_3$ Incorporation. *Surf. Coat. Technol.* **2010**, *205*, 66–75.
32. Nowak, P.; Socha, R. P.; Kaisheva, M.; Fransaer, J.; Celis, J.-P.; Stoinov, Z. Electrochemical Investigation of the Codeposition of SiC and SiO$_2$ Particles with Nickel. *J. Appl. Electrochem.* **2000**, *30*, 429–437.
33. Watson, S. W. Electrochemical Study of SiC Particle Occlusion during Nickel Electrodeposition. *J. Electrochem. Soc.* **1993**, *140*, 2235–2238.
34. Zimmerman, A. F.; Clark, D. G.; Aust, K. T.; Erb, U. Electrodeposition of Ni–SiC Nanocomposite. *Mater. Lett.* **2002**, *52*, 85–90.
35. Hu, F.; Chan, K. C. Electrocodeposition Behavior of Ni–SiC Composite under Different Shaped Waveforms. *Appl. Surf. Sci.* **2004**, *233*, 163–171.
36. Kaisheva, M.; Fransaer, J. Influence of the Surface Properties of SiC Particles on their Codeposition with Nickel. *J. Electrochem. Soc.* **2004**, *151*, 89–96.
37. Stroumbouli, M.; Gyftou, P.; Pavlatou, E. A. Codeposition of Ultrafine WC Particles in Ni Matrix Composite Electrocoatings. *Surf. Coat. Technol.* **2005**, *195*, 325–332.
38. Ramesh Bapu, G. N. Electrocodeposition and Characterization of Nickel–Titanium Carbide Composites. *Surf. Coat. Technol.* **1994**, *67*, 105–110.
39. Lee, E. C.; Choi, J. W. A Study on the Mechanism of Formation of Electrocodeposited Ni–diamond coatings. *Surf. Coat. Technol.* **2001**, *148*, 234–240.
40. Cho, Y.; Choi, G.; Kim, D. A Method to Fabricate Field Emission Tip Arrays by Electrocodeposition of Single-Wall Carbon Nanotubes and Nickel. *Electrochem. Solid State Lett.* **2006**, *9*, 572–575.
41. Abi-Akar, H.; Riley, C.; Maybee, G. Electrocodeposition of Nickel–Diamond and Cobalt–Chromium Carbide in Low Gravity. *Chem. Mater.* **1996**, *8*, 2601–2610.
42. Krishnaveni, K.; Narayanan, S. Electrodeposited Ni–B–Si$_3$N$_4$ Composite Coating: Preparation and Evaluation of its Characteristic Properties. *J. Alloys Compds.* **2008**, *466*, 412–420.
43. Hovestad, A.; Heesen, R. J. C. H. L.; Janssen, L. J. J. Electrochemical Deposition of Zinc–Polystyrene Composites in the Presence of Surfactants. *J. Appl. Electrochem.* **1999**, *29*, 331–338.

44. Ebdon, P. R. The Performance of Electroless Nickel/PTFE Composites. *Plat. Surf. Finish.* **1988,** *75,* 65–68.
45. Mohan, S.; Ramesh Bapu, G. Electrodeposition of Nickel–PTFE Polymer Composites. *Plat. Surf. Fin.* **1995,** *82,* 86–88.
46. Ghouse, M.; Viswanathan, M.; Ramachandran, E. G. Electrocodeposition of Nickel Molybdenum Disulfide and Nickel–Tungsten Disulfide. *Met. Finish.* **1980,** *78,* 44–47.
47. Sofer, Y.; Yarnitzky, Y.; Dirnfeld, S. F. Evaluation and Uses of Composite Ni–Co Matrix Coatings with Diamonds on Steel Applied by Electrodeposition. *Surf. Coat. Technol.* **1990,** *42,* 227–236.
48. Su, Y.; Wang, H.; Ding, G.; Cui, F.; Zhang, W.; Chen, W. Electroplated Hard Magnetic Material and Its Application in Microelectromechanical Systems. *IEEE Trans. Magnet.* **2005,** *41,* 4380–4383.
49. Gomez, E.; Pane, S.; Valles, E. Magnetic Composites CoNi–Barium Ferrite Prepared by Electrodeposition. *Electrochem. Commun.* **2005,** *7,* 1225–1231.
50. Onoda, M.; Shimizu, K.; Tsuchiya, T.; Watanabe, T. Preparation of Amorphous/Crystalloid Soft Magnetic Multilayer Ni–Co–B Alloy Films by Electrodeposition. *J. Magn. Magnet. Mater.* **1993,** *126,* 595–598.
51. Osaka, T.; Takai, M.; Hayashi, K.; Ohashi, K.; Satto, M.; Yamada, K. A Soft Magnetic CoNiFe Film with High Saturation Magnetic Flux Density and Low Coecivity. *Nature* **1998,** *392,* 796–801.
52. Takai, M.; Hayashi, K.; Aoyagi, M.; Osaka, T. Electrochemical Preparation of Soft Magnetic CoNiFeS Film with High Saturation Magnetic Flux Density and High Resistivity. *J. Electrochem. Soc.* **1997,** *144,* 203.
53. Mevrel, R. State of the Art on High-Temperature Corrosion-Resistant Coatings. *Mater. Sci. Eng.* **1989,** *120–121,* 13–24.
54. Malone, G. A. Electrodeposition of Dispersion-Strengthened Alloys. *Plat. Surf. Finish.* **1991,** *78,* 58–62.
55. Pushpavanam, S.; Pushpavanam, M.; Natarajan, S.; Narasimham, K. C.; Chinnasamy, S. High Surface Area Nickel Cathodes from Electrocomposites. *Int. J. Hydrogen Energy* **1993,** *18,* 277–281.
56. Jung, A.; Natter, H.; Hempelmann, R. Nanocrystalline Alumina Dispersed in Nanocrystalline Nickel: Enhanced Mechanical Properties. *J. Mater. Sci.* **2009,** *44,* 2725–2735.
57. Lekka, M.; Koumoulis, D.; Kouloumbi, N.; Bonora, P. L. Mechanical and Anticorrosive Properties of Copper Matrix Micro- and Nano-Composite Coatings. *Electrochem. Acta* **2009,** *54,* 2540–2546.
58. Thiemig, D.; Bund, A.; Talbot, J. B. Influence of Hydrodynamics and Pulse Plating Parameters on the Electrocodeposition of Nickel–Alumina Nanocomposite Films. *Electrochim. Acta* **2009,** *54,* 2491–2498.
59. Maurin, G.; Lavanant, A. Electrodeposition of Nickel/Silicon Carbide Composite Coatings on a Rotating Disc Electrode. *J. Appl. Electrochem.* **1995,** *25,* 1113–1121.
60. Stojak, J. L.; Talbot, J. B. Investigation of Electrocodeposition Using a Rotating Cylinder Electrode. *J. Electrochem. Soc.* **1999,** *146,* 4504–4513.
61. Stojak, J. L.; Talbot, J. B. Effect of Particles on Polarization during Electrocodeposition Using a Rotating Cylinder Electrode. *J. Appl. Electrochem.* **2001,** *31,* 559–564, 2001.
62. Perez, T.; Nav, J. Numerical Simulation of the Primary, Secondary and Tertiary Current Distributions on the Cathode of a Rotating Cylinder Electrode Cell. Influence of Using Plates and a Concentric Cylinder as Counter Electrodes. *J. Electroanal. Chem.* **2014,** *719,* 106–112.

63. Whithers, J. C. Electrodepositing. *Prod. Finish.* **1962**, *26*, 62–68.
64. Williams, R. V.; Martin, P. W. Electrodeposition and Metal Finishing. *Trans. Inst. Met. Finish.* **1964**, *42*, 182–188.
65. Brandes, E. A.; Goldthorpe, D. Electrodeposition of Cermets. *Metallurgia* **1967**, *76*, 195–198.
66. Saifullin, R. C. Combined Electrochemical Coatings and Materials. *M.: Chem.* **1972** [in Russian].
67. Bazzard, R.; Boden, P. J. Nickel–Chromium Alloys by Co-deposition: Part I. Co-deposition of Chromium Particles in a Nickel Matrix. *Trans. Inst. Met. Finish.* **1972**, *50*, 63–69.
68. Guglielmi, N. Kinetics of the Deposition of Inert Particles from Electrolytic Baths. *J. Electrochem. Soc.* **1972**, *119*, 137–146.
69. Celis, J. P.; Roos, J. R. Kinetics of the Deposition of Alumina Particles from Copper Sulfate Plating Baths. *J. Electrochem. Soc.* **1977**, *124*, 1508–1511.
70. Celis, J. P.; Roos, J. R. Kinetics of the Deposition of Alumina Particles from Copper Sulfate Plating Baths. *J. Electrochem. Soc.* **1997**, *124*, 1508–1511.
71. Narayan, R.; Narayana, B. H. Electrodeposited Chromium-Graphite Composite Coatings. *J. Electrochem. Soc.* **1981**, *128*, 1704–1708.
72. Suzuki, Y.; Asai, O. Adsorption–Codeposition Process of Al_2O_3 Particles onto Ag–Al_2O_3 Dispersion Films. *J. Electrochem. Soc.* **1987**, *134*, 1905–1910.
73. Guo, H.; Qin, Q.; Wang, A. Mass Transport Process of Solid Particles in Composite Electrodeposition. *Proc. Electrochem. Soc.* **1988**, *46*, 88–18.
74. Graydon, J. W.; Kirk, D. W. Suspension Electrodeposition of Phosphorus and Copper. *J. Electrochem. Soc.* **1990**, *137*, 2061–2066.
75. Narayan, R.; Chattopadhyay, S. Electrodeposited Cr–Al_2O_3 Composite Coatings. *Surf Technol.* **1982**, *16*, 227–234.
76. Kariapper, A. M.; Foster, J. Further Studies on the Mechanism of Formation of Electrodeposited Composite Coatings. *Trans. Inst. Met. Finish.* **1974**, *52*, 87–91.
77. Buelens, C.; Celis, J. P.; Roos, J. R. Electrochemical Aspects of the Codeposition of Gold and Copper with Inert Particles. *J. Appl. Electrochem.* **1983**, *13*, 541–548.
78. Celis, J. P.; Roos, J. R.; Buelens, C. A Mathematical Model for the Electrolytic Codeposition of Particles with a Metallic Matrix. *J. Electrochem. Soc.* **1987**, *134*, 1402–1408.
79. Valdes, J. L. Electrodeposition of Colloidal Particles. *J. Electrochem. Soc.* **1987**, *134*, 223–228.
80. Eisenberg, M.; Tobias, C. B.; Wilke, C. R. Ionic Mass Transfer and Concentration Polarization at Rotating Electrodes. *J. Electrochem. Soc.* **1954**, *101*, 306–319.
81. Fransaer, J. Analysis of the Electrolytic Codeposition of Non-Brownian Particles with Metals. *J. Electrochem. Soc.* **1992**, *139*, 413–420.
82. Hwang, B. J.; Hwang, J. R. Kinetic Model of Anodic Oxidation of Titanium in Sulphuric Acid. *J. Appl. Electrochem.* **1993**, *23*, 1056–1062.
83. Vereecken, P. M.; Shao, I. Particle Codeposition in Nanocomposite Films. *J. Electrochem. Soc.* **2000**, *147*, 2572–2575.
84. Huerta, M. E.; Pritzker, M. D. EIS and Statistical Analysis of Copper Electrodeposition Accounting for Multi-Component Transport and Reactions. *J. Electroanal. Chem.* **2006**, *594*, 118–132.
85. Lee, J.; Talbot, J. B. A Model of Electrocodeposition on a Rotating Cylinder Electrode. *J. Electrochem. Soc.* **2007**, *154*, 72–78.

86. Eroglu, D.; West, A. C. Mathematical Modeling of Ni/SiC Co-deposition in the Presence of a Cationic Dispersant. *J. Electrochem. Soc.* **2013**, *160*, 354–360.

87. Lee, C. C.; Wan, C. C. A Study of the Composite Electrodeposition of Copper with Alumina Powder. *J. Electrochem. Soc.* **1988**, *135*, 1930–1933.

88. Berkh, O.; Eskin, S.; Zahavi, J. Electrochemical Cr–Ni–Al₂O₃ Composite Coatings, Part II: Mechanical Properties and Morphology. *Plat. Surf. Finish.* **1995**, *82*, 72–78.

89. Takahashi, A.; Miyoshi, Y.; Hada, T. Effect of SiO₂ Colloid on the Electrodeposition of Zinc–Iron Group Metal Alloy Composites. *J. Electrochem. Soc.* **1994**, *141*, 954–957.

90. Meguno, K.; Ushida, T.; Hiraoka, T.; Esumi, K. Effects of Surfactants and Surface Treatment on Aqueous Dispersion of Silicon Carbide. *Bull. Chem. Soc. Jpn.* **1987**, *60*, 89–94.

91. Greco, V. P.; Baldauf, W. Electrodeposition of Ni–Al₂O₃, Ni–TiO₂ and Cr–TiO₂ Dispersion-Hardened Alloys. *Plating* **1968**, *55*, 250–257.

92. Tomaszewski, T. W.; Tomaszewski, L. C.; Brown, H. Composition of Finely Dispersed Particles with Metals. *Plating* **1969**, *56*, 1234–1239.

93. Verelst, M.; Bonino, J. P.; Rousset, A. Electroforming of Metal Matrix Composite: Dispersoid Grain Size Dependence of Thermostructural and Mechanical Properties. *Mater. Sci. Eng.* **1991**, *A135*, 51–59.

94. Ramesh, B.; Mohammed, Y. Electrodeposition of Nickel–Vanadium Pentoxide Composite and its Corrosion Behavior. *Mater. Chem. Phys.* **1993**, *36*, 134–138.

95. Berkh, O.; Bodnevas, A.; Zahavi, J. Electrodeposited Ni–P–SiC Composite Coatings. *Plat. Surf. Finish.* **1995**, *82*, 62–66.

96. Yeh, S. H.; Wan, C. C. Codeposition of SiC Powders with Nickel in a Watts bath. *J. Appl. Electrochem.* **1994**, *24*, 993–1000.

97. Maurin, G.; Lavanant, A. Electrodeposition of Nickel/Silicon Carbide Composite Coatings on a Rotating Disc Electrode. *J. Appl. Electrochem.* **1995**, *25*, 1113–1121.

98. Helle, K. Electroplating with inclusions. In *Proceedings of the 4th International Conference in Organic Coatings Science and Technology*, vol. 2, Athens, 1979; p 264.

99. Hu, X.; Dai, C.; Li, J.; Wang, D. Zeta Potential and Codeposition of PTFE Particles Suspended in Electrolyses Nickel Solution. *J. Plat. Surf. Finish.* **1997**, *84*, 51–53.

100. Lakshminarayanan, G. R.; Chen, E. S.; Sautter, F. K. Electrodeposited Cu and Cu–Al₂O₃ Alloys: Physical and Mechanical Properties. *Plat. Surf. Finish.* **1976**, *63*, 38–46.

101. Bhagwat, M. S.; Celis, J. P. Adsorption of Cations on Alumina in Relation to Co-Deposition with Nikel. *Trans. Inst. Met. Finish.* **1983**, *61*, 72–79.

102. Dobiáš, B. *Coagulation and Flocculation.* Marcel Dekker Inc.: New York, 1993.

103. Probstein, R. F. *Physicochemical Hydrodynamics.* J. Wiley & Sons: New York, 1994.

104. Ghouse, M.; Viswanathan, M.; Ramachandran, E. G. Electrocodeposition of Nickel Molybdenum Disulfide and Nickel–Tungsten Disulfide. *Met. Finish.* **1980**, *78*, 44–47.

105. Tacken, R. A.; Jiskoot, P.; Janssen, L. J. J. Effect of Magnetic Charging of Ni on Electrolytic Codeposition of Zn with Ni Particles. *J. Appl. Electrochem.* **1996**, *26*, 129–134.

106. Bazard, R.; Boden, P. J. Nickel–Chromium Alloys by Codeposition (Part II): Diffusion Heat Treatment of Codeposited Composite. *Trans. Inst. Met. Finish.* **1972**, *50*, 207–210.

107. Suzuki, Y.; Asai, O. Adsorption & Codeposition Process of Al₂O₃ Particles onto Ag/Al₂O₃ Dispersion Films. *J. Electrochem. Soc.* **1987**, *134*, 1905–1910.

108. Graydon, J. W.; Kirk, D. W. Suspension Electrodeposition of Phosphorus and Copper. *J. Electrochem. Soc.* **1990**, *137*, 2061–2066.

109. Bozzini, B.; Giovannelli, G.; Cavallotti, P. L. An Investigation into Microstructure and Particle Distribution of Ni–P/Diamond Composite Thin Films. *J. Microsc.* **1997**, *185*, 283–291.

110. Pashley, R. M.; Israelachvili, J. N. A Comparison of Surface Forces and Interfacial Properties of Mica in Purified Surfactant Solutions. *Colloids Surf.* **1981,** *2,* 169–187.
111. Apachitei, I.; Duszczyk, J.; Katgerman, L.; Overkamp, P. Particles Co-deposition by Electroless Nickel. *Scr. Mater.* **1998,** *38,* 1383–1390.
112. Hayashi, H.; Izumi, S.; Tari, I. Codeposition of α-Alumina Particles from Acid Copper Sulfate Bath. *J. Electrochem. Soc.* **1993,** *140,* 362–365.
113. Yeh, S. H.; Wan, C. C. A Study of SiC/Ni Composite Plating in the Watts Bath. *Plat. Surf. Finish.* **1997,** *84,* 54–58.
114. Podlaha, E. J.; Landolt, D. Pulse-Reverse Plating of Nanocomposite Thin Films. *J. Electrochem. Soc.* **1997,** *144,* 200–202.
115. Qu, N. S.; Chan, K. C.; Zhu, D. Pulse Co-electrodeposition of Nano-Al_2O_3 Whiskers Nickel Composite Coating. *Scr. Mater.* **2004,** *50,* 1131–1134.
116. Shen, Y. F.; Xue, W. Y.; Wang, Y. D.; Liu, Z. Y.; Zuo, L. Mechanical Properties of Nanocrystalline Nickel Films Deposited by Pulse Plating. *Surf. Coat. Technol.* **2008,** *202,* 5140–5145.
117. Gyawali, G.; Cho, S. H.; Woo, D.; Lee, S. W. Effect of Ultrasound on the Mechanical Properties of Electrodeposited Ni–SiC Nano Composite. *Kor. J. Mater. Res.* **2010,** *20,* 235–239.
118. Kim, M.; Sun, F.; Lee, J.; Hyun, Y. K.; Lee, D. Influence of Ultrasonication on the Mechanical Properties of Cu/Al_2O_3 Nanocomposite Thin Films during Electrocodeposition. *Surf. Coat. Technol.* **2010,** *205,* 2362–2368.
119. Vakhrushev, A.; Molchanov, E. Simulation of Nanocomposite Coating Created by Electrocodeposition Method. *Int. J. Mater.* **2016,** *3,* 44–55.
120. Vakhrushev, A. V.; Molchanov, E. K. Hydrodynamic Modeling of Electrocodeposition on a Rotating Cylinder Electrode. *Key Eng. Mater.* **2015,** *654,* 29–33.
121. Gabe, D. R.; Walsh, F. C. The Rotating Cylinder Electrode: A Review of Development. *J. Appl. Electrochem.* **1983,** *13,* 3–22.
122. Wilcox, D. C. *Turbulence Modeling for CFD.* DCW Industries, 1998.
123. Driver, D. M.; Seegmiller, H. L. Features of a Reattaching Turbulent Shear Layer in Diverging Channel Flow. *Am. Inst. Aeronaut. Astronaut. J.* **1985,** *23,* 163–171.
124. Chang, K. C.; Hsieh, W. D.; Chen, C. S. A Modified Low-Reynolds-Number Turbulence Model Applicable to Recirculating Flow in Pipe Expansion. *J. Fluid Eng.* **1995,** *117,* 417–423.
125. Abe, K.; Kondoh, T.; Nagano, Y. A. New Turbulence Model for Predicting Fluid Flow and Heat Transfer in Sepa-Rating and Reattaching Flows—I. Flow Field Calculations. *Int. J. Heat Mass Transf.* **1994,** *37,* 139–151.
126. Yang, Z.; Shih, T. H. New Time Scale Based k–ε Model for Near-Wall Turbulence. *Am. Inst. Aeronaut. Astronaut. J.* **1993,** *31,* 1191–1198.
127. Launder, B.; Sharma, B. Application of the Energy Dissipation Model of Turbulence to the Calculation of Flow near a Spinning Disc. *Lett. Heat Mass Transf.* **1974,** *1,* 131–138.
128. Jagadeesh, P.; Murali, K. Application of Low-Re Turbulence Models for Flow Simulations Past Underwater Vehicle Hull Forms. *J. Naval Architect. Mar. Eng.* **2005,** *1,* 41–54.
129. Fares, E.; Schröder, W. A Differential Equation for Approximate Wall Distance. *Int. J. Numer. Methods Fluids* **2002,** *39,* 743–762.
130. Averill, A. F.; Mahmood, H. S. Determination of Tertiary Current Distribution in Electrodeposition Cell—Part 1. Computational Techniques. *Trans. Inst. Met. Finish.* **1997,** *75,* 228–233.

131. Zienkiewicz, O. C.; Taylor, R. L.; Zhu, J. Z. *The Finite Element Method: Its Basis and Fundamentals*. Butterworth-Heinemann: Oxford, 2005.
132. Andereck, C. D.; Liu, S. S.; Swinney, H. L. Flow Regimes in a Circular Couette System with Independently Rotating Cylinders. *J. Fluid Mech.* **1986,** *164,* 155–183.
133. Taylor, G. I. Stability of a viscous liquid contained between two rotating cylinders The Philosophical Transactions of the Royal Society. **1923,** *223,* 289–343.
134. Eisenberg, M.; Tobias, C. B.; Wilke, C. R. Ionic Mass Transfer and Concentration Polarization at Rotating Electrodes. *J. Electrochem. Soc.* **1954,** *101,* 306–319.
135. Lee, C. C.; Wan, C. C. A Study of the Composite Electrodeposition of Copper with Alumina Powder. *J. Electrochem. Soc.* **1988,** *135,* 1930–1933.

CHAPTER 15

LOCAL ANESTHETICS CLASSIFICATION: ARTIFICIAL INTELLIGENCE INFORMATION ENTROPY

FRANCISCO TORRENS[1*] and GLORIA CASTELLANO[2]

1Institut Universitari de Ciència Molecular, Universitat de València, Edifici d'Instituts de Paterna, PO Box 22085, E-46071 València, Spain

2Departamento de Ciencias Experimentales y Matemáticas, Facultad de Veterinaria y Ciencias Experimentales, Universidad Católica de Valencia San Vicente Mártir, Guillem de Castro-94, E-46001 València, Spain

*Corresponding author. E-mail: Francisco.Torrens@uv.es

CONTENTS

ABSTRACT

In this chapter, local anesthetics classification with particular focus on artificial intelligence information entropy is reviewed in detail.

In 1948, Shannon published a seminal article entitled *A mathematical Theory of Communication*; an essential result of Shannon's work was the expression for the amount of information entropy (IE), which was the first accurate measurement, scientific, and acceptable of the 20th century, but the most fascinating aspect was that Shannon's IE definition was similar to an equation, created in the 19th century, for a peculiar and elusive physics law: *the entropy*; Shannon established an entropy—IE equivalence, entirely applicable to any system or process.[1,2] The IE was applied to different areas of science and technology, which initially were not related to the model, for example, biology, physics, chemistry, linguistics, economy, social/computer sciences, etc. Since the 1990s, IE was used in chemistry to characterize atoms and small molecules; decades later, several studies were focused on the study of chemical reactions: the researchers showed that the tool could give valuable information, with which is possible to complement natural-systems description.

Chemistry presents an important idea elucidated in 400 years of experimental science: the existence and organization of the periodic table of the elements (PTE, the other is bioevolution). In earlier publications, PTE,[3] IE, molecular classification,[4-17] and local anesthetics[18-20] were discussed. A scheme of research reorientation in translational science in a society was presented.[21] Shannon proposed a measure of the amount of IE supplied by a probabilistic experiment. In the present report, his theory is explained and applied to the world of molecules. Even evolutionary processes present thresholds not lacking in the inorganic world, for example, water heats gradually but, at 100°C, it begins to boil suddenly, starting liquid to gas transit. In local anesthetics, the aromatic portion of molecules is primarily responsible for lipid solubility, which allows diffusion across the nerve cell membrane, determining the intrinsic potency of agents.[22-24] The aromatic and amine portions determine protein-binding characteristics, which are the primary determinant of anesthesia duration. The present analysis of local anesthetic (procaine analogues) includes chemical compounds that fit the following general scheme: (lipophilic portion)–(intermediate chain)–(hydrophilic part), since these are the most numerous and present the widest range of uses among the species used in practice of local anesthesia. The lipophilic fragment normally consists of at least one phenyl radical, the hydrophilic function is most usually a secondary/tertiary amine, and the intermediate chain commonly presents an ester/amide linkage. The *structural elements* of a local anesthetic molecule are *ranked* according to their contribution to anesthetic potency, in the following order of decaying importance: lipophilic portion > hydrophilic part > intermediate chain > number of N atoms > number of O atoms.

A molecular classification is defined based on a similarity matrix, which IE is calculated. After the equipartition conjecture (EC), one has a decision criterion between different classifications. The distance between two similarity matrices is computed. Attributing complexity to models/not to natural systems and relativizing the concept to a chosen framework, one arrives at complexity, capturing the idea and comparing formulations illuminating problems. Structure-based subcellular pharmacokinetics models the chemical behavior and effects in biosystems, determined by chemicals and biosystems physicochemical properties and structures: pharmacokinetic properties are predicted via modeling with molecular descriptors via statistics; some approaches capture the complexity of absorption, distribution, metabolism, excretion, and toxicity (ADMET) via the unjustified use of descriptors, rather than focusing on chemicobiological interactions and building mechanistic models, which are nonlinear in optimized coefficients. Certain attributes of molecular structure come from a chemical predisposition. On observing a complex molecule, complexity does not occur from the fact that it is difficult to draw or memorize. That happening in complex molecules could be true for others that seem complicated, but that could be generated from simpler *building blocks*, easy to put together. Self-assembling is the definition of simplicity although, at first sight, an object may seem complex. Synthetic biology (SB) engineering principles of standardization, abstraction, and modularity are applied to biotechnology, to make engineering new functions in life systems less intimidating and more accessible/predictable. Applying all these definitions to local anesthetics, the periodic table (PT) of local anesthetics, anesthetic alcohols, and ice is calculated, presented, and discussed.

15.1 INFORMATION ENTROPY

A connection exists between the entropy notion in thermodynamics and information/uncertainty. Based on the classical Boltzmann's work, Shannon proposed a measure of the amount of information or uncertainty supplied by a probabilistic experiment. Consider an experiment in which event *i* results with probability p_i, which is described by the finite probability distribution:

$$p_i \geq 0, i = 1, \ldots, n; \quad \sum_i p_i = 1 \tag{15.1}$$

Shannon IE associated with this experiment is

$$h_n(p_1, p_2, \ldots, p_n) = -\sum_i p_i \ln p_i \tag{15.2}$$

The function properties give a reasonable measure of uncertainty in a probabilistic experiment, for example:

1. $h_n(p_1, p_2, \ldots, p_n) \geq 0$
2. $h_n(p_1, p_2, \ldots, p_n) = 0$ if $p_i = 1$ for some i and $p_j = 0$ for $i \neq j$ (if only one result is possible, uncertainty is null).
3. iii. $h_{n+1}(p_1, p_2, \ldots, p_n, 0) = h_n(p_1, p_2, \ldots, p_n)$ (adding an impossible event does not change uncertainty).
4. iv. $h_n(p_1, p_2, \ldots, p_n) \leq h_n(1/n, \ldots, 1/n, \ldots, 1/n) = h_{max}$ (greatest uncertainty corresponding to equally likely outcomes, cf. Fig. 15.1).

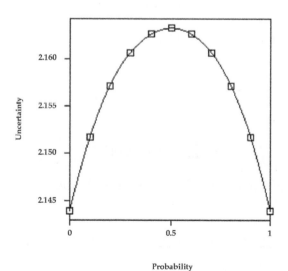

FIGURE 15.1 Variation of uncertainty as a function of probability.

Random events exist, probabilities cannot be directly evaluated; for example, results of the measurements made on microsystems are the mean values of some random variables. Many random distributions exist compatible with a given mean value. The problem is how to decide which is the best distribution. The principle of maximum information entropy (PMIE) can be considered as such a criterion. According to this, systems choose the random distribution maximizing IE. The IE is the primary concept and use the probability distribution that maximizes IE subject to certain constraints for the statistical evolution inference.[25] The PMIE appears to present a subjective character. As long as IE is accepted as being the most suitable measure of uncertainty, the system selects that particular random distribution, which contains the largest amount of uncertainty compatible to the given restraints. Success of PMIE in classical and quantum mechanics suggests extending its range of application. Making inferences on the basis of partial information, it is needed to use that probability distribution that has maximum information entropy, as a measure of uncertainty, subject to whatever is known.[26] Boltzmann's eta (H)-theorem (1872) states the rise in IE as a measure of uncertainty. For a large class of Markovian stochastic evolutions, the H-theorem holds. Denote by $p_{t,t+1}(w', w)$ the probability of transition from state w, at moment t to the state w', at moment $t + 1$. Note that $p_{t,t+1}(w', w) \geq 0$ and:

$$\sum_{w'} p_{t,t+1}(w', w) = 1 \tag{15.3}$$

Markovian evolution is described by

$$p_{t+1}(w') = \sum_w p_t(w) p_{t,t+1}(w', w) \tag{15.4}$$

At every moment, IE is given by

$$h_t = -\sum_w p_t(w) \ln p_t(w) \tag{15.5}$$

The H-theorem proves that if transition stochastic matrix $p_{t,t+1}(w', w)$ is bistochastic, that is, for any (w, w'):

$$\sum_w p_{t,t+1}(w', w) = \sum_{w'} p_{t,t+1}(w', w) = 1 \tag{15.6}$$

then:

$$h_t \leq h_{t+1} \tag{15.7}$$

Non-Archimedean (NA) counterpart of PMIE says that the systems choose the random distribution, which maximizes, in NA frame, IE. However, maxima are not unique in NA frames. Consequently, a variety of acceptable distributions will result, and other decision criteria should be selected to ensure uniqueness. Specific real norms would be used to choose one of them.[27-35] The NA categorization of the H-theorem is discussed. Consider two NA times $T = [n, t], T' = [n', t']$ and corresponding NA entropies: $H(T) = [h_0(n), h_1(n)], H(T') = [h_0(n'), h_1(n')]$. The H-theorem NA counterpart would establish that NA IE H(T) always rises in time but the relation of order is a new one in the NA frame; for example, if $T < T'$ in the NA order, by model categorization, the NA-valid H-theorem implies:

$$H(T) < H(T') \tag{15.8}$$

which should be considered with the same NA order as for time T. The NA inequality shows that it is sufficient that IE rise at level $m = 0$, that is, $h_0(n) < h_0(n')$ for $n < n'$, to assure NA–IE rise, despite IE decay of other levels contribution. It is possible that $h_1(n) > h_1(n')$, that is, on higher levels, the level-associated IE decay. Restricting the analysis to a single level, $m = 1$, may show results apparently contradicting PMIE but they are clarified in the multilevel frame. The multilevel IE rises, while allowing self-organization at the focused level $m = 1$. Consider the use of a qualitative spot test to determine the presence of Fe in a water sample.[36] Without any sample history, the testing analyst must begin assuming that both outcomes are equiprobable with probabilities 1/2. When up to two metals may be present in the sample (e.g., Fe, Ni, both), four possible outcomes exist, ranging from neither (0,0) to both (1,1) being present with probabilities $1/2^2$. Which of these four possibilities turns up is determined via two tests, each having observable states. Similarly, with three elements, eight possibilities exist, every one with a probability of $1/2^3$. Three tests are needed. The IE and uncertainty are defined in terms of the base 2 logarithm of the number of possible analytical outcomes:

$$I = H = \log_2 n \tag{15.9}$$

where I indicates the amount of IE, and H, quantity of uncertainty. Initial uncertainty is defined in terms of the probability of the occurrence of every outcome; for example, the following definition results:

$$I = H = \log_2 n = \log_2 \frac{1}{p} = -\log_2 p \tag{15.10}$$

where I is IE contained in answer, given that n possibilities existed, H, initial uncertainty resulting from need to consider the n possibilities, and p, probability of every outcome if all n possibilities are equiprobable. Equation 15.10 is generalized if each outcome probability is not the same. If one knows

$$\hat{\mathbf{R}}_{\hat{i}\hat{j}} = \max\left(r_{wt}\right); \qquad w \in \hat{i}, \quad t \in \hat{j} \tag{15.14}$$

where w designates any index of species belonging to the class of i and, similarly, t, any index referring to the class of j. To any similarity matrix $\mathbf{R}$, IE $H(\mathbf{R})$ is associated:

$$H(\mathbf{R}) = -\sum r_{ij} \ln r_{ij} - \sum \left(1 - r_{ij}\right) \ln \left(1 - r_{ij}\right) \tag{15.15}$$

which expresses the quantity of IE associated with the matrix of design. The defined IE is a measure of the imprecision in classifying experiments. To compare two similarity matrices $\mathbf{R} = [r_{ij}]$ and $\mathbf{W} = [w_{ij}]$, a distance D is introduced:

$$D(\mathbf{R}, \mathbf{W}) = -\sum r_{ij} \ln \left(r_{ij}/w_{ij}\right) - \sum \left(1 - r_{ij}\right) \ln \left[\left(1 - r_{ij}\right)/\left(1 - w_{ij}\right)\right] \tag{15.16}$$

which measures the discrepancy between two similarity matrices and the associated classification. In the classification algorithm, every *hierarchical tree* corresponds to an IE dependence versus *grouping level b* and an *H–b* diagram is obtained. The EC is a decision criterion between different variants, resulting from the classification between hierarchical trees.[38] According to EC, for a given task, the best configuration of a flow sheet is that in which entropy production (EP) is the most uniformly distributed.[39–43] The EC implies an EP linear dependence versus grouping level, so that EC line is

$$H_{eqp} = H_{max} b \tag{15.17}$$

As the classification is discrete, EC is a regular staircase function. The best variant is chosen to be that minimizing the sum of the squares of the deviations:

$$S = \sum_{b_i} \left(H - H_{eqp}\right)^2 \tag{15.18}$$

15.3   FISABIO/BIMCV: *BIG DATA* STORM IN CANCER/IMAGE DATA/DEVICES

Big data (BD) is a concept that makes reference to the accumulation of great data amounts and procedures to find patterns in them. The features involve a series of obstacles to save: having high-scale data bears noise accumulation and obtaining correlations with low precision predictions. Conventional algorithms for data analysis are insufficient, forcing to use other methods that allow an optimized exploration. It is crucial to dispose of automatic tools that are able to detect and correct data uncertainty sources and homogeneity degree. Salas reported a genetic *advice* in cancer in Valencia Community.[44] A molecular analysis showed that cancer is sporadic 70–80%, family 10–25% and hereditary 5–10%. Hereditary cancer syndromes resulted breast/ovary (60%), hereditary nonpolyposis colorectal cancer (31%), familial adenomatous polyposis, multiple endocrine neoplasia type-2, von Hippel–Lindau syndrome, retinoblastoma, Peutz–Jeghers syndrome, etc. The Foundation for Health and Biomedical Research Promotion of Valencia Region (FISABIO) is a nonprofit organization of scientific and care character, aiming scientific-technical, health and biomedical research promotion, impulse and development in the Valencia region (VR).[45] The Spanish Type Culture Collection of Universitat de València, Barcelone Waters Society and Bosch & Gimpera Foundation develop the Project *Drinking Water Library*, to create the first database for matrix-assisted laser desorption–ionization time-of-flight mass spectrometry (MALDI-TOF MS), of microorganisms isolated from water for human consumption. The identification of this type of aquatic bacteria results a qualitative improvement for evaluating water microbiological quality. On the other hand, imaging databanks (IDs) appear as strategic research infrastructures. Medical Imaging Databank of the Valencia Community (BIMCV) is devised as a knowledge repository oriented to favor medical imaging (MI) technological advances, providing technological-coverage services in support of research and development projects. It has a geographical ambit, including all VR centers. However, the management system of integrated medical-imaging databanks (MIDs) will be later compatible with the Health National System central node and other international nodes. De la Iglesia and Martínez proposed questions (Qs), hypotheses (Hs), and problems (P) on the BIMCV databank.[46]

Q1. How does BD make the present BD storm?

Q2. How is BD used in public health and data analysis?

Q3. How is US responding to anomalies detections in the Department of Defense?

H1. Paralyzed academic investigator's disease syndrome (PAIDS) (Goldstein, 1986): Clinical–scientist–physicists displacement to basic research created a clinical research space that was filled with physicists lacking in minimal and fundamental research skills.

P1. The BD Ps are: classification, prediction, clustering, features selection, and outliers detection.

H2. (Van Horn and Toga, 2014). Human neuroimaging (NI) is a *BD* science.[47,48]

H3. Kryder's law: *There is an exponential growth of data in NI.*

Q4. Is NI a computation storm?

Q5. Why cannot people combine beauty and science?

Q6. How do people interact with the world?

H4. Methodologists do not question only data/samples but also method and use statistics; however, epistemologists question data/samples and statistics-itself value.

H5. Discovery versus H-science research (van Horn, 2004; van Horn and Toga, 2014): With neurobiological-data accumulation into large DBs and compute cluster-enabled-means availability for large-scale data processing, a new form of discovery-oriented neuroscience is on the horizon.

   Perkel looked at data-analysis ins/outs and proposed Qs/H on making sense of BD.[49]

Q7. How big is BD?

Q8. What might the sequencing community learn from the microscopy group of people?

H6. Microscopy community developed software longer than geneticists, but they can learn much from their younger cousins (open-source ethos, collaboration spirit).

15.4 TALKING LIFE: SYNTHETIC BIOLOGY THEORY, EXPERIMENTS AND ETHICS

València Biocampus participated in International Genetically Engineered Machine (iGEM, 2012) competition raising a question.[50]

Q1. What is *SB*?

   They made a brainstorm, proposing the following starting working hypotheses.

H1. To generate a living battery.

H2. To generate bacteria that transform rubbish in fuel (or heat).

H3. To try to talk with common diarrhea-causing bacterium *Escherichia coli*. They proposed the following questions, answers, and hypotheses.

Q2. Is it possible to talk with a bacterial culture and that it answer to one?

Q3. How to talk with bacteria?

A3. Three steps are needed: that they listen to, that they understand, that they answer to you.

H4. It can be made with light.

H5. To generate bacteria that will be able to understand the light message and answer. Five examples.

Q4. Are you hungry [blue fluorescence (FLU)]?

A4. Yes, feed me.

Q5. Can you breath (green FLU)?

Q6. Do you have enough oxygen?

Q7. Do you have enough nitrogen (yellow FLU)?

Q8. Are you hot (red FLU, cf. Fig. 15.2)?

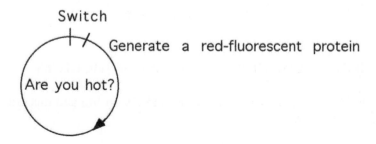

FIGURE 15.2 DNA-modification design of bacterium *E. coli* to answer the question: *Are you hot?*

They proposed the following driving hypotheses.

H6. To generate bacteria that are able to carry out an order, for example, manufacture a protein.

H7. The process (cf. Fig. 15.3) follows.

Ask a question ⟶ Translate to exciting wavelength ⟶ Cells emit light in response ⟶ Light is translated to a voice message

FIGURE 15.3 Scheme of the process

They proposed additional hypotheses, facts, and questions.

H8. Bacteria did not work due to mutation that they did not synthesize FLU protein, saving energy.

F1. In the beginning, a few of such mutants existed but then they proliferated and dominated.

H9. Possible application (PA)1. Bacteria, for example, *E. coli*, are used every day in laboratory; an example.

Q9. How are you?

H10. PA2. Yeasts (*Saccharomyces cerevisiae*) are used every day in industry (bread); two examples.

Q10. How are you?

Q11. Are you producing ethanol?

They performed a study of the ethical aspects of SB and proposed hypotheses and facts.

H11. The bacteria learn to lie and do not answer the questions as expected.

H12. Evolutionary model. Lying bacteria behave to take the place of honest ones with frequency.

F2. Such behavior was observed via experiments.

15.5   LIFE AND DISEASE IN THE PLANET OF THE GENES

López-Otín proposed questions, answers and Hs on life and disease in the planet of the genes.[51]

Q1. What is life?

A1. (Schrödinger, 1944). Life arose from an evolutionary process that became complex.[52]

Q2. What are the laws of life?

Q3. Can one undo life?

A3. Life is: cell to cell, society to death (control of cells working badly: apoptosis, autophagy, etc.).

Q4. What is disease?

A4. Disease arose similarly to life.

Q5. Why does all molecular harmony disappear, do cells weaken, tissues lose functions, and does the organism age/fall ill?

Q6. Will it be possible to control cancer and other diseases for which no adequate responses exist?

Q7. Could people in the near future extend longevity?

Q8. Where does disease come from?

A8. Multicellularity causes risk of disease. Stem cells could be for either good or evil.

Q9. What is the mechanism of division of deoxyribonucleic acid (DNA)?

A9. The DNA is formed by two complementary strands (A...T, C...G); when double strand splits, each chain is completed with complementary nucleotides.

Q10. If everybody is constituted by the same (99.5%) DNA, why are people different?

A10. Because of translocations of nucleotides in DNA.

Q11. What does make people human?

A11. Essential differences exist in genetic-regulation programs.

Q12. Where do diseases come from?

A12. Ones are common and others are hereditary.

Q13. Why ones and not others?

Q14. Were there scientists that looked to the future?

A14. Yes, in Spain, Ramón y Cajal and Severo Ochoa.

Q15. Can a robot think?

A15. (Turing, 1936). Turing's *a* (automatic)-*machine*.

Q16. However, this... what is it for?

H1. Untrue H. *Money is most global planet good and people must take care of it.*

Q17. Why is basic science so important?

A17. Talent is most global planet good and people must take care of it.

Q18. How to keep oneself up to date in epigenomics and metagenomics?

A18. Teaching institutions should continue their work; people must support who could change.

Q19. Is it possible that cells listen to people and carry out their orders?

A19. There are a series of capacities and information.

Q20. Where do they arrive up to?

A20. They arrive far.

Q21. Have you an elixir of eternal youth?

A21. Immortality is unnecessary; error should be avoided: Life has a *molecular language*.

Q22. Can philosophy help?

A22. Philosophy cannot be made without an ideology.

Q23. How will the future be?

A23. Genomic edition, but it could be a moratorium because of the secondary effects.

Q24. What do people know and where are they going to with the commensal bioma?

A24. Bioevolution is a dynamic organisms cooperation: the loss of a bacterium causes *dysbiosis*.

Q25. How must education change?

A25. Education must inquire, make students think.

15.6 MECHANISTIC STUDIES OF DEOXYRIBONUCLEIC ACID REPAIR

Miranda and Lhiaubet-Vallet proposed Qs, As, and Hs on studies of DNA repair mechanisms (DRMs).[53]

Q1. Why is light important?

A1. Because light technologies in 21st century will be as significant as 20th-century electronic ones.

Q2. How stable is DNA?

Q3. Thus, could DNA molecules really be stable for a whole lifetime?

Q4. The DNA stability…. What does it happen?

Q5. *N*-glycosylases: How do they work?

A5. (Lindahl). Base-excision repair (BER) DRM in inferior animals and marsupial mammals.

Q6. Photolyase light DRM in bacteria and plants (Sancar)—How does it work?

A6. (Sancar). Photolyase DRM works by a light-dependent process occurring in a short timescale.

Q7. (6–4) Photolyase was discovered: What is its DRM?

Q8. Does such DRM include an oxetane intermediate?

H1. (Sancar). Nucleotide-excision repair (NER) is other DRM in placentary mammals and man.

Q9. Mismatch repair (MMR) DRM (Modrich)—What is it?

Q10. From the therapeutic point of view, what is DNA repair for?

A10. To design artificial DRMs to avoid damage, for example, sunbath; weaken DRMs, for example, cancer cells.

Q11. Is there photodynamic therapy (PDT) sensitized with porphyrins in the infrared (IR)?

A11. Limit is in singlet $^1O_2$ formation from O_2 (requiring 23 kcal $mol^{-1} \approx 600$ nm in visible (VIS).

Q12. Is there PDT sensitized with phthalocyanines?

A12. There is work in progress but the molecules are not approved for clinical use.

Q13. Is TiO_2 innocuous as used, for example, in solar screens?

A13. TiO_2 could oxide biomolecules but it does not because: state is solid; no skin penetration.

Eritja proposed questions, answers and hypothesis on the tools of DNA repair.[54]

Q14. How much stable is DNA?

A14. It was thought that DNA was immutable.

Q15. (Modrich, 1989). In case of wounds coming from duplication errors, what is the good chain?

H2. Chemotherapy provokes DNA repair; a solution is to administer an inhibitor of DNA repair.

Q16. In paternity tests, do pathologists search for genes?

A16. They do not search for genes but for the number of repetitions in noncodifying positions.

15.7 MAGNETS, THEIR INFLUENCE ON LIFE AND METAL DICHALCOGENIDES MX_2

Is negative the chemistry concept? On one hand, the chemophoby false myth exists: All that *smells* chemistry is bad. Natural/artificial dilemma comes from the times of alchemy.[55] The cause is lack of spreading. Comparing physics and chemistry, is chemistry the future? New trends (nanotechnology, etc.) show decaying differences between physics and chemistry, science and technology, economy and capitalism, etc. On the other hand, chemistry has a positive meaning: *Between two persons there is chemistry*. In this laboratory, Lloret proposed Qs and As on magnets (cf. Fig. 15.4) and their influence on life.[56]

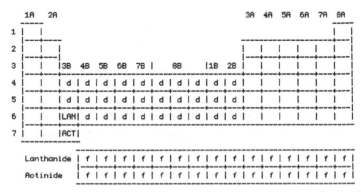

FIGURE 15.4 The PTE of magnetic elements includes those elements of both *d-block* and *f-block*.

Q1. How is detected the magnetic field in which rocks solidified?

Q2. Why are there magnets in the brain?

A2. People do not know.

Q3. Why are there magnets in bacteria?

Q4. Why do bacteria *want* magnets?

Q5. Why do bacteria need to orient themselves?

Q6. Where do bacteria *want* to go?

A6. To know what is up/down, anaerobic bacteria search for minimum oxygen concentration.

Q7. Are there magnetostatic bacteria in Mars?

Q8. Earth magnetic field provides protection, but how to colonize a planet without a magnetic field?

Q9. How to generate a magnetic-field shield?

A9. People generate potent magnetic fields but with small energy compared to the thermal one.

Q10. How does ThyssenKrupp rope-free elevator system moving vertically and horizontally work?

Q11. Does Masical magnetic descaler work?

Q12. Do people use DNA for information storage (electronic memory devices)?

The PTE of two-dimensional (2D)-layered transition metal dichalcogenides (TMDs) MX_2 (M = Ti–Cr, Zr–Tc, Hf–Re, X = S, Se, Te, cf. Fig. 15.5) shows oxidation states (OSs) 4, 5, 6, 7, and 3.[57]

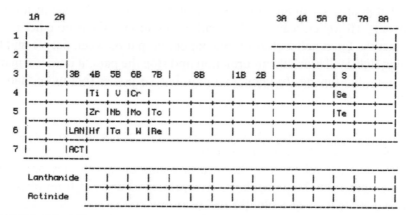

FIGURE 5 The PTE of 2D-layered metal dichalcogenides MX_2.

The PTE of metal oxide nanoparticles (NPs) ZnO, CuO, V_2O_3, Y_2O_3, Bi_2O_3, In_2O_3, Sb_2O_3, Al_2O_3, Fe_2O_3, SiO_2, ZrO_2, SnO_2, TiO_2, CoO, NiO, Cr_2O_3, and La_2O_3 (of Al, Si, Ti–Cr, Fe–Zn, Y, Zr, In–Sb, La and Bi, cf. Fig. 15.6) shows OSs 3, 2, and 4.[58]

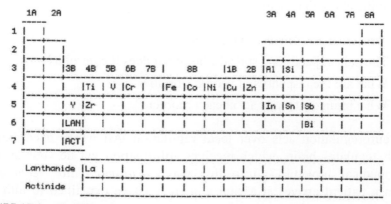

FIGURE 15.6 The PTE of metal oxide NPs.

15.8　PAIN AND PLEASURE, THE BRAIN CONTROLS PAIN AND PAIN IS AVOIDABLE

People does not think about pain frequently, unless it affect them directly, in which case they do not stop thinking about it.[59] To try to avoid or ease pain leads people to separate the hands from an alight stove or start costly research programs in search of more efficacious new analgesic drugs (i.e., more efficacious molecules). Pain is only information. It is like a flashing light that people see in the distance; the more rapidly it blinks, the stronger the pain is. However, pain is only a call to attention and has no other utility than that the brain notice it. A simple paper leaf put between people and the flashing light will block the information and stop the pain, it does not matter the strong that this be. To know that pain has no power (not more than the one that people's mind want grant it) does not help people when they suffer, but it implies that drugs must be neither excessively strong nor rough to stop it. They must be only intelligent. Many analgesics that people use are extracts processed from natural plants, their synthetic copies or chemical compounds close relatives of the natural plants on which they are based. Plants do not try to help people. Active principles derived from plants are toxic and serve the vegetable like defense; in fact, that is what makes them efficacious as drugs. A substance that blocks nerves action can kill people, if it blocks the nerves that reach the heart, or can ease the pain derived from an operation if it blocks the nerves that communicate the brain with the incision place. The field researchers that develop new drugs get excited when they find a new poisonous specimen of a plant, insect, frog, bacterium, or fungus. Grisolía and de Andrés organized XI Yearly Day: Pain Is Avoidable.[60] De Andrés reported that brain controls abdominal pain and raised Qs on brain controlling abdominal pain.

Q1.　How does people's nervous system work?

Q2.　The pain expands and spreads, but what does it control it?

Q3.　What are the responses of people's brain?

Q4.　What brain centers do act at each moment?

Neuroregulation acts over local regulation. All works according to which regulates brain, which receives information via nervous system, and medical novelties go in this way: *neuromodulator techniques*. When a patient arrives at a pain unit, it must be explained to him that his abdominal-pain problem is because of how his nervous system works, which is sensitized by an appendicitis suffered in childhood, which generated a plastic change in his colon form, which sensitized a zone of his nerves, which, in turn, send a signal to his spinal cord and from this to his brain, and the problem must be solved in his cord, putting him a control system. The patient must understand that a regulation exists and, sometimes, the solution is not yet in the stomach but in the same brain. The pain expands/spreads and what controls it is the nervous system. It is analyzed what people's brain responses are, what brain centers act at each moment and people know that in determined circumstances of intestinal pathologies, they have reflexes of their organs that travel information from trunk to brain, which act as symptomatology perpetuators (*visceral hypersensitivity*), that is, problems maintenance not by their viscera itself but enervation; *that is, it is the brain and nervous-character structures, which, in the end, perpetuate patient symptomatology.* Díaz Insa organized a day and raised questions on migraines.[61]

Q5. What does the new international classification of migraines change?

Q6. What does such classification provide?

Q7. Why has one got a headache?

Q8. What must physicians know about migraines physiopathology to improve their patients?

Q9. What clinical tests to carry out?

Q10. What patients with migraines to carry out to?

Q11. Are chronic migraines a paradigm of how pain is perpetuated?

Q12. What drugs do physicians use?

Q13. What patients to treat?

Q14. Is neurostimulation already a reality in migraines?

Q15. What is botulinum toxin (BTX) in migraines for?

Q16. How is BTX action in migraines?

Q17. Why is BTX useful in migraines?

Q18. Is calcitonin gene-related peptide (CGRP) a road to future treatments?

Local anesthetics (procaine analogues), alcohols and ice PT is shown in Table 15.1.

TABLE 15.1 Table of Periodic Properties for Local Anaesthetics, Anaesthetic Alcohols, and Ice.

Period	g000	g010	g100	g101	g110	g111
p00	Ice		Benzyl alcohol		Diperodon	Cocaine
			Propofol		Pramoxine	Cyclomethy-
			4-Iodopropofol		Mexiletine	caine
p01		Dibucaine		Benzocaine	Dyclonine	Hexylcaine
		Propanolol		Butamben		Piperocaine
p10		Dimethisoquin			Bupivacaine	Benoxinate
					Etidocaine	Proparacaine
					Lidocaine	Propoxycaine
					Mepivacaine	
					Prilocaine	
					Tocainide	
					(S)-ropivacaine	
p11			Phenytoin			Butacaine
						2-Chloropro- caine
						Procaine
						Tetracaine

In *reference* to procaine (cf. Fig. 15.7a), the lipophilic portion is a phenyl, hydrophilic function is an amine, intermediate chain is an ester and its group is g111. There are two N and two O atoms, and its period is p11. In benoxinate, there are not two O atoms (period p10). In benzocaine, the hydrophilic fragment is not an amine (group g101), and there are not two N atoms (period p01).

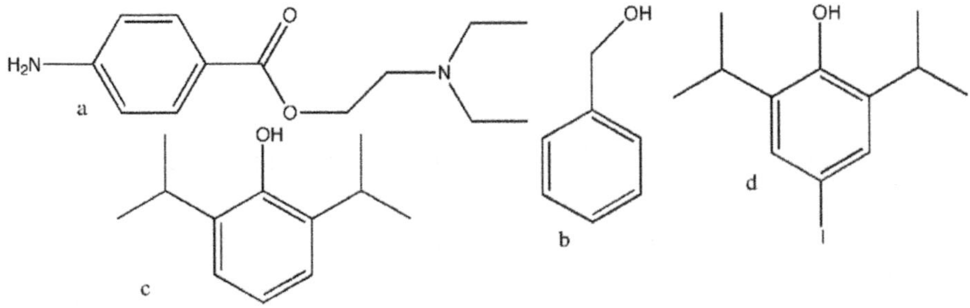

FIGURE 15.7 The PTE of metal oxide NPs.

15.9 *TRANSLATING INNOVATION* INTO USEFUL PRODUCTS FOR HEALTHCARE

Many initiatives and research institutes, for example, the Institute for Molecular Science of the University of Valencia, worldwide stated their mission to harness multidisciplinary advances in underlying sciences (improved understanding of disease's molecular basis, materials sciences/engineering advances, new approaches to clinical trial design) with the aim of bringing quickly, safely, and cost-effectively innovative medical products to patients. A need exists for addressing modern healthcare challenges in society. However, translation from laboratory to clinic is poorly efficient. Duncan raised Qs on *translating innovation* into useful products for improved healthcare and understanding pathophysiology/quality by design.[62]

Q1. Does a miracle occur in translation from laboratory to clinic?

Q2. Why is translation from laboratory to clinic still poorly efficient?

Q3. Have researchers an interest to solve the problems associated with the black box of failure?

Q4. Basic research … products?

Q5. Who is responsible (the researchers, industry, the funding strategy, regulatory agencies)?

Q6. What is responsible?

Q7. Why is translation poorly efficient?

Q8. How can one resolve the problem?

Q9. Will one never make a product if he does not know what he wants to make?

Q10. What is a report?

Q11. What is a product in clinical trial?

Q12. Still no products in the market … understanding why?

Q13. What does it take to produce a breakthrough drug?

Q14. High-throughput (HT) screening (HTS), *omics* … have made it worse?

Q15. Why is the drug failure rate so high?

Q16. Poor research?

Q17. Bad model for development?

Q18. Biomarkers … are they clinically relevant?

Q19. A reductionism approach to *the omics*?

Q20. How does the human machine work in real time in health and disease?

Q21. Fundamental question. Can a drug cross the biological barriers?

Q22. Receptor level (down regulation?).

Q23. What can one do to help?

Q24. *Yes, we can pass all day in the details, but is not generality more interesting?*

Q25. Do you think that one of the most important problems to do translational work is endocytosis?

Q26. If one does not know the biological problems, how to do?

Q27. Is the role of the regulatory agencies the problem?

Pérez-Alonso raised Qs on translating to society biomedical-research knowledge.[63]

Q28. How to translate to society the knowledge generated in biomedical research?

Q29. Why do scientists research?

Scientists are uncomfortable because they do not see wholly clear A to the following Qs.

Q30. How do benefits arrive to society?

Q31. What do taxpayers expect from scientists?

Q32. Does all this worry scientists?

Pérez-Alonso proposed additional questions and answers.

Q33. Is the solution entrepreneuring?

Q34. In addition, then, is there something more?

A34. To create a spin-off enterprise.

Q35. How have scientists achieved to reach here?

Q36. How to promote that researchers show interest in developing applications from work results?

Q37. Why to have a Spanish association of scientific-technician entrepreneurs?

A37. To promote *in first person* entrepreneuring, technology transfer and investment in scientific enterprises.

Q38. Do you know that biotechnology already forms part of your life?

Cremades proposed Qs/As on innovation/entrepreneuring initiatives in health research.[64]

Q39. What is *innovation*?

A39. It is the opposite process to research.

Q40. What is *innovating* in health sciences?

Q41. Why *to innovate* in health sciences?

A41. (1) *Opportunity*: enterprises increasing interest in collaborating with research centers; (2) *necessity*: incomes to research and hospital/ National Health sustainability; (3) *obligation*: results exploitation in Horizon 2020 and national official announcements.

Q42. Can one do something?

A42. To facilitate the whole process.

Research reorientation in translational (marketable) science is needed. In chemistry and physics, reorientation is performed by bench-guided reverse engineering (cf. Fig. 15.8). Ideas in biology are valued by the number of questions that they generate. On one hand, in biology (science), reorientation is performed by clinic-guided reverse engineering. The same applies to medicine (technology).

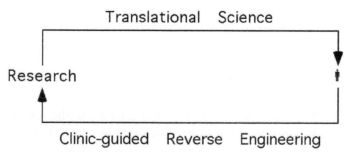

FIGURE 15.8 Scheme of the reorientation of research in translational science.

On the other hand, research reorientation in translational science in a society is done by culture-guided reverse engineering (cf. Fig. 15.9).

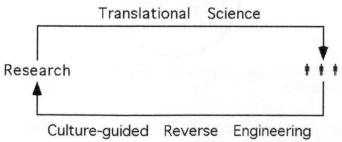

FIGURE 15.9 Scheme of the reorientation of research in translational science in a society.

Translating science from researchers of different fields is applied to cancer (cf. Fig. 15.10).

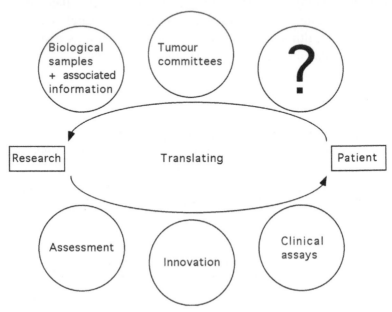

FIGURE 15.10 Translating science from researchers of different fields is applied to cancer.

Health literacy (cf. Fig. 15.11) requires improving the communication abilities and reducing the complexity of information about health.

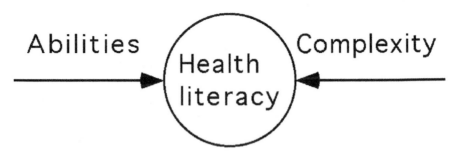

FIGURE 15.11 Health literacy needs improving communication abilities/reducing information complexity.

Research categories suggested by Stokes[65] were modified by Chalmers et al. (cf. Fig. 15.12).[66]

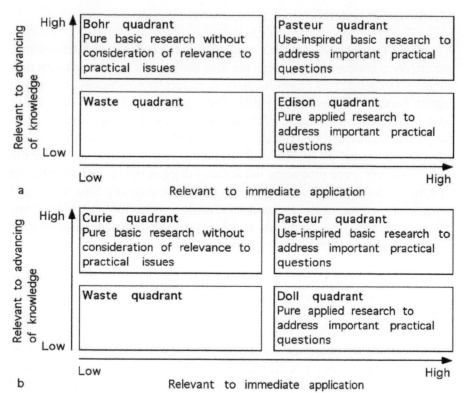

FIGURE 15.12   Classification of different categories of research: (a) after Stokes; (b) after Chalmers *et al.*

There was a change from pregenomic to postgenomic research (cf. Fig. 15.13).[67]

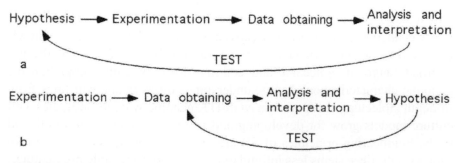

FIGURE 15.13   From (a) pregenomic research to (b) postgenomic research.

15.10 DISCUSSION

Attributing complexity to models and relativizing concept to a framework, one arrives at complexity, capturing the idea and comparing formulations illuminating problems. This report focused on molecular complexity measure to capture decomposing-expression difficulties and sketched formal structure to relate different formal languages to source analysis; the study forms part of *complexity science* and describes experience with approach assessing molecular diversity based on IE; the method shows trends to oversample remote areas of feature space producing unbalanced designs because of information type, which definition is inappropriate for molecular diversity. Structure-based subcellular pharmacokinetics models chemical behavior and biosystems effects, determined by chemicals and biosystems physicochemical properties and structures: pharmacokinetic properties are predicted modeling with molecular descriptors via statistics; approaches capture ADMET complexity with unjustified descriptors, rather than chemicobiological interactions and nonlinear mechanistic models. Implicit code was that simple systems behave simply and complex behavior was complex-cause result; however, after chaos theory, results could be interchanged. Molecular-structure attributes come from a chemical predisposition. Observing a complex molecule, complexity does not occur from fact that it is difficult to draw or memorize. That happening, in complex molecules, could be true for others that seem complicated but that could be generated from simpler *building blocks*, easy to put together. Self-assembling is simplicity definition although an object seem complex.

Just as Darwin's theory of evolution, Pross' one simply hints how simple, fragile replicating systems complexified into intricate living chemical systems; however, understanding the process waits until ongoing studies reveal chemical-materials classes and -pathway kinds, which simple replicating systems are capable, in drive to complexity and replicative stability. Given maturation of next-generation sequencing (NGS), it is not surprising to see challenges, which bring effectiveness for translating genomic knowledge into therapeutics via characterized tumor cell lines, from sets offering a broad range of molecular and cellular diversity with comprehensive genomic characterization. As one understands diseases and cell lines from in-depth molecular standpoint via NGS/HT, genome-wide technologies, cell culture models grow for developing and testing treatments before clinical trials. Engineering principles of SB are applied to biotechnology to make functions in life systems less intimidating and more accessible/predictable.

In anesthetics PT, the first three components of properties vector mark the group. Components subordination order is lipophilic > hydrophilic > ester/

amide, meaning that anesthetics in the same group and period are similar. Classification criteria, selected to reduce the analysis to a manageable quantity of structures, refer to the structural parameters related to lipophilic portion, etc. Many classification algorithms are based on IE. For sets of moderate size, an excessive number of results appear compatible with the data, suffering a combinatorial explosion. However, after EC, one has a decision criterion between different variants, resulting from classification between hierarchical trees. According to EC, the best configuration of a flow sheet is that in which EP is most uniformly distributed. The method avoids the problem of other continuum-variables ones because, for compounds with constant vector, the null standard deviation causes a Pearson's correlation coefficient of one. The lower level classification processes show lower entropy, being more parsimonious (Occam's razor). The present method for molecular classification based on EP EC is a simple, reliable, efficient, and a fast procedure. It was written not only to analyze EP EC but also to explore the world of molecular classification. The good comparison of the present classification results with other taken as *good* confirms adequacy of the vector selected for anesthetics. The IE classes anesthetics agreeing with principal component analyses. Ester/amide anesthetics are classed separately. Low-potency and short-duration agents result apart from those with high potency and long duration. Final classification is shown precise and with low bias. The present classification model improves the signal-to-noise contribution.

15.11 FINAL REMARKS

From the preceding results and discussion, the following final remarks can be drawn.

First, *topical anesthetics advance for minimizing pain during cutaneous procedures*. Eutectic mixture of similar local anesthetics lidocaine/prilocaine 2.5/2.5 wt.% (EMLA®) is most widely used topical anesthetic given efficacy/safety. Present classification is useful for design of mixtures of dissimilar anesthetics (low potency + long duration with high potency + short duration) [lidocaine/prilocaine/dibucaine (betacaine-LA), 1:1 lidocaine/tetracaine eutectic mixture (S-caine)] and applies to difficult cases (ice, alcohols). Ice/EMLA decay discomfort associated with needle injection: EMLA surpasses in pain control but ice, in ease of use, fast action, and inexpensiveness.

Second, benzyl alcohol (Fig. 15.7b, bacteriostatic, solvent)/phenylalcohols are percutaneous enhancers for transdermal drugs (DNA-synthesis inhibitor fluorouracil). It is an efficient anesthetic for intact mucous membranes, surpassing procaine, ranking with alypine/β-eucaine and weaker than holocaine/cocaine. Its action is not as lasting as cocaine and 1% solutions produce smarting. Although benzyl-alcohol 0.9% anesthesia duration is limited, advantages as a local anesthetic in minor plastic surgery are inexpensiveness and less adverse reactions. Bacteriostatic-saline preinjection decreases severity of propofol (Fig. 15.7c, 2,6-diisopropylphenol) intravenous-injection pain, comparable to mixing lidocaine/propofol. Propofol/sevoflurane protocols are used in anesthesia maintenance. However, propofol is recommended neither in a neonatal/pediatric population nor at home, because of narrow therapeutic index: overdose causes cardiac/respiratory arrests (Michael Jackson's death), requiring machines only available in hospitals.

Further work will deal with anesthetic alcohols/propofol family, for example, 4-iodopropofol (Fig. 15.7d, 2,6-diisopropyl-4-iodophenol), which has similar effects to propofol on isolated receptors, acting as γ-aminobutyric-acid type-A positive modulator/Na^+-channel blocker but, in animals, it has anxiolytic/anticonvulsant effects, lacking propofol sedative/hypnotic profile.

ACKNOWLEDGMENTS

Francisco Torrens belongs to the Institut Universitari de Ciència Molecular, Universitat de València. Gloria Castellano belongs to the Departamento de Ciencias Experimentales y Matemáticas, Facultad de Veterinaria y Ciencias Experimentales, Universidad Católica de Valencia *San Vicente Mártir*. The authors thank support from the Spanish Ministerio de Economía y Competitividad (Project No. BFU2013-41648-P), EU ERDF and Universidad Católica de Valencia *San Vicente Mártir* (Project No. PRUCV/2015/617).

KEYWORDS

- equivalence
- random distribution
- real norms
- equiprobable
- hydrophilicity

REFERENCES

1. Shannon, C. E. A Mathematical Theory of Communication: Part I. Discrete Noiseless Systems. *Bell Syst. Tech. J.* **1948,** *27,* 379–423.
2. Shannon, C. E. A Mathematical Theory of Communication: Part II. The Discrete Channel with Noise. *Bell Syst. Tech. J.* **1948,** *27,* 623–656.
3. Torrens, F.; Castellano, G. Reflections on the Nature of the Periodic Table of the Elements: Implications in Chemical Education. In *Synthetic Organic Chemistry*; Seijas, J. A., Vázquez Tato, M. P., Lin, S.-K., Eds.; MDPI: Basel, 2015; vol. 18, pp 1–15.
4. Torrens, F.; Castellano, G. Molecular Classification of Thiocarbamates with Cytoprotection Activity against Human Immunodeficiency Virus. *Int. J. Chem. Model.* **2011,** *3,* 269–296.
5. Castellano, G.; Tena, J.; Torrens, F. Classification of Polyphenolic Compounds by Chemical Structural Indicators and Its Relation to Antioxidant Properties of *Posidonia oceanica* (L.) Delile. *MATCH Commun. Math. Comput. Chem.* **2012,** *67,* 231–250.
6. Torrens, F.; Castellano, G. Molecular Diversity Classification *via* Information Theory: A Review. *ICST Trans. Complex Syst.* **2012,** *12* (10–12), e4.1–8.
7. Torrens, F.; Castellano, G. Information Theoretic Entropy for Molecular Classification: Oxadiazolamines as Potential Therapeutic Agents. *Curr. Comput.-Aided Drug Des.* **2013,** *9,* 241–253.
8. Torrens, F.; Castellano, G. Molecular Classification of 5-Amino-2-aroylquinolines and 4-Aroyl-6,7,8-Trimethoxyquinolines as Highly Potent Tubulin Polymerization Inhibitors. *Int. J. Chemoinf. Chem. Eng.* **2013,** *3* (2), 1–26.
9. Castellano, G.; González-Santander, J. L.; Lara, A.; Torrens, F. Classification of Flavonoid Compounds by Using Entropy of Information Theory. *Phytochemistry* **2013,** *93,* 182–191.
10. Castellano, G.; Lara, A.; Torrens, F. Classification of Stilbenoid Compounds by Entropy of Artificial Intelligence. *Phytochemistry* **2014,** *97,* 62–69.
11. Torrens Zaragozá, F. Classification of Lactic Acid Bacteria against Cytokine Immune Modulation. *Nereis* **2014,** (6), 27–37.
12. Torrens, F.; Castellano, G. Molecular Classification of Pesticides Including Persistent Organic Pollutants, Phenylurea and Sulphonylurea Herbicides. *Molecules* **2014,** *19,* 7388–7414.
13. Torrens, F.; Castellano, G. Molecular Classification of Styrylquinolines as Human Immunodeficiency Virus Integrase Inhibitors. *Int. J. Chem. Model.* **2014,** *6,* 347–376.
14. Torrens Zaragozá, F. Classification of Fruits Proximate and Mineral Content: Principal Component, Cluster, Meta-analyses. *Nereis* **2015,** *7,* 39–50.
15. Castellano, G.; Torrens, F. Information Entropy-Based Classification of Triterpenoids and Steroids from *Ganoderma*. *Phytochemistry* **2015,** *116,* 305–313.
16. Torrens Zaragozá, F. Classification of Food Spices by Proximate Content: Principal Component, Cluster, Meta-Analyses. NEREIS **2016,** *8,* 23–33. ISSN: 1888-8550.
17. Urdă A.; Marcu I.-C. Catalysis. *The Explicative Dictionary of Nanochemistry*; Putz, M. V., Ed.; Apple Academic Press & CRC Press of Taylor & Francis: Waretown, 2015; pp 17–19.
18. Torrens, F.; Castellano, G. Periodic Classification of Local Anaesthetics (Procaine Analogues). *Int. J. Mol. Sci.* **2006,** *7,* 12–34.
19. Castellano Estornell, G.; Torrens Zaragozá, F. Local Anaesthetics Classified Using Chemical Structural Indicators. *Nereis* **2009,** *2,* 7–17.

20. Torrens, F.; Castellano, G. Using Chemical Structural Indicators for Periodic Classification of Local Anaesthetics. *Int. J. Chemoinf. Chem. Eng.* **2011,** *1* (2), 15–35.
21. Torrens, F.; Castellano, G. AIDS destroys Immune Defences: Hypothesis. *New Front. Chem.* **2014,** *23*, 11–20.
22. Covino, B. G. Local Anesthesia. *N. Engl. J. Med.* **1972,** *286*, 975–983.
23. Covino, B. G. Local Anesthetic Agents for Peripheral Nerve Blocks. *Anaesthesist* **1980,** *29* (7), 33–37.
24. Covino, B. G. Pharmacology of Local Anaesthetic Agents. *Br. J. Anaesth.* **1986,** *58*, 701–716.
25. Jaynes, E. T. Information Theory and Statistical Mechanics. *Phys. Rev.* **1957,** *106*, 620–630.
26. Dewar, R. C. Information Theory Explanation of the Fluctuation Theorem, Maximum Entropy Production, and Self-Organized Criticality in Non-Equilibrium Stationary States. *J. Phys. A* **2003,** *36*, 631–641.
27. Iordache, O. M.; Maria, G. C.; Pop, G. L. Lumping Analysis for the Methanol Conversion to Olefins Kinetic Model. *Ind. Eng. Chem. Res.* **1988,** 27, 2218–2224.
28. Iordache, O.; Bucurescu, I.; Pascu, A. Lumpability in Compartmental Models. *J. Math. Anal. Appl.* **1990,** 146, 306–317.
29. Corriou, J. P.; Iordache, O.; Tondeur, D. Classification of Biomolecules by Information Entropy. *J. Chim. Phys. Phys.—Chim. Biol.* **1991,** 88, 2645–2652.
30. Iordache, O.; Corriou, J. P.; Tondeur, D. Neural Network for Systemic Classification and Process Fault Detection. *Hung. J. Ind. Chem. Veszprém* **1991,** *199*, 265–274.
31. Iordache, O.; Corriou, J. P.; Garrido-Sanchez, L.; Fonteix, C.; Tondeur, D. Neural Network Frames: Application to Biochemical Kinetic Diagnosis. *Comput. Chem. Eng.* **1993,** *17*, 1101–1113.
32. Iordache, O.; Corriou, J. P.; Tondeur, D. Separation Sequencing: Use of Information Distance. *Can. J. Chem. Eng.* **1993,** *71*, 955–966.
33. Iordache, O. *Modeling Multi-level Systems*; Springer: Berlin, 2011.
34. Iordache, O. *Self-Evolvable Systems: Machine Learning in Social Media*; Springer: Berlin, 2012.
35. Iordache, O. *Polytope Projects*; CRC: Boca Raton, FL, 2014.
36. Clegg, D. E.; Massart, D. L. Information Theory and its Application to Analytical Chemistry. *J. Chem. Educ.* **1993,** *70*, 19–24.
37. Willett, P. Chemical Similarity Searching. *J. Chem. Inf. Comput. Sci.* **1998,** *38*, 983–996.
38. Tondeur, D.; Kvaalen, E. Equipartition of Entropy Production: An Optimality Criterion for Transfer and Separation Processes. *Ind. Eng. Chem. Res.* **1987,** *26*, 50–56.
39. Glansdorff, P.; Prigogine, I. *Thermodynamic Theory of Structure, Stability, and Fluctuations*. Wiley: London, 1971.
40. Varmuza, K. *Pattern Recognition in Chemistry*. Springer: New York, 1980.
41. Pierce, T. H.; Hohne, B. A. *Artificial Intelligence: Applications in Chemistry*; ACS Symposium Series 306. ACS: Washington, 1986.
42. Hohne, B. A.; Pierce, T. H. *Expert Systems: Applications in Chemistry*; ACS Symposium Series 408. ACS: Washington, 1989.
43. Benzécri, J. P. *L'Analyse des Données I. La Taxonomie*; Dunod: Paris, 1980.
44. Salas, M. D. Personal communication.
45. Cortina, B.; Blasco, C.; Fajardo, S.; Vazquez, M.; Beltran, A.; Benavent, A.; Ferrús, J. *Book of Abstracts, Encuentro de Investigadores en Cáncer: Dando la Cara por la Sociedad*, Alcoi (Alacant), Spain, March 15, 2016. FISABIO: València, Spain, 2016; p 1.

46. De la Iglesia, M.; Martínez, C. Personal communication.
47. Van Horn, J. D.; Grafton, S. T.; Rockmore, D.; Gazzaniga, M. S. Sharing Neuroimaging Studies of Human Cognition. *Nat. Neurosci.* **2004,** *7,* 473–481.
48. Van Horn, J. D.; Toga, A. W. Human Neuroimaging as a *Big Data* Science. *Brain Imag. Behav.* **2014,** *8,* 323–331.
49. Perkel, J. Making Sense of Big Data. *BioTechniques* **2016,** *60,* 108–112.
50. Collantes, J. M. Personal communication.
51. López-Otín, C. Personal communication.
52. Schrödinger, E. *What is life?* Cambridge University: Cambridge, 1944.
53. Miranda, M. A.; Lhiaubet-Vallet, V. Personal communication.
54. Eritja, R. Personal communication.
55. Bensuade-Vincent, B.; Simon, J. *Chemistry: The Impure Science.* Imperial College Press: London, 2012.
56. Lloret, F. Personal communication.
57. Geim, A. K.; Grigorieva, I. V. Van der Waals Heterostructures. *Nature (London)* **2013,** 499, 419–425.
58. Puzyn, T.; Rasulev, B.; Gajewicz, A.; Hu, X.; Dasari, T. P.; Michalkova, A.; Hwang, H. M.; Toropov, A.; Leszczynska, D.; Leszczynski, J. Using nano-QSAR to Predict the Cytotoxicity of Metal Oxide Nanoparticles. *Nat. Nanotechnol.* **2011,** *6,* 175–178.
59. Gray, T. *Molecules: The Elements and the Architecture of Everything*; Black Dog & Leventhal: New York, 2014.
60. Grisolía, S.; de Andrés, J. *XI Jornada Anual: El Dolor Es Evitable. El Sistema Gastrointestinal,* Fundación Valenciana de Estudios Avanzados: València, 2015.
61. Díaz Insa, S. *Jornada sobre Cefaleas y Migrañas,* Fundación Valenciana de Estudios Avanzados: València, 2016.
62. Duncan, R. Personal communication.
63. Pérez-Alonso, M. Personal communication.
64. Cremades, E. Personal communication.
65. Stokes, D. E. *Pasteur's Quadrant—Basic Science and Technological Innovation.* Brookings Institution Press: Washington, 1997.
66. Chalmers, I.; Bracken, M. B.; Djulbegovic, B.; Garattini, S.; Grant, J.; Gülmezoglu, A. M.; Howells, D. W.; Ioannidis, J. P. A.; Oliver, S. Research: Increasing Value, Reducing Waste. 1. How to Increase Value and Reduce Waste When Research Priorities Are Set. *Lancet* **2014,** *383,* 156–165.
67. Martínez González, L. J. Personal communication.

CHAPTER 16

ENTROPIC FACTORS CONFORMATIONAL INTERACTIONS

G. A. KORABLEV[1], V. I. KODOLOV[2], YU. G. VASILIEV[3,4],
N. N. NOVYKH[5], and G. E. ZAIKOV[6*]

[1]Department of Chemistry, Izhevsk State Agricultural Academy, Izhevsk, Udmurtskaja Respublika, Russia

[2]Department of Chemistry and Chemical Engineering, Izhevsk State Agricultural Academy, Izhevsk, Udmurtskaja Respublika, Russia

[3]Department of Physiology and Animal Sanitation, Izhevsk State Agricultural Academy, Izhevsk, Udmurtskaja Respublika, Russia

[4]Department of Histology, Cytology and Embryology, Izhevsk State Technical University, Izhevsk, Udmurtskaja Respublika, Russia

[5]Department of Anatomy and Biology, Izhevsk State Agricultural Academy, Izhevsk, Udmurtskaja Respublika, Russia

[6]Department of Chemistry, Emmanuel Institute of Biochemical Physics, RAS, Moscow, Russia

*Corresponding author. E-mail: GEZAIKOV@Yahoo.com

CONTENTS

ABSTRACT

The concept of the entropy of spatial-energy interactions is used similarly to the ideas of thermodynamics on the static entropy. The idea of entropy appeared based on the second law of thermodynamics and ideas of the adduced quantity of heat. Such rules are general assertions independent of microscopic models. Therefore, their application and consideration can result in a large number of consequences which are most fruitfully used in statistic thermodynamics. In this research, we are trying to apply such regularities to self-organization assess the degree of spatial-energy interactions using their graphic dependence in the form of S-lines. The nomogram to assess the entropy of different processes is obtained. The variability of entropic S-line demonstrations is discussed in biophysical processes.

16.1 INTRODUCTION

Over than 50 years ago at the contest of witty graphs, a future academician and director of Institute of Biochemical Physics Prof. N. M. Emanuel drew an S-curve and wrote down only one word—"foster-mother." Such curve can be frequently found in many processes of chemical kinetics in general regularities of biosystem formation, as well as at the development stages of engineering systems. At the same time, the curves with one or more steps can be seen[5,6] and each subsystem corresponds to common progressive development of the initial system (Fig. 16.1).

In this figure, the period of "technological gap" corresponds to the transition to a new more rational organizational system. Thus, the process of system self-organization is reflected, which is the main process of structure formation by time. Such systems can comprise live systems at the cellular level or structures formed due to interatomic interactions, for example, in the process of crystal formation.

"Joint (cooperative) movement of large groups of molecules is common in all phenomena of ordered structure formation during irreversible processes in strongly non-equilibrium systems."[1] The areas of particles statistically different by their characteristics appear and disappear, that is, the fluctuation takes place. With such collective interactions, the resonance interactions of two or more particles called "bifurcation" are possible. "This process is no longer microscopic and results in macroscopic effect—self-organization."[2]

According to the evolution criterion of Glensdorf–Prigozhin, the speed of change in entropy production, conditioned by the changes in thermodynamic forces, decreases and tends to zero, thus leading to the formation of ordered structures.[1]

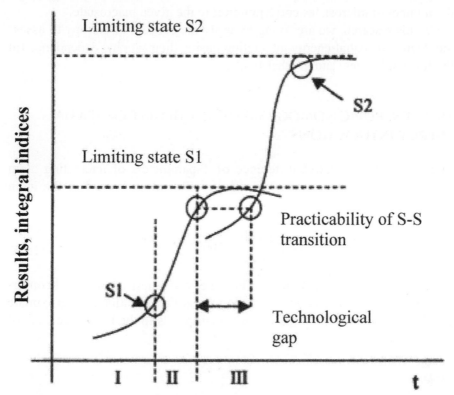

FIGURE 16.1   System transitions by time t.

Consequently, the concept of entropy can be a quantitative and functional characteristic of S-lines.

In statistic thermodynamics, the entropy (S) of the closed and equilibrious system equals the logarithm of the probability of its definite macrostate:

$$S = k \ln W \tag{16.1}$$

where W is the number of available states of the system or degree of the degradation of microstates; k is the Boltzmann's constant.

This correlation is a general assertion of macroscopic character and does not contain any references to the structure elements of the systems considered and it is completely independent of microscopic models.[2]

Therefore, the application and consideration of these laws can result in a large number of consequences. The thermodynamic probability W is the main characteristic of the process. The thermodynamic probability equals the number of microstates corresponding to the given macrostate.

In this research, we are trying to apply the concept of entropy to assess the degree of spatial-energy interactions using their graphic dependence (in the form of S-lines) and in other fields.

16.2 ENTROPIC NOMOGRAM OF THE DEGREE OF SPATIAL-ENERGY INTERACTIONS

The value of the relative difference of P-parameters of interacting atom components—the structural interaction coefficient α is used as the main quantitative characteristic of structural interactions in condensed media[9]:

$$\alpha = \frac{P_1 - P_2}{(P_1 + P_2)/2} 100\% \tag{16.2}$$

Based on this equation, the maximum effectiveness of structural conformations takes place at paired interaction under the condition of approximate equality of parameters P_1 and P_2 that corresponds to the resonance state of this process.

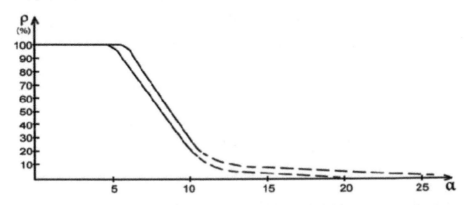

FIGURE 16.2 Nomogram of structural interaction degree dependence (ρ) on coefficient α.

Applying the reliable experimental data, we obtain the nomogram of structural interaction degree dependence (ρ) on coefficient α, the same for a wide range of structures (Fig. 16.1). This approach gives the possibility to evaluate the degree and direction of the structural interactions of phase formation, isomorphism and solubility processes in multiple systems, including molecular ones.

Such nomogram can be demonstrated[9] as a logarithmic dependence:

$$\alpha = \beta \, (\ln \rho)^{-1}, \tag{16.3}$$

where coefficient β is the constant value for the given class of structures. β can structurally change mainly within ±5% from the average value. Thus, coefficient α is reversely proportional to the logarithm of the degree of structural interactions and therefore can be characterized as the entropy of spatial-energy interactions of atomic-molecular structures.

Actually, the more the ρ is, the more probable is the formation of stable ordered structures (e.g., the formation of solid solutions), that is, the less is the process entropy, but also the less is coefficient α.

Thus, the relative difference of spatial-energy parameters of the interacting structures can be a quantitative characteristic of the interaction entropy: $\alpha = S$ (in Fig. 16.2).

In the chapter,[3] thus, the carbonization rate, as well as the functions of many other physical–chemical structural interactions, can be assessed via the values of the calculated coefficient α and entropic nomogram.

16.3 NOMOGRAMS OF BIOPHYSICAL PROCESSES

1. On the kinetics of fermentative processes

For ferments with stoichiometric coefficient n not equal 1, the type of graphical dependence of the reaction product performance rate (μ) depending on the substrate concentration (c) has[7] a sigmoid character with the specific bending point (in Fig. 16.1).

The graph of the dependence of electron transport rate in biostructures on the diffusion time period of ions is similar (p 278).[7]

2. Dependence of biophysical criteria on their frequency characteristics

a) The passing of alternating current through live tissues is characterized by the dispersive curve of electrical conductivity—this is the graphical dependence of the tissue total resistance (z-impedance) on the alternating current frequency logarithm (log ω). Normally, such curve, on which the impedance is plotted on the coordinate axis, and log ω—on the abscissa axis, formally, completely corresponds to the entropic nomogram (Fig. 16.2).

b) The fluctuations of biomembrane conductivity (conditioned by random processes) "have the form of Lorentz curve"[8] (p. 99). In this graph, the fluctuation spectral density (ρ) is plotted on the coordinate axis, and the frequency logarithm function (log ω)—on the abscissa axis. The type of such curve also corresponds to the entropic nomogram in Figure 16.2.

16.4   S-CURVES ("LIFE LINES")

Already in the last century, some general regularities in the development of some biological systems depending on time (growth in the number of bacteria colonies, population of insects, weight of the developing fetus, etc.) were found.[5,6] The curves reflecting this growth were similar, first of all, by the fact that three successive stages could be rather vividly emphasized on each of them: slow increase, fast burst-type growth, and stabilization (sometimes decrease) of number (or another characteristic). Later, it was demonstrated that engineering systems go through similar stages during their development. The curves drawn up in coordinate system where the numerical values of one of the most important operational characteristics (e.g., aircraft speed, electric generator power, etc.) were indicated along the vertical and the "age" of the engineering system or costs of its development along the horizontal were called S-curves (by the curve appearance), and they are sometimes also called "life lines."

16.5   ENTROPIC TRANSITIONS IN PHYSIOLOGICAL SYSTEMS

In actual processes, the resources often come to an end not because the system spent them, but because the new system appeared which starts to more effectively perform the similar function and attracts the resources to itself (Fig. 16.1).[6]

This is characteristic not only for short-term acting processes but also for general development of interstructural and cellular interactions.

It is much more complicated to consider and mathematically analyze heteromorphic and rather dynamic intercellular and cellular–cellular interactions.

Neurogenesis can serve as an example of such structural transformations in the organization of biological system.[4] In particular, let us consider the neurogenesis on the example of motor nucleus of trifacial nerve. It is known that on the 10–12th day of embryogenesis, rats are characterized by high mitotic activity. In this period, the after-brain is composed of externally isomorphic populations of medulloblasts different on molecular and submolecular levels. During the indicated period, the size and shape of cells are approximately the same and variety of sizes of cellular populations in mantle layer is limited. From the 12th day, the processes of neuroblast immigration become activated that are followed by the end of their proliferative activity and activation of axon growth.

At this moment, the entropy is demonstrated in dividing the cell groups into mitotically active populations and neuroblasts with different number of arms, which stopped splitting. As a result, at the moment of birth the variety of sizes of nerve cells reach considerable values from 7 to 8 mcm in diameter in small neurons and neuroblasts and up to 25–30 mcm in large neurons. The differences are also revealed in the number of arms, degree of morphological maturity of nerve cells. By the age of 9 months, a pubertal animal demonstrates the stagnation processes with externally rich variety of neurons inside the nucleus. This tendency of anti-entropy growth proceeds in accordance with Figure 16.2 and is demonstrated in Table 16.1. The factors are compared with a pubertal rat. At the same time, there are vividly expressed methods inside each population with the distribution curve close to a normal one.

TABLE 16.1 Ratio of Types of Neuroblastic Cells in Motor Nucleus of Trifacial Nerve for Medium Rats ($M \pm m$).

Development periods	Percentage of cell type ratio
Newly born	59.0 ± 2.6
1 week	67.4 ± 3.2
1 month	40.8 ± 3.2
Pubertal	26.6 ± 2.3

Thus, in a complex biological object—a mammal brain—the moments in the development of cellular populations are found when the significant dispersion is observed followed by temporary manifestations of entropy increase, which is in compliance with node points of accelerated development and transition to new qualitative change in the population composition (Fig. 16.1).

It is the increase in the structure variety, and thus, controllable transient enhancement of the system chaotic character that can be the basis initiating the transition to new states, to the development of certain cells, cell populations in general (in accordance with Fig. 16.1).

Apparently, the self-organizing processes generally follow the same principle: slow development from the structural variety, fast growth and stabilization of the renewed biosystem. In such way, the nature is struggling with entropy development in organism, maintaining it on the constant level as the main condition of stationary state.

16.6 GENERAL CONCLUSION

Entropic S-lines have diversified manifestation in conformational interactions biophysical systems.

KEYWORDS

- self-organization
- S-lines
- entropy
- spatial-energy parameter
- biophysical processes

REFERENCES

1. Bochkarev, A. I.; Bochkareva, T. S.; Saksonov, S. V. *Concept of Modern Natural Science.* http://allrefs.net/c12/4e3ae/p94/.
2. Gribov, L. A.; Prokofyeva, N. I. Basics of Physics. *M.: Vysshaya shkola*; 1992; 430 p.
3. Kodolov, V. I.; Khokhriakov, N. V.; Trineeva, V. V.; Blagodatskikh, I. I. Activity of Nanostructures and its Manifestation in Nanoreactors of Polymeric Matrixes and Active Media. *Chem. Phys. Mesosc.* **2008,** *10* (4), 448–460.
4. Kuzin, A. V.; Vasiliev, Yu. G.; Chuchkov, V. M.; Shorokhova, T. G. *Ensemble Interactions in Central Nerve System*; ANK Publishers: Izhevsk-Berlin, 2004; 160 p.
5. Kynin, A. T.; Lenyashin, V. A. *Evaluation of Parameters of Engineering Systems with Growth Curves.* http://www.metodolog.ru/01428/01428.html.
6. Lyubomirsky, A.; Litvin, S. *Laws of Engineering System Development.* http://www.metodolog.ru/00767/00767.html.

7. Rubin, A. B. Biophysics. Book 1. Theoretical Biophysics. *M.: Vysshaya shkola*; 1987; 319 p.
8. Rubin, A. B. Biophysics. Book 2. Biophysics of Cell Processes. *M.: Vysshaya shkola*; 1987; 303 p.
9. Korablev, G. A. *Spatial-Energy Principles of Complex Structures Formation*; Brill Academic Publishers and VSP: Netherlands, 2005; 426 pp (monograph).

CHAPTER 17

APPLICATION OF ARTIFICIAL NEURAL NETWORKS AND A METAHEURISTIC ALGORITHM IN APPLIED CHEMISTRY AND CHEMICAL ENGINEERING

SHAHRIAR GHAMMAMY[1*] and MAHDI GHAVAMI[2]

[1]*Department of Chemistry, Faculty of Science, Imam Khomeini International University, Ghazvin, Iran*

[2]*Department of Mechanical Engineering, Isfahan University of Technology, Isfahan, Iran*

Corresponding author. E-mail: shghamami@yahoo.com

CONTENTS

ABSTRACT

This chapter presents a framework that aims to facilitate the modeling, design, and optimization of phosphoric acid production processes to propose solutions that improve the performance of the existing production technologies.

A neural network or artificial neural network (ANN) is a powerful computational data model that is able to capture and represent complex input/output relationships. The idea of ANNs is based on the belief that working of human brain by making the right connections. The human brain is composed of 100 billion nerve cells called neurons. They are connected to other thousand cells by axons. Inputs electrical impulses can travel through the neural network. A neuron can send the message to other neuron to handle the issue or does not send it forward. ANNs are composed of multiple nodes, which imitate biological neurons of human brain. The neurons are connected by links and they interact with each other. They come in many shapes and sizes. Their shapes and connections help them carry out specialized functions, such as storing memories or controlling muscles. The nodes can take input data and perform simple operations on the data (Fig. 17.1).

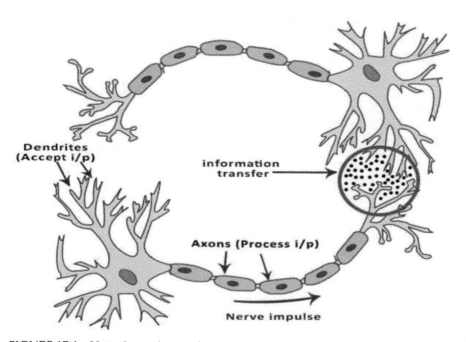

FIGURE 17.1   Natural neural network.

The result of these operations is passed to other neurons. The output at each node is called its activation or node value. Each link is associated with weight.

Computer scientists have long been inspired by the human brain. In 1943, Warren S. McCulloch, a neuroscientist, and Walter Pitts, a logician, developed the first conceptual model of an ANN. In their paper, "A logical calculus of the ideas imminent in nervous activity," they described the concept of a neuron, a single cell living in a network of cells that receives inputs, processes those inputs, and generates an output.[1]

The simplest definition of an ANN is provided by the inventor of one of the first neurocomputers, Dr. Robert Hecht-Nielsen. He defines a neural network as:

> ...a computing system made up of a number of simple, highly interconnected processing elements, which process information by their dynamic state response to external inputs.

Although the mathematics involved with neural networking is not a trivial matter, a user can rather easily gain at least an operational understanding of their structure and function.

To better understand artificial neural computing, it is important to know first how a conventional "serial" computer and its software process information. A serial computer has a central processor that can address an array of memory locations where data and instructions are stored. Computations are made by the processor reading an instruction as well as any data the instruction requires from memory addresses, the instruction is then executed and the results are saved in a specified memory location as required. In a serial system (and a standard parallel one as well), the computational steps are deterministic, sequential, and logical, and the state of a given variable can be tracked from one operation to another.

In comparison, ANNs are not sequential or necessarily deterministic. There are no complex central processors, rather there are many simple ones which generally do nothing more than take the weighted sum of their inputs from other processors. ANNs do not execute programed instructions, they respond in parallel (either simulated or actual) to the pattern of inputs presented to it. There are also no separate memory addresses for storing data. Instead, information is contained in the overall activation "state" of the network. "Knowledge" is thus represented by the network itself, which is quite literally more than the sum of its individual components.

17.1   ARCHITECTURE OF NEURAL NETWORKS

Neural networks are typically organized in layers. Layers are made up of a number of interconnected "nodes" which contain an "activation function." Patterns are presented to the network via the "input layer," which communicates to one or more "hidden layers" where the actual processing is done via a system of weighted "connections." The hidden layers then link to an "output layer" where the answer is output. More graphically, the process looks something like as presented in Figure 17.2.

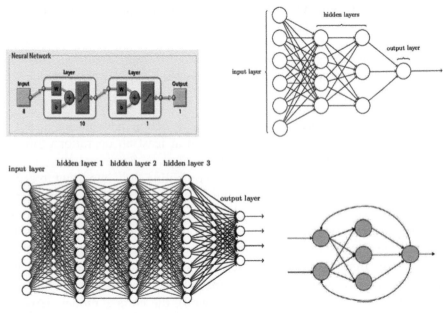

FIGURE 17.2   Neural network structure.[2]

In the topology diagrams shown, each arrow represents a connection between two neurons and indicates the pathway for the flow of information. Each connection has a weight, an integer number that controls the signal between the two neurons. Most neural networks are fully connected, which means each hidden unit and each output unit is connected to every unit in the layers either side.

ANNs are capable of learning, which takes place by altering weight values. The weights in most ANN can be both negative and positive, therefore, providing excitatory or inhibitory influences to each input. As each input enters the neuron, it is multiplied by its weight. The neuron then sums all these new input values which gives us the activation (a floating point number which can be negative or positive). If the activation is greater than a threshold, the neuron outputs a signal. If the activation is less than the threshold, the neuron outputs zero (Fig. 17.3).

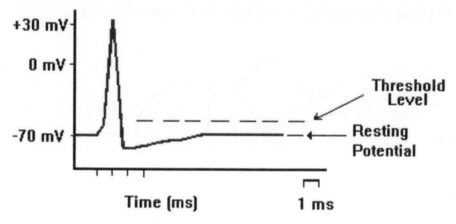

FIGURE 17.3 Neuron threshold level.[1]

ANNs learn by example as do their biological counter parts. Although there are many different kinds of learning rules used by neural networks, this demonstration is concerned only with one: the delta rule. The delta rule is often utilized by the most common class of ANNs called "back-propagational neural networks". Back-propagation is an abbreviation for the backward propagation of error.[3]

With the delta rule, as with other types of back-propagation, "learning" is a supervised process that occurs with each cycle or "epoch" (each time the network is presented with a new input pattern) through a forward activation flow of outputs, and the backward error propagation of weight adjustments. More simply, when a neural network is initially presented with a pattern, it makes a random "guess" as to what it might be, it then sees how far its answer was from the actual one and makes an appropriate adjustment to its connection weights.

If the network generates a "good or desired" output, there is no need to adjust the weights. However, if the network generates a "poor or undesired" output or an error, then the system alters the weights to improve subsequent results.

The inputs are fed into the input layer and get multiplied by interconnection weights as they are passed from the input layer to the first hidden layer.

Within the first hidden layer, they get summed then processed by a nonlinear function (usually the hyperbolic tangent).[3] As the processed data leaves the first hidden layer, again it gets multiplied by interconnection weights, then summed and processed by the second hidden layer. Finally, the data are multiplied by interconnection weights then processed one last time within the output layer to produce the neural network output (Fig. 17.4).

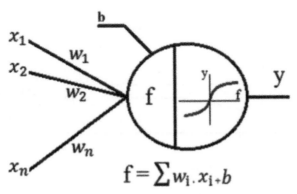

FIGURE 17.4 Neuron function.[2]

A typical neural network has anything from a few dozen to hundreds, thousands, or even millions of artificial neurons. Neural network analysis often requires a large number of individual runs to determine the best solution. Most learning rules have built-in mathematical terms to assist in this process which control the "speed" (beta-coefficient) and the "momentum" of the learning. The speed of learning is actually the rate of convergence between the current solution and the global minimum. Momentum helps the network to overcome obstacles (local minimum) in the error surface and settle down at or near the global minimum.

The input data are repeatedly presented to the neural network. With each presentation the output of the neural network is compared to the desired output and an error is computed. This error is then fed back (back-propagated) to the neural network and used to adjust the weights such that the error decreases with each iteration and the neural model gets closer and closer to produce the desired output. This process is known as "training." The training data set includes a number of cases, each containing values for a range of input and output variables. The number of cases required for neural network training frequently presents difficulties. There are some heuristic guidelines, which relate the number of cases needed to the size of the network (the simplest of these says that there should be 10 times as many cases as connections in the network). Actually, the number needed is also related to the complexity of the underlying function which the network is trying to model and to the variance of the additive noise. As the number of variables increases, the number of cases required increases nonlinearly, so that with even a fairly small number of variables, a huge number of cases are required.

For most practical problem domains, the number of cases required will be hundreds or thousands. For very complex problems more may be required, but it would be a rare (even trivial) problem which required less· than a hundred cases.

Once a neural network is "trained" to a satisfactory level, it may be used as an analytical tool on other data. To do this, the user no longer specifies any training runs and instead allows the network to work in forward propagation mode only. New inputs are presented to the input pattern where they filter into and are processed by the middle layers as though training were taking place, however, at this point, the output is retained and no back-propagation occurs. The output of a forward propagation run is the predicted model for the data which can then be used for further analysis and interpretation.

- Capturing associations or discovering regularities within a set of patterns.
- Where the volume, number of variables or diversity of the data are very great.
- The relationships between variables are vaguely understood.
- The relationships are difficult to describe adequately with conventional approaches.

There are many advantages and limitations to neural network analysis and to discuss this subject properly we would have to look at each individual type of network, which isn't necessary for this general discussion. Neural networks are in a sense the ultimate "black boxes." Apart from defining the general structure of a network and perhaps initially seeding it with a random numbers, the user has no other role than to feed it input and watch it train and await the output. In fact, it has been said that "you almost don't know what you're doing." The final product of this activity is a trained network that provides no equations or coefficients defining a relationship. The only available information at the end of the training process are the updated weights matrixes that could help define the mathematical relationship between input and output by knowing of network structure.[1]

There are several strategies for learning, such as

- **Supervised learning:** Essentially, a strategy that involves a teacher that is smarter than the network itself. For example, let's take the facial recognition example. The teacher shows the network a bunch of faces, and the teacher already knows the name associated with each face. The network makes its guesses; then, the teacher provides the network with the answers. The network can then compare its answers to the known "correct" ones and make adjustments according to its errors. Our first neural network in the next section will follow this model.

- **Unsupervised learning:** It is required when there isn't an example data set with known answers. Imagine searching for a hidden pattern in a data set. An application of this is clustering, dividing a set of elements into groups according to some unknown pattern. We won't be looking at any examples of unsupervised learning in this chapter, as this strategy is less relevant for our examples.

- **Reinforcement learning:** A strategy built on observation. Think of a little mouse running through a maze. If it turns left, it gets a piece of cheese; if it turns right, it receives a little shock. Presumably, the mouse will learn over time to turn left. Its neural network makes a decision with an outcome (turn left or right) and observes its environment. If the observation is negative, the network can adjust its weights to make a different decision the next time. Reinforcement learning is common in robotics; the robot performs a task and observes the results. Did it crash into a wall or fall off a table? Or is it unharmed? We'll look at reinforcement learning in the context of our simulated steering vehicles.

This ability of a neural network to learn, to make adjustments to its structure over time, is what makes it so useful in the field of artificial intelligence.

Overtraining is the most typical problem when we create neural-network-based models. One way to avoid overfitting is to use a lot of data. The main reason overfitting happens is because you have a small data set (training data) and you try to learn from it. The algorithm will have greater control over the data and it will make sure it satisfies all the data points. But if you have a million data points, then the algorithm is forced to generalize and come up with a good model that suits all the points (Fig. 17.5).

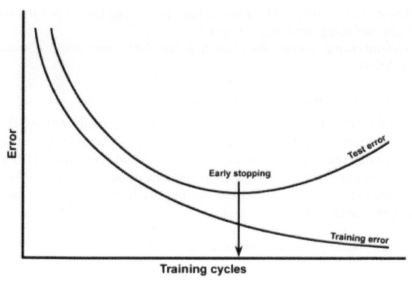

FIGURE 17.5   Error convergence and divergence.

There are several methods available to check the degree of generalization and/or to detect overfitting:

- cross-validation,
- noise addition, and
- validation set.

We don't have the luxury of gathering a large database all the time. Sometimes we are limited to a small database and we are forced to come up with a model based on that. In these situations, a technique is used called cross validation. Only the data points in the training data set are used to come up with the model and the testing data set is used to test how good the model is. This is repeated with different partitions of training and testing data sets. This method gives a fairly good estimate of the underlying model. Another method is validation set, when monitoring the performance of the neural network on an independent data set while it is trained, the training process can be stopped when its performance on the independent data set is optimal. This validation set has to be an independent data set.

The true power and advantage of neural networks lies in their ability to represent both linear and nonlinear relationships and in their ability to learn these relationships directly from the data being modeled. Traditional linear models are simply inadequate when it comes to modeling data that contains nonlinear characteristics. Of course, character recognition is not the only problem that neural networks can solve.

The following paragraphs contain some commonly asked questions about ANNs.

When should neural networks be used?

To avoid solving simple tasks with complex models, traditional statistical methods dealing with linear mappings should be exploited first. To check whether nonlinear methods provide better results and reveal more information, experiments with simple standard versions of neural networks should be conducted. Later, the model can be improved by adding more layers, more units, or feedback loops.

How many data points are necessary to train a neural network?

Since neural networks are trained through examples, large data sets are required. Before starting the experiments, try to collect as many examples as possible. Especially, models with many degrees of freedom (many connections in the network) require a large number of examples. There exist heuristics for finding out the maximum number of degrees of freedom for a given number of examples, or the minimum number of required examples for a given number of degrees of freedom. However, this criterion is hardly ever met in practice. When the available data set is not large enough, the results are not reliable.

What about special cases in the training data set?

Usually, the neural network does not learn to treat the special cases correctly, because they are not presented to the network often enough. The neural network takes the statistical distribution of the data into account and tends to neglect outliers.

Can ANNs be retrained with new examples?

Whether this is possible depends on the neural network model, the multi-layer perceptron is not well suited to retraining. Starting from scratch is usually faster and provides better results. Other models may be applicable to the additional integration of examples.

Can a neural network handle several tasks at once?

Whether this is reasonable depends on the tasks. If the tasks are closely related, this can improve the performance, because the weights leading to the hidden units prestructure the task appropriately. However, tasks which are too different usually interfere. In general, use separate neural networks with single output units for each task. This provides a better overview and allows smaller networks to be used.

How many hidden units should be used?

The hidden units prestructure the inputs so that they are useful for solving the task. The lowest number of hidden layer should be used units as possible. When there are many hidden units, the network tends to adapt too well to the training set. Thus, it is less suited to generalizing. The removal of a single hidden unit considerably reduces the size of the network (and thus the number of degrees of freedom), because a single hidden unit is connected with all the input units and all the output units.

What kind of real-world problems can neural networks solve?

Neural networks have been successfully applied to broad spectrum of data-intensive applications, such as

- **Character recognition**—The idea of character recognition has become very important as handheld devices like the Palm Pilot are becoming increasingly popular. Neural networks can be used to recognize handwritten characters.
- **Process modeling and control**—Creating a neural network model for a physical plant then using that model to determine the best control settings for the plant.
- **Machine diagnostics**—Detect when a machine has failed so that the system can automatically shut down the machine when this occurs.
- **Portfolio management**—Allocate the assets in a portfolio in a way that maximizes return and minimizes risk.
- **Image compression**—Neural networks can receive and process vast amounts of information at once, making them useful in image compression. With the Internet explosion and more sites using more images on their sites, using neural networks for image compression is worth a look.
- **Target recognition**—Military application which uses video and/or infrared image data to determine if an enemy target is present.
- **Medical diagnosis**—Assisting doctors with their diagnosis by analyzing the reported symptoms and/or image data such as MRIs or X-rays.

- **Credit rating**—Automatically assigning a company's or individual's credit rating based on their financial condition.
- **Targeted marketing**—Finding the set of demographics which have the highest response rate for a particular marketing campaign.
- **Voice recognition**—Transcribing spoken words into ASCII text.
- **Financial forecasting**—Using the historical data of a security to predict the future movement of that security.
- **Quality control**—Attaching a camera or sensor to the end of a production process to automatically inspect for defects.
- **Intelligent searching**—An internet search engine that provides the most relevant content and banner ads based on the users past behavior.
- **Fraud detection**—Detect fraudulent credit card transactions and automatically decline the charge.

To facilitate understanding of the neural network, a practical example of chemical engineering is expressed.

17.2   CASE STUDY: PREDICTION OF PHOSPHORIC ACID PRODUCTION (WET PROCESS)

Phosphoric acid production due to its different uses and applications in various industrial sectors has been in a lot of attention and there are many different ways to produce it.[5] Available research efforts focus on phosphoric acid process modeling issues, while overlooking the prospect for improvements in the existing processes through optimization approaches. In this study, we have used mine soil phosphate for phosphoric acid production. Several experiments were performed and the percentage of purity achieved is different,[6-11] but one of the recent critical issues that have been of interest to many researchers, the production of phosphoric acid, is a high purity. There are many prediction methods to reduce the number of experiments to achieve the best results in the shortest time. This work has been performed by software and production design of phosphoric acid modeling and parameters and their effects have been studied. This work proposes a way that facilitates the modeling and design of phosphoric acid production processes. The result of this study makes a framework that aims to facilitate the modeling, design, and optimization of phosphoric acid production processes to propose solutions that improve the performance of the existing production technologies (Fig. 17.6).[4]

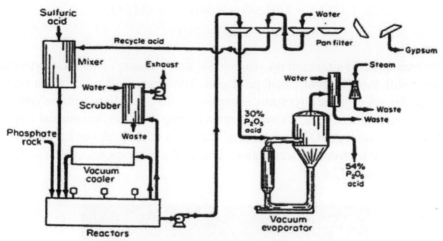

FIGURE 17.6   Phosphoric acid production process (wet process).[4]

The amount of sulfuric acid, water, soil and the amount of phosphate mining soil are parameters of the process. Of course, this isn't completely in software, but phosphoric acid produced in the laboratory and the information obtained give to the software for estimation by ANN. At the end of the prediction process, the amount of the selected parameters and the results are estimated successfully.

17.2.1   PHOSPHORIC ACID PRODUCTION

Phosphoric acid is produced in a reactor that facilitates the mixing and contact of phosphate rock with an aqueous solution of sulfuric and phosphoric acid. The phenomenon can be described by the following two-stage reaction. The phenomena associated with the kinetics of the reactor involve the dissolution of the phosphate rock into the reactor and the nucleation and crystal growth of the gypsum crystals. The equations used for the mathematical representation of the above phenomena are taken from Abu-Eishah and Nizar.[12]

$$Ca_{10}F_2(PO_4)_6 + 14H_3PO_4 \rightarrow 10Ca(H_2PO_4)_2 + 2HF\uparrow \qquad (17.1)$$

$$10Ca(H_2PO_4)_2 + 10H_2SO_4 + 20H_2O \rightarrow 20H_3PO_4 + 10CaSO_4 \cdot 2H_2O \quad (17.2)$$

Since the substances participating in the second reaction are the same materials produced in the first reaction, production of phosphoric acid process has expressed in an equation:

$$Ca_{10}F_2(PO_4)_6 + 10H_2SO_4 + 10nH_2O \rightarrow 6H_3PO_4$$
$$+ 10CaSO_4 \cdot nH_2O + 2HF\uparrow \qquad (17.3)$$

The main product stream of the reactor is a slurry stream containing gypsum mixed with the phosphoric acid produced. The slurry stream leads to a filter where the solid phase (gypsum) is separated from the liquid phase (phosphoric acid, sulfuric acid, water). Streams exiting the filter unit consist either of solid gypsum ($CaSO_4$) with a certain fraction of moisture or liquid phosphoric acid at a certain concentration. The production process can be seen in Figures 17.6 and 17.7.

FIGURE 17.7   An example of phosphoric acid and solid phase (gypsum) in lab.[4]

17.2.2   CHEMICALS AND REAGENTS

Estimation process is carried out with the help of ANNs. To achieve neural network training data, chemical reaction has been done by different input (water, soil, sulfuric acid, and mixing time). In laboratory, we mixed different rate of water, soil, and sulfuric acid by different time. After several minutes, reaction completed (Figs. 17.8 and 17.9).

Number	Water	Soil	Sulfuric acid	Time	Value of phosphoric acid	Purity percent
1	300	30	30	480	235	0.2
2	300	30	60	480	213	0.33
3	300	30	90	480	290	0.39
4	300	30	120	480	290	0.45
5	300	30	150	480	325	0.54
6	465	40	30	480	415	0.17
7	465	50	30	480	424	0.165
8	465	60	30	480	236	0.161
9	465	70	30	480	320	0.18
10	465	80	30	480	300	0.1722

FIGURE 17.8   Effective parameters of chemical reaction and outputs.[4]

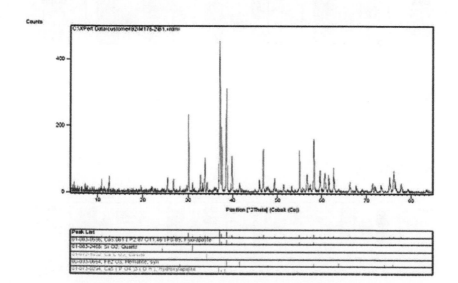

FIGURE 17.9   XRD results of phosphate concentrate raw material.

Compound	LOI	P_2O_5	CaO	MgO	Al_2O_3	SiO_2	Fe_2O_3	Y_2O_3	CeO_2	Cl
Percentage	0.56	40.97	48.55	1.69	0.29	3.19	3.28	0.19	1.01	0.27

17.2.3   MODELING WITH ARTIFICIAL NEURAL NETWORK

Considering four effective parameters including water rate, acid rate, soil rate, and time, which have more effects on phosphoric acid volume and purity percent, to recognize the first objective function (phosphoric acid volume), ANN is designed with four input neuron layers, two hidden layers with six and four neurons, and output layer with one neuron. Furthermore, to recognize the second objective function (purity percent), another neural network is designed with four input neuron layers, three hidden layers with five, seven, and four neurons, and output layer with one neuron. Figures 17.10 and 17.11 show the modeling of ANN for the first and the second objective functions.

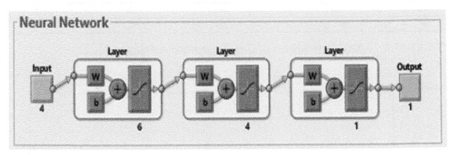

FIGURE 17.10　Topology of ANN for the first objective function.

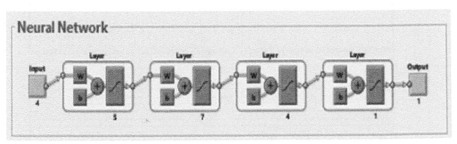

FIGURE 17.11　Topology of ANN for the second objective function.

Network topology is feed-forward type and the neuron activity function is of tangent sigmoid type. In addition, Levenberg–Marquart algorithm is used for synchronization of neural network weights.[2] More than 179 data (experimental test) are generated. Data are including amount of water, soil, and sulfuric acid; mixing time; the volume; and purity of phosphoric acid. Eighty percent of data are considered for network training and the remained data are used to test the trained network. First neural network estimate phosphoric acid volume.

The output of the trained neural network and experimental result for test data are shown in Figure 17.12.

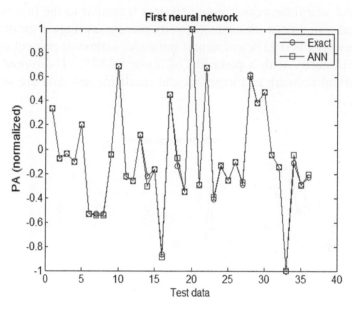

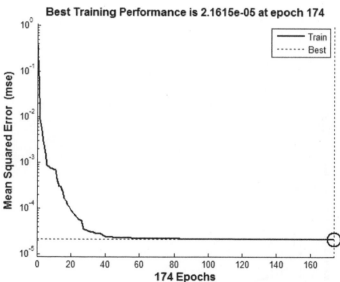

FIGURE 17.12 (a) training error diagram and (b) compliance between ANN and experiments for the first objective function (test data).

Average error of first neural networks is 0.49% and neural networks estimates mass and natural frequencies with a precision of about 99.5%.

The second neural network training process is similar to the first network. For second neural network estimate purity percent, average error of second neural networks is 0.23% and neural networks estimates percent of phosphoric acid purity with a precision of about 99.77%. The output of the trained neural network and experimental result for test data are shown in Figure 17.13.

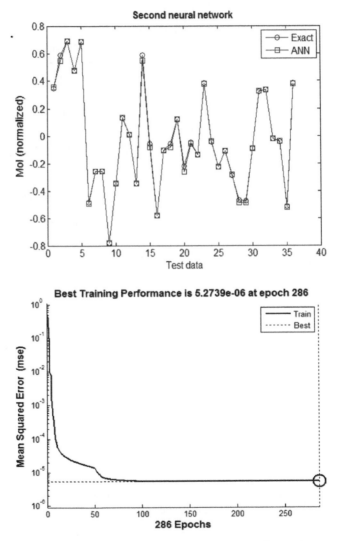

FIGURE 17.13 (a) Training error diagram and (b) compliance between ANN and experiments for the second objective function (test data).

By knowing of the neural network weights and activation function neurons, the output of the neural network can be expressed based on its input by an explicit mathematical relationship.[13] The activation function of network's neurons is tansigmoid (tansig) and first and second trained neural network has three and four layers. Therefore, output of trained neural network can be defined by using mathematical relationships presented in Ref. [13].

After this step, the problem can be mentioned as a direct mathematical expression. The mathematical relationship can be used for optimization problem. In optimization problem, not only the objective function exists but also constraints can be mentioned. Neural network can estimate both cases. There are several ways to convert the constrained optimization problem into unconstrained optimization problem. The most common method is penalty function method. A constrained optimization problem that can be defined to minimize is expressed as eq 17.4. $h(x)$ is equality constraint and $g(x)$ is inequality constraint.[3]

$$\min f(x_i)$$
$$g_j(\{x_i\}) \le 0 \qquad (17.4)$$
$$h_k(\{x_i\}) = 0$$

Objective function of optimization problems is a criterion for comparing different designs and selects the best plan. Minimizing the objective function should not be adversely effect on the behavior and performance of the optimization problem. To convert the constrained optimization problem into unconstrained optimization problem,[3] the exterior penalty function is defined the following equation:

$$\min \Phi(x_i, r) = f(\{x_i\}) + r\left[p(\{x_i\})\right] \qquad (17.5)$$
$$P(\{x_i\}) = \sum_{j=1}^{n}\left\{\max\left[0, g_j(\{x_i\})\right]^2\right\} + \sum_{k=1}^{m}\left\{\left[h_k(\{x_i\})\right]^2\right\}$$

$\Phi(x,r)$ is secondary objective function or artificial objective function, $f(x)$ is the main objective function, r is penalty function coefficient, penalty function coefficient is usually constant and large value. Using a constant and large value for the penalty function coefficient makes it easier to search in the search space. However, the optimization problem can be solved with most optimization algorithms such as genetic algorithms.

17.3   ANOTHER CASE STUDY: PURIFICATION OF MATERIALS

17.3.1   *ALUMINUM PURIFICATION APPROXIMATION IN ZONE MELTING METHOD*

High-purity aluminum is one of the most widely used metal alloys. If the zone melting method is used to achieve high-purity aluminum, one of the important points in this context is how to achieve high purity. In other words, how many repetitions of zone-melting process are needed to achieve the required pureness? One way to respond to this issue is using of neural networks. This subject has been studied by use of experimental data and design of ANN.

High-purity aluminum is used for a wide range of electronics applications, including anode foils for aluminum electrolytic capacitors, hard disk substrates, bonding wires, and wiring materials for semiconductors and liquid crystal display panels, due to the following reasons: Oxide films having excellent permittivity and insulation properties can be obtained through surface treatment, high-purity aluminum contains only a small amount of impurity elements, precipitates and inclusions, and it has high electrical and thermal conductivities. In recent years, it has been used in stabilized superconductors and thermal conductors, making the most of its outstanding properties, which can be demonstrated at low temperatures.

Conventional forming methods are ineffective in achieving of favorable properties of produced parts because ultragrained (UFG) structure is not formed. Moreover, only limited levels of structural and strength-plastic characteristics can be obtained through conventional forming methods. An application of nonconventional forming methods as well as severe plastic deformation (SPD), such as the most preferred methods of equal-channel angular pressing (ECAP) and equal-channel angular rolling technologies could be the solution.[3] They provide the possibility to obtain the UFG structure. UFG polycrystalline metals prepared by SPD methods have suitable combination of strength and ductility. It is unique and it indeed represents interesting cases from the point of view of mechanical properties. In recent years, the scientific research has been focused on the increase of strength in Al alloys by SPD. The observation of the material structure through transmission electron microscopy (TEM) is useful and it is used to confirm the various theories about material behavior during the ECAP processing.

The conduction properties of aluminum change significantly in the temperature zone ranging from low up to room temperature. Particularly, in high-purity aluminum the conductivity can become 10,000 times greater. To understand and make the most of such conductivity over a broad temperature range, as well as the electrical resistivity, which is its inverse, it is essential to understand the factors of electrical resistivity. The electrical resistivity factors include phonon, impurity elements (chemical impurities), surface scattering, point defects, line defects (dislocations), and plane defects (grain boundaries and stacking faults), and it has been known as Mathieson's Rule that the resistivity components of each factor are countable. Generally, the effects of phonon and impurity elements are large.

17.3.2 EXPERIMENTAL PROCEDURES

The zone melting method used in this experimental that has been created from Ovens rotary, engine, and quartz-tube graphics. In this experiment, argon has been used as vacuum because the atmosphere creates impurity in our product. After melting one side of a long, slender raw material, if the melting part is slowly moved toward the other end of the material either by moving the material itself or moving the heating mechanism, the impurity elements will move toward the same end based on the same principle of the segregation process. This technique for refining other zones of the raw material (except for the melting end) is called the zone refining process. The melting operation can be performed only once, or it can be repeated to enhance the purification effect. The raw material can be positioned horizontally or vertically. For the heating mechanism, several techniques such as resistance heating, high-frequency induction heating, and optical heating can be utilized. Although the zone refining process is suitable for small quantity production due to its lengthy refinement time, high purity exceeding 5 N (99.999%) can be achieved by using high-purity aluminum obtained through the three layer process as the raw material. The zone refining was performed by moving the melting part by three or seven passes. The purified sample was then cut up at even intervals.

17.3.3 EXPERIMENTAL RESULTS

Aluminum has other properties that can be improved by improving the purity. Thus, its physical properties are being evaluated and its applications are being developed. Furthermore, due to the high-purity aluminum-based alloying, uniquely special properties could be obtained. Quantum-meter set has been used to obtain percent elements. Regarding Mg, the result of impurity analysis was significantly smaller than the simulation result. Because those results indicate that the purification behaviors of a large number of elements (except for Mg) can be successfully evaluated and that the effects of experimental conditions such as the number of passes can be examined through this simulation, it can be concluded that this simulation technique is an effective tool. In this experiment, most impurities are related elements Sn and Fe. Mg sits on quartz glass after evaporation. Graphic have been used in experiment for no contact aluminum with quartz glass since quartz glass create impurities Sn (Figs. 17.14 and 17.15).

FIGURE 17.14   Experiment apparatus design in software (solid work).

FIGURE 17.15   Experimental apparatus in laboratory.

After ensuring the accuracy of the experimental set up, the required number of tests was carried out in lab. After all the experiments were carried out, the approximation process begins. To speed up data retrieval, the ANN is used as an alternative to experimental model. If the design and training of the neural network was successful, neural network can estimate the experimental model and its results.

17.3.4 ANN SIMULATION

A typical ANN has three layers: the input layer, the hidden layer, and the output layer. Each neuron in the input layer represents the value of one independent variable.[3] The neurons in the hidden layer are only for computational purposes. Each of the output neurons computes one dependent variable. Signals are received at the input layer, pass through the hidden layer, and reach the output layer. The appropriate number of neurons in each layer depends on the type of problem. These layers can have a diverse number of neurons and activation functions, such as sigmoid and linear functions. All neurons are linked to the neurons in the next layer through their connectivity weights.[3] A schematic of ANN architecture with three hidden layers is shown in Figure 17.16.

The first input variable represents the temperature of furnace, the second variable represents the speed of furnace movement, and third variable represents the flow of argon gas in quartz pipe. Variable (Y) is the percent aluminum purity that is measured by quantum meter in laboratory.

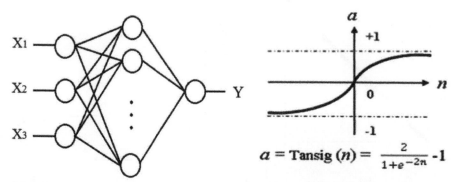

FIGURE 17.16 A schematic of neural network architecture and activation function of neurons.[2]

ANN has been designed with 10 neurons in the first hidden layer and 9 and 1 neurons in second hidden layer and output layer, respectively. Process of training neural network is shown in Figures 17.17 and 17.18.

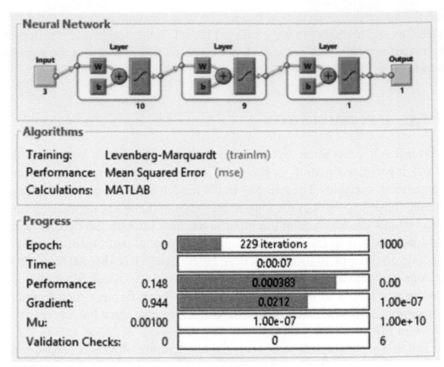

FIGURE 17.17   Neural network training in software.

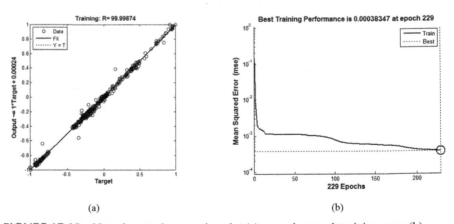

| (a) | (b) |

FIGURE 17.18   Neural network regression plot (a); neural network training error (b).

Using experiments, 560 data are generated that data are including inputs (x) and output (y). Eighty percent data are considered for network training and the rest of the data are used to test the trained network. It is observed that

trained neural network estimates output with acceptable error. The output of
the trained neural network and experimental data are shown in Figure 17.19.

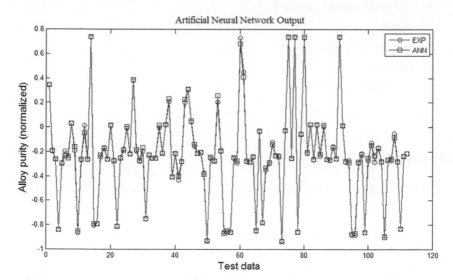

FIGURE 17.19   The output of the artificial neural network and experimental set up for data
test.

Neural network predicts the objective function, by knowing of the neural
network weights and activation function neurons, the output of the neural
network can be expressed based on its input by an explicit mathematical
relationship. The activation function of network's neurons is tansig (tansigmoid) and trained neural network has three layers; therefore, output of
trained neural network can be defined by using mathematical relationships
already presented and explained in detail in reference.[2]

This explicit mathematical relationship provides a feature that without
testing in the lab only by the determination of purity required for aluminum,
the number of required repetitions (round oven) specified.

As specified in the regression plot, the accuracy of designed neural
network is around 99.998 that favorable conditions have been achieved
for purity of $4n$ aluminum. To achieve higher purity of aluminum, greater
number of repetitions is required for neural network training. The results
show that network is able to estimate the purity aluminum by sufficient data.
The designed three-layer network can support prediction of aluminum purity
by suitable data to $4n$.

KEYWORDS

- artificial neural network
- nodes
- neural networking
- neurons
- learning

REFERENCES

1. Menhaj, M. B. *Fundamentals of Neural Networks*. Amir Kabir University Publication: Tehran, Iran, 2009.
2. Ghamami, M.; Shariat Panahi, M.; Rezaei, M. Optimization of Locomotive Body Structures by Using Imperialist Competitive Algorithm. *J. Comput. Appl. Res. Mech. Eng.* **2014**, *3* (2), 105–113.
3. Ghadiri, M.; Mosavi Mashhadi, M.; Ghamami, M. Study of Effective Parameters of Parallel Tubular Channel Angular Pressing (PTCAP). *Modares Mech. Eng.* **2015**, *14* (16), 27–33.
4. Grivani, G.; Ghammamy, S.; Yousefi, F.; Ghammamy, M.; Ghahremani, R. Simulation, Design and Optimization of Industrial Phosphoric Acid Production Processes by Genetic Algorithm. *J. Res. Appl. Inorg. Chem.* **2015**, *1*, 55–67.
5. Papadopoulos, A. I.; Theodosiadis, K.; Seferlis, P. Modeling, Design and Optimization of Industrial Phosphoric Acid Production Processes. *Chem. Eng. Trans.* **2007**, *12*, 477–482.
6. Papadopoulos, A. I.; Theodosiadis, K.; Seferlis, P. A Generic Framework for Modeling, Design and Optimization of Industrial Phosphoric Acid Production Processes. In *18th European Symposium on Computer Aided Process Engineering—ESCAPE 18*; Braunschweig, B., Joulia, X., Eds.; Elsevier BV./Ltd.: Amsterdam, 2008 (all rights reserved).
7. Azaroual, M.; Kervevan, C.; Lassin, A.; André, L.; Amalhay, M.; Khamar, L.; El Guendouzi, M. Thermo-kinetic and Physico-Chemical Modeling of Processes Generating Scaling Problems in Phosphoric Acid and Fertilizers Production Industries. *Proc. Eng.* **2012**, *46*, 68–75.
8. Boulkroune, N.; Meniai, A. H. Modeling Purification of Phosphoric Acid Contaminated with Cadmium by Liquid–Liquid Extraction. *Energy Proc.* **2012**, *18*, 1189–1198.
9. Bharathi Kannamma, G.; Prabhakaran, D.; Kannadasan, T. Analysis and Simulation of Dihydrate Process for the Production of Phosphoric Acid (Reactor Section). *Am. J. Eng. Res.* **2013**, *2*, 1–8.
10. Cameron, J. E. Pollution Control in Fertilizer Production, Chapter 19. *Phosphoric Acid by Wet Process: Pond Water Management*; Marcel Dekker Inc.: New York, 1994; pp 225–236.

11. Mathias, P. M.; Mendez, M. Simulation of Phosphoric Acid Production by the Dihydrate Process. In *22nd Clearwater Convention on Phosphate Fertilizer & Sulfuric Acid Technology Sheraton Sand Key Resort*, Clearwater Beach, FL, May 22–23, 1998.
12. Abu-Eishah, S. I.; Abu-Jabal, N. M. Parametric Study on the Production of Phosphoric Acid by the Dihydrate Process. *Chem. Eng. J.* **2001,** *81* (1), 231–250.
13. Gan, C.; Limsombunchai, V.; Clemes, M.; Weng, A. Artificial Neural Networks versus Logistic Model. *J. Soc. Sci.* **2005,** *1*, 211–219.

INDEX

Printed and bound by CPI Group (UK) Ltd, Croydon, CR0 4YY

23/10/2024

01777704-0008